RENEWALS 458-4574

# SUPERCONDUCTING MATERIALS

THE INTERNATIONAL CRYOGENICS MONOGRAPH SERIES

General Editors

Dr. K. Mendelssohn, F. R. S.
The Clarendon Laboratory
Oxford, England

Dr. K. D. Timmerhaus
Engineering Research Center
University of Colorado, Boulder, Colorado
and
Engineering Division
National Science Foundation, Washington, D.C.

H. J. Goldsmid
 Thermoelectric Refrigeration, 1964
G. T. Meaden
 Electrical Resistance of Metals, 1965
E. S. R. Gopal
 Specific Heats at Low Temperatures, 1966
M. G. Zabetakis
 Safety with Cryogenic Fluids, 1967
D. H. Parkinson and B. E. Mulhall
 The Generation of High Magnetic Fields, 1967
W. E. Keller
 Helium-3 and Helium-4, 1969
A. J. Croft
 Cryogenic Laboratory Equipment, 1970
A. U. Smith
 Current Trends in Cryobiology, 1970
C. A. Bailey
 Advanced Cryogenics, 1971
D. A. Wigley
 Mechanical Properties of Materials at Low Temperatures, 1971
C. M. Hurd
 The Hall Effect in Metals and Alloys, 1972
E. M. Savitskii, V. V. Baron, Yu. V. Efimov,
M. I. Bychkova, and L. F. Myzenkova
 Superconducting Materials, 1973

SUPERCONDUCTING MATERIALS

E. M. Savitskii, V. V. Baron, Yu. V. Efimov,
M. I. Bychkova, and L. F. Myzenkova

*A. A. Baikov Institute of Metallurgy
Academy of Sciences of the USSR
Moscow, USSR*

Translated from Russian by
G. D. Archard

Translation Editor
K. D. Timmerhaus
Engineering Research Center
University of Colorado
Boulder, Colorado
and
Engineering Division
National Science Foundation
Washington, D.C.

PLENUM PRESS • NEW YORK - LONDON

Evgenii Mikhailovich Savitskii was born in 1912 and completed his work at the Institute of Nonferrous Metals and Gold in 1936, specializing in physical metallurgy. From 1937 to 1953 he worked in the Institute of General and Inorganic Chemistry, Academy of Sciences of the USSR, where in 1940 he became the Director of the Mechanical-Testing Laboratory. Since 1953 he has worked in the A. A. Baikov Institute of Metallurgy as the Director of the Laboratory of Rare Metals and Alloys. In 1953 E. M. Savitskii defended his dissertation and obtained the degree of Doctor of Physical Metallurgy. In 1966 he was selected as a Corresponding Member of the Academy of Sciences of the USSR.

Veronika Vladimirovna Baron, Candidate of Technical Sciences, Senior Scientific Fellow, was born in 1914. In 1939 she finished work at the Technological Faculty of the Moscow Institute of Nonferrous Metals and Gold. Since 1944 she has been working in the field of metallurgy in the Institute of General and Inorganic Chemistry, Academy of Sciences of the USSR, and since 1953 in the A. A. Baikov Institute of Metallurgy, Academy of Sciences of the USSR. In 1954 she defended her Candidate's Dissertation.

Yurii Vladimirovich Efimov, Candidate of Technical Sciences, Senior Scientific Fellow, was born in 1931. In 1956 he finished his work at the Physico-Chemical Faculty of the Moscow Institute of Steel and Alloys. Since 1956 he has worked in the A. A. Baikov Institute of Metallurgy, Academy of Sciences of the USSR. In 1967 he defended his Candidate's Dissertation.

Margarita Ivanovna Bychkova, Candidate of Technical Sciences, Junior Scientific Fellow, was born in 1932. In 1956 she finished work at the Moscow Physical-Engineering Institute, specializing in physical metallurgy. Since 1957 she has worked in the A. A. Baikov Institute of Metallurgy, Academy of Sciences of the USSR. In 1969 she defended her Candidate's Dissertation.

Larisa Filippovna Myzenkova, Diploma Engineer, Junior Scientific Fellow, was born in 1935. In 1958 she finished work at the Metallurgical Faculty of Leningrad Mining Institute, and since 1959 has worked in the A. A. Baikov Institute of Metallurgy, Academy of Sciences of the USSR.

The original Russian text, published by Nauka Press in Moscow in 1969, has been corrected by the authors for this edition. This translation is published under an agreement with Mezhdunarodnaya Kniga, the Soviet book export agency.

METALLOVEDENIE SVERKHPROVODYASHCHIKH MATERIALOV

E. M. Savitskii, V. V. Baron, Yu. V. Efimov, M. I. Bychkova, and
L. F. Myzenkova

Металловедение сверхпроводящих материалов

Е. М. Савицкий, В. В. Барон, Ю. В. Ефимов, М. И. Бычкова, Л. Ф. Мызенкова

Library of Congress Catalog Card Number 72-91517

ISBN 0-306-30586-0

© 1973 Plenum Press, New York
A Division of Plenum Publishing Corporation
227 West 17th Street, New York, N.Y. 10011

United Kingdom edition published by Plenum Press, London
A Division of Plenum Publishing Company, Ltd.
Davis House (4th Floor), 8 Scrubs Lane, Harlesden, London NW10 6SE, England

All rights reserved

No part of this publication may be reproduced in any form without written permission from the publisher

Printed in the United States of America

Foreword

With the increased interest in superconductivity applications throughout the world and the necessity of obtaining a firmer understanding of the basic concepts of superconductivity, the editors of the International Cryogenics Monograph series are extremely grateful for the opportunity to add Superconducting Materials to this series. This comprehensive review and summary of superconducting materials was originally prepared by the Russian authors in 1969 and has been specifically updated for this series. It is the most thorough review of the literature on this subject that has been made to date. Since advances in the development and use of new superconducting materials are largely associated with the general state and level in the development of the physical theory of superconductivity, the physical chemistry of metals, metallography, metal physics, technical physics, and manufacturing techniques, it is hoped that this monograph will provide the stimulus for further advances in all aspects of this exciting field.

The editors express their appreciation to the authors, the translators, and Plenum Publishing Corporation for their assistance and continued interest in making this worthy addition to the series possible.

Washington, D. C. K. Mendelssohn
June 1973 K. D. Timmerhaus

Foreword to the American Edition

This monograph was published in Russian in 1969 by the Nauka Publishing House on behalf of the Academy of Sciences of the USSR in Moscow. The reasons inspiring the authors to produce this monograph, as well as the leading characteristics of the material presented, are indicated in the foreword to the Russian edition. The scientific views expressed in the monograph and the experimental data presented have never encountered any objections in Soviet or other scientific and technical literature.

In preparing the American edition we have introduced certain additions to every section of the monograph. These include the results of various recent publications on the magnetic structure, Debye temperature, and electron specific heat of superconductors, the structure of binary and multicomponent superconducting alloy systems, together with their composition–T_c diagrams, information relating to new superconducting elements, alloys, and compounds, the effect of metallurgical factors and interstitial impurities on their properties, new methods of prediction, study, and production (composite superconductors), and new fields of application of superconducting materials. In view of the limited amount of additions which may conveniently be made to the text in the course of translation, we have had to give a certain preference to our own experimental material, since the book is of a monograph nature. Among recent investigations carried out in our laboratory in the field of superconducting materials, mention may here be made of our initial attempts at predicting the existence of superconducting compounds and their critical temperatures by computer (using basic data relating to the electron structure of the atoms comprising the compounds), the plotting of the composition–

T_c diagrams of ternary and multicomponent systems by experiment-planning techniques, the measurement of the critical superconducting temperature of binary alloys in samples of variable composition, the creation of composite superconductors from two-phase alloys by replacing the low-m.p. phase with another superconductor, semiconductor, or insulator in the solid-liquid state, and also work on the production of a multiple-filament superconducting cable [1, 1a]. Research into the construction of phase diagrams for new superconducting systems and the discovery of experimental laws relating the T_c to the composition and constitution of alloys is continuing uninterruptedly. Investigations have started into the effects of extremal conditions (high cooling rates etc.) on the superconducting properties of alloys and compounds and the creation of metastable superconducting phases. Many of the Russian experimental investigations have been published in collections relating to the research and development of superconducting materials produced by the Nauka Publishing House in 1965, 1967, 1969, and 1970 [2-5]. The fifth collection (Superconducting Alloys and Compounds) was published in 1971. The second collection was translated into English by Consultants Bureau, New York–London (1970) as: Physical Metallurgy of Superconductors, Edited by E. M. Savitskii and V. V. Baron. In relation to individual metals, data regarding superconducting materials have been assembled in a number of monographs written by members of our laboratory [6-11].

In addition to the purely physical theory of the superconducting state of solids, it is essential to emphasize the importance of the physical chemistry of metals and physical metallurgy as a theoretical basis for the development of new metallic materials possessing special physical properties, including superconductivity. The English reader may acquaint himself with the present state of this problem and with the work which has been carried out in the Soviet Union by considering the monograph of E. M. Savitskii and G. S. Burkhanov: Physical Metallurgy of Refractory Metals and Alloys, Plenum Publishing Corporation, New York–London (1970).

As physical chemists, the authors of this monograph adhere to the point of view that all properties of matter and materials depend on their internal structure and the technological and service conditions. Many research workers (including ourselves) are daily convinced of the vast amount of information which may be extracted by an intelligent use of the Mendeleev Periodic Table, phase diagrams, composition-property diagrams, x-ray data, and electron-microscope and metallographic investigations at ordinary and cryogenic temperatures in order to secure scientific generalizations and develop new superconducting materials, and also to establish a reasonable technology for the production of these materials, together with a deeper understanding of the complex quantum phenomena underlying the superconducting state of matter. The problem of the creation and technological application of superconducting materials is as yet young; it is developing very rapidly, and it may well be that the most important discoveries are still to come. Nevertheless, it is our own opinion that physico-chemical and metallographic considerations will have as vital a part to play in the future development of this problem as they have hitherto, and that they will contribute fundamentally to its creative merits.

The mutual exchange of information is vital in such a vigorously-developing field as the study and application of superconductors. Unfortunately, the language barrier often constitutes one of the major difficulties impeding correlation between the work of scientists in different countries, as well as up-to-date information regarding their research, achievements, and new ideas. Translations of monographs and articles help in reducing this gap. For this reason many important publications by scientists in other countries relating to the problem of superconductivity and superconducting materials are being translated into Russian in the Soviet Union.

We are grateful to Plenum Publishing Corporation, the Scientific Editor, and the translators for their initiative and interest in our monograph and the considerable work involved in its publication in English; we regard this as a friendly act toward our country and ourselves personally.

E. M. Savitskii
V. V. Baron
Yu. V. Efimov
M. I. Bychkova
L. F. Myzenkova

LITERATURE CITED

1. E. M. Savitskii, Vestnik Akad. Nauk SSSR, 7:44 (1970).
1a. E. M. Savitskii, Izv. Akad. Nauk SSSR, Metally, No. 2 (1970).
2. Metallography and Physical Metallurgy of Superconductors, Izd. Nauka, Moscow (1965).
3. Metallography, Physical Chemistry, and Physical Metallurgy of Superconductors, Izd. Nauka, Moscow (1967).
4. Physical Chemistry, Metallography, and Physical Metallurgy of Superconductors, Izd. Nauka, Moscow (1969).
5. Problems of Superconducting Materials, Izd. Nauka, Moscow (1970).
6. E. M. Savitskii and G. S. Burkhanov, Metallography of the Alloys of Refractory and Rare Metals (second revised and supplemented edition), Izd. Nauka (1971).
7. V. F. Terekhova and E. M. Savitskii, Yttrium, Izd. Nauka, Moscow (1967).
8. E. M. Savitskii, M. A. Tylkina, and K. B. Povarova, Rhenium Alloys, Izd. Nauka, Moscow (1965).
9. E. M. Savitskii, V. P. Polyakova, and M. A. Tylkina, Palladium Alloys, Izd. Nauka, Moscow (1967).
10. Yu. V. Efimov, V. V. Baron, and E. M. Savitskii, Vanadium and Its Alloys, Izd. Nauka, Moscow (1969).
11. E. M. Savitskii, V. F. Terekhova, I. V. Burov, I. A. Markova, and O. P. Naumkin, Alloys of Rare-Earth Metals, Izd. AN SSSR (1962).

Contents

INTRODUCTION 1

CHAPTER I
The Superconducting State of Materials and Methods of
Estimating It................................ 9

    The Phenomenon of Superconductivity 3
        1. History of the Discovery 9
        2. Superconductors of the First Group........ 11
        3. Superconductors of the Second Group 12
        4. Hard Superconductors................. 13
        5. The BCS (Bardeen–Cooper–Schrieffer)
           Theory............................ 14
        6. The GLAG (Ginzburg–Landau–Abrikosov–
           Gor'kov) Theory 18
        7. The Anderson Model.................. 22
        8. The Filament (Sponge) Model of a Hard
           Superconductor 25

    Empirical Rules 27

    Methods of Measuring the Critical Superconducting
    Characteristics of Metals and Alloys.............. 37
        1. Measuring the Temperature of the Transition
           into the Superconducting State 37
        2. Measurement of Critical Magnetic Fields.... 41
        3. Measurement of the Critical Current....... 44

    Low-Temperature Technique 53

Metallography of Superconducting Alloys 59
    1. Preparation of Microsections 59
    2. Etching of the Microsections 62
    3. Study of the Microstructure and Properties of Alloys 66

Literature Cited 73

CHAPTER II
Superconducting Elements 81

    Properties of Superconducting Elements 81

    Effect of Deformation and Interstitial Impurities on the Superconducting Properties of the Elements 90

    Literature Cited 102

CHAPTER III
Superconducting Compounds 107

    Compounds with the Cr_3Si Structure............. 108

    Interstitial Phases and Certain Other Compounds of Metals with Nonmetals 120

    Sigma and Laves Phases and Similar Compounds 135

    Superconducting Compounds with Other Types of Structures 150

    Effect of Alloying Elements and Impurities on the Structure and Properties of Compounds 158
        1. Effect of Transition Metals on the Properties of Cr_3Si-Type Compounds 159
        2. Effect of B Subgroup Elements on the Properties of Cr_3Si-Type Compounds 168
        3. Influence of the Simultaneous Replacement of the A and B Components on the Properties of Compounds of the Cr_3Si-Type 179
        3. Effect of Interstitial Impurities on the Properties of Cr_3Si Compounds 179
        5. Effect of Alloying on the Properties of Compounds with Other Types of Crystal Structure . 184

Effect of Heat Treatment and Other Factors on the
Superconducting Characteristics of Compounds 194

Literature Cited 203

CHAPTER IV
Physicochemical Analysis of Superconducting Systems 215

Binary Superconducting Systems 221
    1. Systems with Unlimited Solubility in the
       Liquid and Solid States 221
    2. Systems with Unlimited Solubility and a Poly-
       morphic Transformation of the Components . . 228
    3. Systems of the Eutectic, Peritectic, and
       Monotectic Types. 252
    4. Systems Involving the Formation of Interme-
       diate Phases 262

Ternary and More Complex Superconducting Systems . . . 307
    1. Ternary Systems 307
    2. Quaternary System 351
    3. Pseudoternary Superconducting Systems 353
    4. Pseudoquaternary System 358

Literature Cited 361

CHAPTER V
Production of Superconducting Materials 373

Effect of Composition, Deformation, and Heat Treatment
on the Critical Current of Superconducting Alloys 374

Technology of the Production of Superconducting Alloys . 387

Properties and Production Technology of Parts Made
from Superconducting Compounds 391
    1. Production of Vanadium–Gallium Wire by
       Working the Quenched Solid Solution 392
    2. Production of Superconducting Coatings 393
    3. Production of Superconducting Wire from
       Compounds by Working a Mixture of the Orig-
       inal Components in a Soft Sheath with Subse-
       quent Heat Treatment (Kunzler Method) 401

4. Production of Superconducting Coatings by
   Hydrogen Reduction 407
5. Production of Composite Superconductors
   from Compounds of the Cr_3Si-Type 409
6. Production of Large Superconducting Parts ... 411

Literature Cited 414

CHAPTER VI
Applications 419

Superconducting Magnets...................... 420

Computing Technology......................... 430
   1. Cryotrons......................... 430
   2. Memory Devices 431

Electronics and Measuring Technology............. 433
   1. Bolometers – Receivers of Thermal
      Radiation 433
   2. Superconducting Magnetic Lenses 433
   3. Masers........................... 434

Nuclear Power and Space 436
   1. Magnets for Thermonuclear Reactions...... 436
   2. Elementary-Particle Accelerators 439
   3. Bubble Chambers................... 440
   4. Resonance Pump 441
   5. Gyroscopes....................... 441
   6. "Zero" Magnetic Field 442
   7. Magnetohydrodynamic (MHD) Generators 442
   8. Protection of Astronauts from Radiation..... 444
   9. Hydromagnetic Braking............... 445
   10. Energy Stores 445

Electrical Machines 445

Conclusion 447

Literature Cited 451

Index 457

Introduction

Metal science constitutes an ever-widening field at a time of scientific and technical revolution such as the present. Metallography, or the science of the structure of metals, dates from the time of D. K. Chernov, who in 1868 discovered the existence of certain special or "critical" points in the heating of steel, these being associated with the polymorphic transformations of iron. Metallography accordingly celebrated its centenary in 1968. The present-day science of metals, however, covers a wider field than elementary metallography. Metal science is concerned with the relationship between the electron characteristics and the structures of metals and alloys and their composition, as well as with their physical, chemical, technological, and practical characteristics under various thermodynamic and kinetic conditions. In certain other countries the term "metal science" has been replaced by the term "physical metallurgy," which has a certain justification in view of the fact that it emphasizes the necessity of a physical approach to the study of metallurgical substances and processes. Metal science and physical metallurgy have a great deal in common, but in metal science the greatest stress is laid upon the physicochemical aspect (study of phase diagrams, changes in composition and properties, and so on). In the last few years the Soviet Union (particularly the Academy of Sciences) has started using the term "physicochemistry of metals," which underlines the fundamental basis of metal science. Soviet literature has also adopted the terminology of the "physicochemical analysis of metal systems," thus emphasizing the fact that this aspect of research belongs to the N. S. Kurnakov tradition, which is of course chiefly concerned with the study of physical properties in relation to changes in the composition of physicochemical systems.

Physical chemists and metal scientists base their research on the fundamental principles of modern science: the structure of matter, quantum mechanics, the periodic table, thermodynamics, physical chemistry, and solid-state physics. The science of metals and alloys incorporates all the achievements of the exact, fundamental sciences, particularly physics and chemistry.

Subjects studied in (theoretical) "metal science," the "physicochemistry of metals," or "physical metallurgy" include metals and alloys, particularly solid solutions, eutectic mixtures, and intermetallic compounds. The general aim of this science is that of creating a consistent theory and deriving experimental relationships such as will enable alloys of predetermined parameters and properties to be developed for every possible need. Substantial advances have already been made in this direction.

As regards methods of studying metals and alloys, "metal science" can and should use every possible theoretical and experimental technique which will help in solving the problems of developing new metallic materials and using these to best advantage. Metal science incorporates a vital technical branch, concerned with the servicing and monitoring of metallurgical and engineering production. As a result of the ever-increasing demands of technology, particularly those of its latest ramifications, the post-war development of the sciences of metals and alloys has been exceedingly vigorous. By virtue of the many rare metals and alloys now coming into production, the field of metal science is continually expanding; completely new branches, such as the metal science of fissile materials, have been created.

During the 1960s another new aspect of metal science unfolded before us – the metal science of superconductors; this is concerned with the laws governing the changes in the structure and properties of superconducting metals, alloys, and compounds, the development of superconducting materials of specified properties, and the scientific and technological aspects of their production. This branch of metal science has not yet been fully established; it is still in the course of a dynamic development. The present monograph is a reflection of this continuing development.

The general plan of the book, the choice of material, and the development of the arguments concerning this material are based

INTRODUCTION

on physicochemical principles; experience shows that at the present time this approach to the problem of superconducting materials is one of the most fruitful.

The monograph has been prepared by a number of authors working together in the Laboratory of Refractory and Rare Metals and Alloys in the A. A. Baikov Institute of Metallurgy, Academy of Sciences of the USSR, on the physicochemical and metallographical study of superconductors. The particular scientific interests of the authors have primarily determined the nature of the book.

The importance of the physical phenomena taking place during the passage of an electric current through superconductors at low temperatures is obvious; equally obvious at the present time is the vital influence of metallurgical factors on the characteristics of superconducting metals, alloys, and intermetallic compounds.

It is the very nature of the superconductor, its composition, internal structure, purity, macrostructure, microstructure, and fine structure, its defects, stresses, and cold hardening, which determine the characteristics of superconducting materials, magnets, and other devices, the stability of their behavior, and the economic viability of employing them in the latest physical instruments; the production of superconducting wire for solenoids only became possible after the development of methods of producing, melting, and thermomechanically treating rare refractory metals and their alloys in vacuum or in protective media.

The application of superconductivity in numerous fields of modern technology is developing with astonishing rapidity. Even so, no monograph discussing the problem of superconductivity has yet appeared in the world's literature.

This book constitutes our own attempt at filling this gap. The reader is left to judge how far this has been successful. The authors will be grateful for any comments and will take these into account in future work.

The monograph is intended for metal scientists, physicochemists, and metallophysicists as well as other specialists in industry, research institutes, and engineering organizations concerned with studying, producing, and using superconducting materials.

The authors wish to thank their colleague N. D. Kozlov for reading through various chapters in the manuscript and also G. S. Kosenko and M. I. Beloborodov for help in setting out the latter.

* * *

The phenomenon of superconductivity was only discovered in the present century, and for a considerable period enjoyed no practical applications at all.

The vanishing of electrical resistance to a steady current at a temperature close to absolute zero, first noted for mercury, and the vanishing of magnetic induction inside a superconductor (the Meissner effect), established much later in tin samples and subsequently in various other metals, constitute fundamental properties of superconductors which, until recently, have hardly been used at all.

Only the creation of materials capable of retaining their superconducting properties in strong magnetic fields and at high current densities opened the possibility of a wide practical application of superconductors.

The development since the war of metallurgy and metal science in relation to rare and refractory metals, the use of vacuum electric-arc and electron-beam methods of melting and refining, improved methods of analyzing impurities, and also improvements to the methods of working these metals and alloys, have created favorable conditions for discovering superconductivity in a large number of materials and developing superconductors of advanced parameters.

Particularly promising is the use of superconducting materials capable of working with high currents and powerful magnetic fields, chiefly in the manufacture of superconducting magnets; these include solenoids and electromagnets with ferromagnetic cores for intensifying and smoothing the magnetic flux created by the superconducting coils.

The elimination of cumbersome supply and water-cooling systems and the possibility of passing high currents (many times greater than would have been possible, for example, in copper solenoids) without disrupting the superconductivity enable us to re-

INTRODUCTION

duce the weight and size of superconducting magnetic systems very substantially by comparison with ordinary systems. Comparative calculations of the economic effect resulting from the use of superconducting and ordinary solenoids carried out in various American organizations in the 1960s showed that the operating costs of the former would be 1000 times lower. On improving the technological processes involved in the manufacture of superconducting materials and broadening the use of these in science and technology, the economical and technical advantages of their use should increase continuously. The development of low-temperature technology and the creation, for example, of helium cryostats working in a closed cycle, and so on, should also be favorably influenced in this respect.

Superconducting magnet systems are already successfully operating in a number of instruments of radio and electronic technology, improving existing devices, and in certain cases giving rise to fundamentally new ones.

The use of superconductors may also prove valuable in other, very different fields of modern technology. In addition to the possibility of using superconductors for the manufacture of transformers, generators, motors, and, perhaps, power lines, special prospects exist in relation to the creation of superconducting coils for magnetohydrodynamic (MHD) generators, which constitute a completely new field of power. Calculations [1] show that the conversion of thermal into electrical power by means of MHD generators can only be economically viable if the magnetic field is created by superconducting coils. The same applies to future thermonuclear reactors.

It is hard to list and still harder to predict all the possible fields of application of superconductors.

Requirements for new technology will present an ever-increasing demand for superconducting materials with specific and stable superconducting and technological parameters.

The development of superconducting devices, particularly magnets, has occupied the attention of dozens of commercial concerns and government scientific organizations in the United States, Japan, Britain, West Germany, France, Holland, Switzerland, and other countries. Similar investigations have been vigorously pursued in the socialist countries. In the USSR, compositions and production technology of superconducting alloys and compounds, superconducting magnets, and other devices have been avidly developed. Starting in 1964, annual scientific conferences have been regularly held on the physicochemistry, metal physics, and metal science of superconductors, and the transactions of these conferences have been published. The problem of superconductors is a complicated one. By its very nature it demands the cooperation of physicists and metals scientists as well as specialists in electronic instruments and power and cryogenic technology.

The properties of superconducting metals and alloys, their so-called critical parameters – the temperature of transition into the superconducting state (T_c), the critical magnetic field (H_c), and the critical current (I_c) – largely depend on the "metallurgical history" of their manufacture. The purity of the original material, the methods employed in melting, the mechanical processing and heat treatment, all have a considerable effect on the level of these properties. This applies particularly to the transition metals and their alloys, which have the highest superconducting properties.

Since the time at which superconductors with a "high" critical temperature, retaining their superconductivity in strong magnetic fields (alloys and compounds of rare and refractory metals), were discovered, no material has ever been found with a T_c of over 18°K. Very recently one communication has appeared regarding an Nb_3Al-base alloy with a critical temperature of 20°K [2]; the critical magnetic field of this reaches 200-250 kOe and the current density about 10^5 A/cm^2. Characteristics of this order are found in the compounds Nb_3Sn, Nb_3Al, V_3Ga, etc. In a number of alloys of the refractory metals (Nb and V) with other elements T_c and H_c reach 10-11°K and 100-120 Oe, respectively.

However, in addition to their chemical composition and interstitial impurity content, the properties of superconductors also depend on phase composition, structure, homogeneity, degree of equilibrium, and ordering. The superconducting characteristics and the other physical properties vary regularly with the chemical composition and structure of the alloys.

In this book we shall attempt to generalize the principal results of Soviet and non-Soviet research workers as to the constitution and properties of superconducting metals, alloys, and compounds.

The authors' own experimental data relating to a variety of superconducting systems will also be presented.

Existing information regarding the superconducting properties of alloys and compounds will be systematized and discussed from the point of view of physicochemical analysis, thus yielding a number of important laws regarding changes in the properties of superconducting systems. Information relating to the fundamental physical concepts of superconductivity will also be presented, together with methods of determining the properties of superconductors and manufacturing them.

Chapter VI is devoted to the applications of superconducting materials in science and technology.

LITERATURE CITED

1. G. W. Wilson and D. G. Roberts, Symposium on MHD Power Generation, Newcastle (September, 1962), p. 14.
2. Science News, 91(20):475 (1967).
3. E. M. Savitskii and V. V. Baron, Izv. Akad. Nauk SSSR, Metallurgiya i Gornoe Delo, No. 5, p. 4 (1963).
4. E. M. Savitskii, Zh. Neorg. Khim., 12(7):1726 (1967).

Chapter I

The Superconducting State of Materials and Methods of Estimating It

THE PHENOMENON OF SUPERCONDUCTIVITY

1. History of the Discovery

Superconductivity was discovered by Kamerlingh Onnes in 1911 [1], who found that the electrical conductivity of metals vanished at low temperatures (Fig. 1). The temperature below which this effect occurs is called the temperature of the transition (transformation) into the superconducting state (T_c) or the critical or transition temperature. In addition to this, any external magnetic field is completely unable to penetrate into the inside of the superconductor, and if the transition into the superconducting state takes place in a magnetic field, then the field is expelled from the superconductor (Meissner effect) [2].

Twenty-seven metals are superconductors; superconductivity also exists in more than a thousand compounds and alloys, where temperatures of transition into the superconducting state range from 0.01 to 21°K [3,39]. The number of superconducting metals, alloys, and compounds is increasing continuously as research advances.

Immediately after the discovery of superconductivity, its possible use in creating very strong magnetic fields was investigated. The first research into the properties of superconductors was encouraging (because of the comparatively high critical currents [4]); however it was subsequently found that a magnetic field of a few oersteds destroyed the superconductivity of mercury, tin,

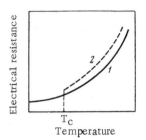

Fig. 1. Dependence of the electrical resistance on the measuring temperature (schematic): 1) normal metal; 2) superconductor.

and lead [5]. Although it was later (1930) found that a lead–bismuth alloy remained superconducting in moderate fields, and the possibility of making superconducting magnets with a strength of 20 kOe was proposed [6], this idea was eventually rejected [7]. Thus, further research into superconducting materials was delayed until 1961, when it was found that the compound Nb_3Sn had high critical currents in magnetic fields up to 70 kOe [8]. This discovery stimulated research into the development of new, hard superconducting materials and the technology of making and processing these. It was in this period that the new branch of science known as "the metal science of superconductors" started developing. All this taken together evoked the vigorous development of cryogenic technology and other branches of science associated with the use of superconductivity.

The microscopic theory of superconductivity appeared early in the 1950s. The effect of the isotopic composition of a metal on its critical temperature was discovered in 1950 [9, 10]. On the basis of this discovery, a theory of the special attractive forces acting between electrons was constructed [11, 12]. In 1956 Cooper showed that the electrons in certain metals existed not as individual particles but as coupled pairs [13]. Bardeen, Cooper, and Schrieffer in the United States [14] and Bogolyubov in the Soviet Union [15] in turn, using the earlier work as a basis, constructed a microscopic theory of superconductivity providing a reasonable explanation for this phenomenon (the BCS theory).

It should also be noted that at an earlier period many physical properties of superconductors were observed and explained as a result of the experimental work of L. V. Shubnikov, A. I. Shal'nikov, Yu. V. Sharvin, N. V. Zavaritskii, and M. S. Khaikin in the USSR, and K. Mendelssohn, E. Appleyard, and A. Pippard in other countries, and the theoretical work of A. Ritgers, C. J. Gorter, the brothers F. and H. London, R. Peierls, and the Soviet physicists L. D. Landau and V. L. Ginzburg.

The properties of superconductors with high critical fields (superconductors of the second group) were outlined by Ginzburg, Landau, Abrikosov, and Gor'kov (the GLAG theory) [16-18] and also by Goodman [19, 20].

2. Superconductors of the First Group

All homogeneous superconductors (containing inhomogeneities no larger than the atomic dimensions) may be separated into two groups: superconductors of the first group, which include all pure metals except niobium and vanadium; superconductors of the second group, which include niobium, vanadium, and all other superconducting alloys and compounds.

Superconductors of the first group (ideal) are characterized by, apart from the temperature of the transition into the superconducting state (which we shall discuss later), a critical magnetic field H_c which constitutes the minimum magnetic field required to destroy the superconductivity in the sample at a specified measuring temperature. The value of this field is a function of temperature. The field reaches its greatest value at absolute zero (a few hundreds or thousands of oersteds) for pure metals. On raising the temperature, the critical magnetic field diminishes and then vanishes [5]; its temperature dependence may be fairly accurately expressed by the formula $H_c = H_0[1 - (T/T_c)^2]$.

The value of the critical magnetic field for superconductors of the first group is fairly low, of the order of a few hundred oersteds. The current flows along the surface of superconductors of the first kind at a depth of δ_0. The superconductivity of a wire or film carrying a current may be destroyed when the current reaches a critical value (I_c). For fairly thick samples, the critical current equals the current which would create a magnetic field equal to H_c on the sample surface (Silsbee rule [21]).

The critical current (I_c) is a very important parameter characterizing superconductors of the first group.

For superconductors of the first group the transition into the superconducting state in the presence of a magnetic field is a phase

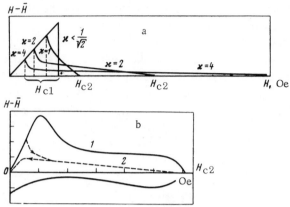

Fig. 2. Magnetization curves of superconductors in relation to the external magnetic field: a) superconductors of the first group ($\varkappa < 1/\sqrt{2}$) and the second group ($\varkappa > 1/\sqrt{2}$); b) hard superconductors (1 – cold-worked, 2 – after annealing).

transformation of the first kind. However, if such a transition occurs in the absence of a magnetic field, then it constitutes a phase transformation of the second kind, at which no latent heat is evolved, and no jump takes place in the specific heat.

The next important property of superconductors of the first group is the Meissner effect [2]. The essence of this lies in the fact that an external magnetic field never penetrates into the inside of a bulk superconductor. If a transition into the superconducting state takes place when the sample is lying in a magnetic field, then the field is driven out of the superconductor.

The magnetization curve of superconductors of the first group is shown in Fig. 2a.

Apart from the properties indicated (specific heat, electrical resistance, magnetic characteristics) the following also occur at the transformation point (T_c): the thermo-emf vanishes, the Hall effect changes, the absorption of ultrasound changes, infrared radiation is absorbed, and so on [22].

3. Superconductors of the Second Group

Superconductors of the second group were first discovered in 1930 [6]; as in the case of those of the first group, they are characterized by the temperature of the transition into the superconducting state (T_c). The special feature of superconductors of this group as compared with the former is the existence of two critical fields (Fig. 2a): lower (H_{c1}) and upper (H_{c2}). Before reaching H_{c1}, these superconductors behave in the same way as those of the first group.

The region between the lower and upper critical fields is called the region of the "mixed" state. A great contribution to the study of this state of metals was made by the well-known Soviet scientist A. V. Shubnikov, and his colleagues.

In the mixed state, superconductors of the second kind have no Meissner effect; the transition from the superconducting to the normal state takes place as a phase transformation of the second kind, even in the presence of a magnetic field. Ideal superconductors of the second group situated in a magnetic field perpendicular to the transport current, having a value greater than H_{c1} but less than H_{c2}, cannot pass a current without destroying the superconductivity. The theory of the mixed state of superconductors of the second group was developed by the Soviet physicists Ginzburg, Landau, Abrikosov, and Gor'kov [16-18] (GLAG theory).

For certain materials, the upper critical magnetic field reaches several hundreds of thousands of oersteds (Nb_3Sn, V_3Si, V_3Ga [23]).

4. Hard Superconductors

Hard superconductors are nonideal superconductors of the second group, i.e., superconductors having chemical and physical inhomogeneities exceeding atomic dimensions. In practice, this means all superconductors of the second group prepared by ordinary metallurgical methods. Hard superconductors possess all the properties characteristic of superconductors of the second group: an upper and lower critical field, a mixed state, a phase transformation of the second kind in the mixed state, and an absence of the Meissner effect. In addition to this, hard superconductors have the following properties peculiar to themselves.

1. Hard superconductors are able to pass a current in the presence of a transverse magnetic field. In some cases, when the structure of the material takes a certain form, the critical current

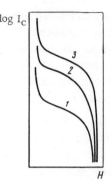

Fig. 3. Critical current as a function of the external magnetic field for a hard superconductor with various degrees of plastic deformation: a) annealed sample; 2-3) cold-worked samples (the degree of deformation of sample 3 is greater than that of sample 2).

may be very high. Figure 3 shows some schematic curves relating the logarithm of the critical current of hard superconductors to the external magnetic field; the critical current is the greater, the higher the degree of plastic deformation of the superconductor.

2. The critical current in the superconductors is proportional to the sample cross section.

3. The magnetization curves of hard superconductors have a hysteresis loop which distinguishes them from ideal superconductors of the second group (Fig. 2b).

In view of these differences, hard superconductors are sometimes called superconductors of the third group.

The possibility of passing large currents through hard superconductors is explained in many papers (for example, that of Anderson [24]) as being due to the "pinning" of the lines of magnetic flux by the defects in the material. In order to explain this phenomenon, the filament (sponge) model of Mendelssohn may be used [25].

5. The BCS (Bardeen–Cooper–Schrieffer) Theory

The discovery of the isotopic effect gave an impetus to the concept of the microscopic theory of superconductivity. It was found [8] that the critical temperature of mercury isotopes was related to the mass M of the isotopes by the equation $T_c M^\alpha = $ const, where α is a coefficient equal to $1/2$. Later this effect was observed in a number of other elements.

The fact that the value of T_c depends on the mass of the isotope suggests a relationship between the development of superconductivity (which is evidently an electronic process) and the mass of the isotope, which only affects the lattice phonon spectrum. This in turn should mean that superconductivity is largely due to a strong interaction between the electrons and the lattice. This mechanism was proposed by Fröhlich [11] independently of the experimental data.

Bardeen, Cooper, and Schrieffer in the United States and Bogolyubov in the USSR proposed a microscopic theory of superconductivity on the basis of this phenomenon [14, 15]. It was shown that the phenomenon of superconductivity was associated with the formation of pairs of electrons (sometimes called C o o p e r p a i r s). The formation of Cooper pairs is associated with the fact that special attractive forces act between the electrons in metals. According to the most modern theory, the magnetic or electric forces acting between bodies may be regarded as the result of an exchange of electromagnetic quanta or p h o t o n s. Photons are continuously emitted by one body and absorbed by the other, and, as a result of this, they interact. The crystal lattice of the ions in a metal by itself forms a system of interacting particles. When energy is given to the lattice, it vibrates, and these vibrations correspond to special quasi particles or p h o n o n s. The electrons in the metal may be exchanged by the phonons. This gives rise to the attractive forces responsible for the formation of the Cooper pairs. Of course, apart from this, the electrons, being similarly charged bodies, repel each other. In those metals in which this repulsive force exceeds the phonon attraction, superconductivity fails to appear. It is an interesting fact that not only do the two electrons forming the pair take part in this phenomenon, but so also does the whole collection of electrons in the metal; it may be shown, in particular, that even a weak attraction is sufficient for the formation of a pair. In order to impart energy to the electron fluid in the superconductor, the pair must be broken. This means that the energy to be imparted cannot be infinitely small but must necessarily be greater than the binding energy of the pair. This restriction on the amount of transferrable energy is known as the g a p in the energy spectrum. The existence of an energy gap has the effect that, up to a certain velocity, the fluid of electron pairs moves without friction; in other words, the electric current flows without any resistance. Since the formation of pairs is a collective effect and is associated with the state of the whole system, the rupture of a few pairs (by raising the temperature) in turn affects the binding energy of the others. The binding energy therefore starts falling, and at the critical temperature it vanishes. No more pairs exist, and the metal is in all respects normal [26, 27].

The BCS theory yields the following relationships.

1. The binding energy of the pairs equals $2\Delta(T)$.

$$\Delta(0) = 2\hbar\omega_D e^{-1/\eta},$$

where ω_D is the Debye frequency (in metals $\hbar\omega_D$ equals 300-400°K); η is a dimensionless constant determining the interaction of the electrons with the lattice (in all known cases $\eta < \frac{1}{2}$).

2. The critical temperature (in energy units)

$$T_c = \frac{\Delta(0)}{1.76}.$$

It is quite possible that the low values usually obtained for T_c are associated with the existence of certain restrictions on η.

3. The electron specific heat*:

a) In the normal state $C_n = \gamma T$, $\gamma \approx 10^2$-10^4 ergs/cm^3·deg;

b) for $T \approx 0$

$$C_s(T)/C_n(T) = 1.35\,(\Delta(0)/T)^{3/2} e^{-\Delta(0)/T};$$

c) for $T = T_c$

$$C_s(T)/C_n(T_c) = 2.43 + 3.77\,(T/T_c - 1).$$

From this we see that the specific heat experiences a jump at $T = T_c$ (Fig. 4).

4. The temperature dependence of the critical field: a) for $T = 0$

$$H_c(T) = H_c(0)\,[1 - 1.06\,(T/T_c)^2].$$

H_c is proportional to T_c and lies between hundreds and thousands of oersteds. b) for $T \approx T_c$

$$H_c(T) = 1.73 H_c(0)\,(1 - T/T_c).$$

---
*This definition for γ is used throughout this monograph.

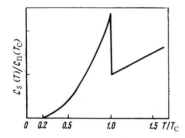

Fig. 4. Dependence of the electron specific heat on the ratio T/T_c.

The critical temperature of the transition into the superconducting state is expressed by the formula

$$kT_c = 1.14\hbar\omega_D \exp\left[-\frac{1}{N(0)V}\right],$$

where $N(0)$ is the density of states at the Fermi surface, V is the interaction parameter [28], and c is Boltzmann's constant.

On comparing these and other theoretical formulas with experimental data for pure metals, good agreement is obtained. The new theory satisfactorily describes the behavior of superconductors in steady and alternating electromagnetic fields. The so-called "superconducting correlation" is responsible for this. Cooper electron pairs are fairly large formations; their radii are of the order of 10^{-4}-10^{-5} cm, i.e., several thousand times greater than the interatomic distances. It follows that the motion of the electrons at various points of the metal is "correlated" over a distance of the order of the dimensions of a pair (this is called the correlation length). Hence the electric current at a specified point of the metal is determined by the electromagnetic field existing, not simply at that particular point, but also over the whole region covered by the correlation length. An integrated, or nonlocalized, relationship develops between the current and the field. If the electromagnetic field varies little at a distance of the order of the correlation length, the effect of correlation is only slight, and the current is determined by the actual field at the same point. In this way we obtain the so-called London equations [29]. In the opposite case we obtain a special type of integrated relation between the current and the field, as proposed by Pippard before the appearance of the microscopic theory of superconductors. As a parameter characterizing the distance within which the field changes substantially, we may take the depth of penetration; it is at this distance that the magnetic field falls to zero (i.e., its value diminishes on passing from the outside into the superconductor). We are, of course, concerned with the relation between the depth of penetration and the correlation length. Experiment shows that all pure metals belong either to the Pippard case or else to an intermediate one, i.e., the correlation length is much greater than the depth of penetration (as in the case of aluminum, for example), or of the same order (as in the case of tin).

6. The GLAG (Ginzburg – Landau – Abrikosov – Gor'kov) Theory

The BCS theory was derived for superconductors of the first group, the critical magnetic fields of these not exceeding a few thousand oersteds. However it has been known for some decades [5] that in certain alloys superconductivity is preserved to some extent up to fields of several tens of thousands of oersteds. When the remarkable properties of the compound Nb_3Sn were discovered in 1961, these were readily explained by the means of the GLAG theory.

The GLAG theory explains the properties of superconductors of the second group. The concept of superconductors of the second group was introduced in 1952 (before the appearance of the complete theory of superconductivity) by Abrikosov [30]. The reason for the development of the properties characteristic of superconductors of the second group is the fact that these contain defects which scatter electrons; this reduces the electron mean free path (l). When the defect concentration is high enough for the free path to become smaller than the dimensions of a Cooper pair, l itself takes up the role of the dimensions of the pair. Any further reduction in l, resulting from an increase in defect or impurity concentration, has the effect that the dimension of the pair become smaller than the depth of penetration of the magnetic field. As a result of this, the energy of the boundary between the superconducting and normal phases becomes negative $\sigma_{ns} < 0$ (for superconductors of the first group σ_{ns} is positive).

The reason for the development of the surface tension (energy) between the normal and superconducting phases is illustrated in Fig. 5. Curve Δ corresponds to a pair-binding energy having a

THE SUPERCONDUCTING STATE 19

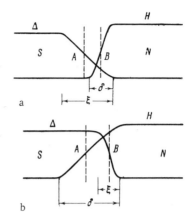

Fig. 5. Development of surface energy between the normal and superconducting phases: a) superconductors of the first group; b) superconductors of the second group.

certain value in the superconducting phase S and a zero value in the normal phase N. As a result of the correlation mentioned earlier, this energy cannot fall suddenly to zero; it falls gradually over a distance of the order of the pair size. Equilibrium between the superconducting and normal phases is only possible if there is a magnetic field equal to the critical value in the normal phase. The magnitude of this field (continuous line H) falls to zero within the superconducting phase at a distance equal to the depth of penetration δ. Figure 5a shows the usual position for superconductors of the first group, when the pair size ξ is greater than the depth of penetration δ. In order to simplify the picture, we may consider an idealized scheme in which the continuous curves are replaced by sharp boundaries. To the left of the broken vertical line A we shall consider the pair-binding energy to be the same as in the main body of the superconducting phase, and to the right of this boundary equal to zero. Analogously the magnetic field is regarded as being critical to the right of the straight line B and zero to the left.

In this way we obtain two boundaries: one relating to the field, the other to the binding energy. In the interval between these (AB) there is no binding energy, but at the same time there is no field. In view of this the layer AB has an excess energy corresponding to the rupture of all the pairs. The excess energy appearing at the boundary brings the body into an unstable state.

In the case of superconductors of the second group, on the other hand, the depth of penetration of the field is greater than the size of the Cooper pair. The boundaries thus change places (Fig. 5b). As a result of this, some of the electrons will be linked into pairs in the layer BA in the presence of a field. This leads to a fall in energy in this region as compared with the normal metal, i.e., to a negative surface tension. This state is energetically favorable for the body, so that superconductors of the second group (with a negative surface tension between the boundaries of the superconducting and normal phases) may be separated into a large number of regions of alternating superconducting and normal phases [26].

It is clear from the foregoing that any superconductor of the first group having an impurity content sufficient to shorten the mean free path appreciably becomes a superconductor of the second group.

A critical parameter ($\varkappa$) determining whether superconductors belong to the first or second group was proposed earlier [16]. For superconductors of the first group $\varkappa = \sqrt{2} \cdot 2e/\hbar c \cdot H_{cm} \delta_0^2 < 1/\sqrt{2}$ (in this case $\sigma_{ns} > 0$), for superconductors of the second group $\varkappa > 1/\sqrt{2}$ ($\sigma_{ns} < 0$); here e is the charge on the electron, H_{cm} is the thermodynamic magnetic field of the bulk superconductor, c is the velocity of light, and δ_0 is the depth of penetration.

In considering the magnetic properties of a superconducting cylinder constituting a superconductor of the second group ($\varkappa > 1/\sqrt{2}$) it was found [17] that for magnetic fields lower than the first critical field (H_{c1}), a cylinder of this kind behaved as a superconductor of the first group; the external magnetic field only penetrated to a depth δ_0. When the field reached a value of H_{c1}, individual quantum filaments of magnetic flux started penetrating into the cylinder [31]. As the external magnetic field increased further, the concentration of magnetic filaments also increased; they became denser and arranged themselves parallel to one another, forming a two-dimensional square configuration. At the corners of the configuration (along the axes of the magnetic tubes) there was no superconductivity; the field equalled the external magnetic field. In the space not occupied by the magnetic filaments, the internal field was smaller than the external, and the concentration of the superconducting electrons reached a maximum in the center of the unit cell; this had a period of $\sqrt{2}\pi\delta/\varkappa$. Further increasing the external field made no difference to this period, but only led to a reduction in the difference between the minimum value of the internal field in the center of the cell and value of the external magnetic field. Finally, for a field equal to the second critical value H_{c2}, this difference fell to zero, a phase transformation of the second kind took place, and the cylinder acquired an electrical resistance.

After refining Abrikosov's numerical calculations, it was later concluded [32] that the minimum free energy of the superconductor in the mixed state corresponded to a triangular rather than a square configuration.

Goodman [33] obtained results qualitatively similar to Abrikosov's using the phenomenological London theory [29]; he proposed a model according to which a superconductor in the mixed state had a laminar structure, with alternating regions of normal and superconducting phases.

The existence of a periodic magnetic structure in a superconductor in the mixed state was confirmed later [34]: In a magnetic field of 1620 Oe a small-angle neutron scattering peak was found at 10', implying a periodic structure with a period of 1500 Å. A study of the magnetic structure by depositing fine particles of iron on the surface of a superconductor showed that the magnetic flux lines in the superconducting material usually formed a triangular lattice. Under certain crystallographic conditions a square magnetic lattice may also exist. The magnetic structure is sensitive to the various types of defects in alloys. A review of the latest data regarding this problem was presented by Selger [34b].

The physical essence of a superconductor of the second group lies in the fact that, for a negative value of σ_{ns}, a state in which the superconductor is, as it were, split up into a large number of normal and superconducting regions is energetically the most favorable state in an external magnetic field; this assertion is based on the principle that as great a volume as possible should be occupied by the actual boundary between the normal and superconducting phases. This situation is realized in the mixed state [35].

The second critical magnetic field determining the level of the fields which the material is capable of providing depends on the temperature of the superconducting transition and other physical characteristics of the sample [36]: $H_{c2} = 2.58 \cdot 10^{-2} \gamma T_c \rho_n$, where

γ is the electron specific-heat coefficient and ρ_n is the residual electrical resistance.

This formula was derived by Gor'kov for dilute solid solutions. However, calculations of the upper critical fields in the niobium–zirconium and niobium–titanium systems [35a] and a comparison of the resultant data with experiment show that this formula may also be used for calculating H_{c2} in systems of continuous solid solutions. In order to estimate H_{c2} we must therefore study such properties as the residual electrical resistance and the specific heat at low temperatures. In certain cases the superconductivity of the samples remains intact in thin surface layers in fields greater than the second critical field. The field up to which this surface superconductivity is preserved is called the t h i r d c r i t i c a l m a g n e t i c f i e l d (H_{c3}) and equals $1.965 H_{c2}$ [39].

7. The Anderson Model

A superconductor of the second group existing in the mixed state is entirely unstable with respect to a superconducting transport current flowing in a direction perpendicular to the magnetic field [36-38]. This instability is associated with the appearance of Lorentz forces, acting when a current flows along the quantized magnetic lines (fluxoids) in a superconductor of the second group.

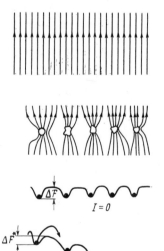

Fig. 6. Attachment of fluxoid associates to inhomogeneities in a hard superconductor (according to Anderson).

THE SUPERCONDUCTING STATE

The Lorentz forces lead to motion of the fluxoids and destroy the superconductivity [36].

In a hard superconductor, in contrast to a superconductor of the second group, physical and chemical inhomogeneities occur. If the inhomogeneity is such that the magnetic flux passes more easily through the material of the inhomogeneity than through the superconducting material, then the whole group of fluxoids forms an association attached to the nearest inhomogeneity (Fig. 6) [40].

Let the size of an inhomogeneity be d. Then the maximum possible depth of the potential well (if the inhomogeneity is a nonsuperconducting material) in which the associate is situated will be $\Delta F_{max} = H_{cm}^2/8\pi d^3$. In the general case we introduce a coefficient p less than unity into this formula. Then the formula takes the form

$$\Delta F = p \frac{H_{cm}^2}{8\pi} d^3 .$$

A transport current of density j perpendicular to the magnetic field acts on the associate of fluxoids with a force $f = j\Phi_B d/c$ (Φ_B is the magnetic flux in the associate). This leads to a potential field $j\Phi_B \, dx/c$ (x is the current coordinate in a direction perpendicular to the field and the current). The associate is situated in a potential field

$$\Delta F^* = p \frac{H_{cm}}{8\pi} d^3 - \frac{1}{c} j \, \Phi_B \, dx .$$

Under the influence of thermal excitations the fluxoid associates will diffuse in the direction of action of the Lorentz forces. The jump frequency is determined by the well-known formula

$$R = R_0 e^{-\Delta F^*/kT} ,$$

where k is Boltzmann's constant; R_0 has the sense of the frequency of the vibrations of the associate in the potential well (according to Anderson $R_0 \sim 10^{10}$ sec^{-1}). When the product $j\Phi_B$ reaches a certain value, the so-called critical mode sets in.

In an alloy of niobium with 25% zirconium an electrical resistance of 10^{-12}-10^{-13} $\Omega \cdot$ cm was found to exist, this varying considerably with the current flowing and the magnetic field [41].

For comparison, we may add that, in a superconductor of the first group, the electrical resistance is generally no greater than 10^{-25} $\Omega \cdot$ cm [42].

We may conclude from the foregoing that the Anderson mechanism on the whole gives a correct representation of the processes taking place in hard superconductors. Anderson himself, however, noted that he had omitted a large number of other factors, such as the dependence of the free energy on the currents and field, the changes taking place in δ_0 with field and temperature, the effect of the shape of the associates, details of the interaction between the flux lines, and so on.

Despite these omissions, Anderson's mechanism is used by a large number of authors to explain the considerable increase in the critical current of samples subjected to heat treatment. These authors consider that the increase in the critical current takes place as a result of the precipitation of phases with low T_c values (as a result of decomposition), these then serving as barriers interrupting the lines of associates. This relates both to Nb–Zr alloys, in which decomposition leads to the precipitation of the α- and ω-phases [43, 44] and also to Nb–Ti alloys, in which similar phases precipitate [45, 46], as well as three-component Nb–Zr–Ti alloys in the presence of interdendritic liquation [48]. The considerable increase taking place in the critical current of Nb–Zr alloys as a result of its decomposition into two solid solutions with high superconducting transition temperatures [49] is probably associated with the appearance of a branched lattice of boundaries as a result of this process.

The mechanism underlying the interaction of dislocations with lines of magnetic flux was studied elsewhere [50]. Thus a superconducting filament 10^{-5} cm thick lying along a screw dislocation interacts with the latter with a force of 10^{-3} dyne/cm. The critical current of the superconductor in an external magnetic field is determined by equating this force to the Lorentz interaction, which gives a value of 10^{-6} A for the critical current in a field of 10 kOe. For a density of filaments equal to 10^{10} cm^{-2} the critical current density is 10^4 A/cm^2. The interaction of an infinitely long filament of quantized magnetic flux with a screw dislocation perpendicular to it was considered at the same time. On equating the interaction to the Lorentz force, the critical current density for niobium is equal to 10^5 A/cm^2.

Elsewhere it was asserted [51] that the dislocation distribution had an effect on the critical current. The authors consider that a two-dimensional dislocation distribution holds the magnetic flux lines together most effectively; a cellular distribution holds them to a slighter extent, while a uniform distribution is the least effective.

It was later found [52] that the rise in critical current observed in annealed (500°C) Nb−54.3at.% Ti (containing 1% oxygen) was associated with the appearance of a cellular structure in the dislocation distribution. The positive effect of cellular structure has been repeatedly noted [52a].

Thus the Anderson theory enables us (to a certain extent) to explain the effect of various kinds of inhomogeneities (chemical and physical) on the rise in critical current.

8. The Filament (Sponge) Model of a Hard Superconductor

In order to explain the retention of superconductivity in strong magnetic fields, it has been suggested [25, 53] that in such fields the greater part of the superconductor (having low superconducting properties) is permeated by a random network of intersecting filaments with a far higher critical magnetic field than the rest. This model explains the retention of superconductivity in fields greater than critical if we suppose that the thickness of the filaments is much smaller than the depth of penetration of the magnetic field (δ_0). The critical field of such a filament is about $H_{cm} \times \delta_0/r_0$ (r_0 is the radius of cross section of the filament). It is considered that the material is initially all superconducting, but that after reaching a certain external magnetic field value the current only flows through the superconducting filaments, while the rest of the material becomes normal.

This model of a hard superconductor explains many of its special features: the high second magnetic field, the existence of a hysteresis loop on the magnetization curve (the capture of magnetic flux as a result of the mutually-coupled nature of the randomly-distributed superconducting intersecting filaments), the phase transformation of the second kind (according to the Ginzburg−Landau theory, a superconductor with a transverse dimension smaller than the depth of penetration transforms into the normal state in an external magnetic field as a result of a phase transformation of the second kind), and the proportionality of the critical current to the cross-sectional area.

This model may be applied to artificially-created spongy superconductors (lead–bismuth alloy in porous glass [54]) and to a severely-worked eutectic niobium–thorium mixture [55]. The mechanism in question may be employed in order to explain the increase in critical current taking place as a result of heat treatment in an alloy of zirconium with 4% Nb [56]. In order to explain the high critical currents in ordinary hard superconducting alloys after working and heat treatment, however, a number of authors propose regarding the dislocation lattice as a set of hypothetical filaments [57-61]. The "effective diameter" of the dislocations is estimated to be 10-100 Å, i.e., the dimensions are just right for such filaments. The dislocations are not limited to the individual grains, but are capable of forming a cellular structure. It is of course well known that increasing the degree of deformation increases the critical current; this may be explained by the rise in the number of dislocations [61]. Furthermore the dislocations are distributed quite evenly over the sample cross section. Thus, in worked alloys the dislocations may very well play the part of the lattice or sponge proposed by Mendelssohn.

The question as to why the dislocations retain superconductivity in high magnetic fields may be resolved in the following manner [61]: The atoms in the centers of the dislocations are displaced from their equilibrium positions, and thus we may expect that the phonon–electron interaction in this region will differ substantially from the interaction in a dislocation-free lattice. Since the phonon–electron interaction is a fundamental aspect of the BCS theory in determining the superconducting properties, we should expect a severe change in the superconducting properties in the neighborhood of dislocations. It is possible that the development of superconductivity in high magnetic fields demands a specific size of dislocation. This may in particular explain the absence of superconductivity in fields above H_c in such superconductors as lead, indium, mercury, etc. On the basis of Abrikosov and Goodman's treatments it was later suggested [36, 62] that a negative surface energy might be essential for the dislocations to act as superconducting filaments.

Thus the two theories supplement one another and may reasonably be used when considering hard superconductors [61].

It may well be that in hard superconductors the current properties are largely governed by dipoles [63] rather than single lines; the dipoles arise as screw dislocations move through the crystal, causing plastic deformation. If this is in fact so, then the presence of the dipoles may explain why the dislocations in lead, indium, and other metals of low melting point fail to act as superconducting filaments. Dipoles are annealed much more rapidly than dislocations, and cease to exist long before single lines of dislocations vanish from the metal. For this reason the sample cannot carry a current in fields above H_c. This requires experimental verification.

A rather different mechanism was proposed later [64] for the formation of the spongy structure in severely-worked Nb–Zr and Nb–Ti alloys. It was suggested that the high superconducting characteristics in worked nonequilibrium Nb–Zr alloys arose as a result of the formation of a well-developed system of thin layers and filaments consisting of two body-centered cubic phases. A similar mechanism for the rise in current was indicated in the case of Nb–Ti alloys [64].

On the basis of the existing model of a filament superconductor the relationship between the critical current, the external magnetic field, and the density of the filaments was also derived quite independently [65]. Consideration of the results showed that the filament model excellently represented the plastic deformation of a hard superconductor.

As yet there are insufficient data available to assign final preference to one model or the other. It is quite probable that different mechanisms of current flow are realized for different structures of the materials.

EMPIRICAL RULES

The modern physical theory of superconductivity is insufficient to predict the properties of real alloys. As yet we cannot say in advance what elements and what proportions should be used to create superconducting alloys with specified critical parameters or produce superconductors for hydrogen, neon, and nitrogen temperatures. In studying the superconductivity of materials, experimental investigations must therefore play a leading part, as in other branches of solid-state physics. There is now a fair amount of experimental material available in order to establish the general relationships between superconductivity and the number of valence electrons per atom, the crystal structure, the atomic masses and volumes, and so on. Of course these empirical rules have many exceptions, but nevertheless they are of considerable practical interest.

Superconductivity is only found in metals with a valence not less than 2 and not greater than 8. For compounds the rule of minimum valence has some exceptions. For example, the compound Ag_3Ga is a superconductor, although its valence is only 1.5 electrons/atom [66].

For transition metals the superconducting transition temperature depends on the number of valence electrons. The maximum transition temperatures occur for transition metals with a mean valence electron concentration of 5 and 7 (Fig. 7) [66a-68].

The dependence of T_c on the number of valence electrons may be explained on the basis of the theoretical model which considers superconductivity as arising as a result of Coulomb interaction [69].

Superconducting metals have a smaller atom volume than other elements in their period [70].

Certain laws are also clearly associated with the crystal structure of the material. Thus the structure of all presently-known superconductors has a center of inversion. There are no superconducting crystals with certain specific symmetry groups, such as those of the $[D_4]$ type.

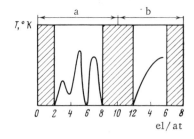

Fig. 7. Dependence of the superconducting transition temperature on the number of valence electrons for transition (a) and nontransition (b) metals.

In the case of alloys forming solid solutions, on alloying a transition metal with another transition metal the T_c of the alloy depends on the number of valence electrons per atom. On alloying a transition metal with a nontransition metal T_c diminishes [71]. In binary alloys formed by elements of groups IV, V, and VI, the optimum electron concentration (calculated on the weighted-mean principle) corresponding to the compositions with the highest values of T_c is lower than in the case of pure metals; it is by no means constant, and varies over a certain range on passing from one system to another. In Nb–Ti alloys the optimum electron concentration equals 4.5 electrons/atom; in Nb–Zr and Nb–Hf the corresponding figures are 4.75 and 4.9 [72].

This difference in the values of the optimum electron concentration is associated with the fact that the superconducting transition temperature is largely governed by a steric factor, determined by the degree of misfit between the solvent and dissolved atoms. A dissolved atom differing in size from the solvent atoms creates a stress field in the lattice and thus influences the electron properties and the electron density on the Fermi surface [73]. If the dissolved atom is larger than the solvent atom, the lattice expands and a local decrease in the charge of the ion occurs. This imparts an effectively lower valence to the dissolved substance than it would have in the unstressed state. Thus, in calculating the valence of the dissolved substance we must introduce a correction for the size of the atom.

The number of electrons per atom may be calculated, with due allowance for the stress introduced by the dissolved atom, from the following formula [7]

$$N = [A] Z_0 + [B] Z^*,$$

where [A] is the atomic proportion of the solvent, [B] is the atomic proportion of the dissolved substance, Z_0 is the valence of the solvent, Z^* is the effective valence of the dissolved substance, $Z^* = Z - (\delta V/V)Z_0$, where Z is the original valence of the dissolved substance, and V is the volume of the solvent atom [74, 75].

When studying the effect of an alloying component on the T_c of solid solutions of transition metals, it was later found [73] that the critical temperature depended on the diameter of the alloying element: If the alloying element has a greater atomic diameter than the base, then the lattice parameter and critical temperature increase with increasing content of the dissolved element; if the diameter is smaller than in the case of the solvent, then the values of these two parameters diminish.

The superconducting transition temperature is substantially affected by the stress field created by the dissolved atoms [76]. In determining the changes in critical temperature taking place in dilute solid solutions we must accordingly allow for the atomic radius of the dissolved substance (or rather for a parameter representing the stress created by the dissolved atom), together with the number of free electrons:

$$\Delta T \sim T_c \, \xi(2r_0) \, N_s^{1/3} \exp\{\alpha - (N_s^{1/3} r_0)^{-1}\},$$

where $\xi(2r_0)$ is the parameter representing the stresses, r_0 is the radius of the dissolved atom, N_s is the number of free electrons, and α is a constant.

Some of the papers [73, 74, 77] relating to the effect of the volume of the elements on changes in T_c are based on early works by Matthias in which he indicated a relationship between T_c and the volume as well as the valence.

In studying the effect of various alloying elements on the superconducting transition temperatures of tin, indium, and aluminum (up to the solubility limit), it was found that for a low additive content (order of several tenths of an atomic percent) T_c decreased in inverse proportion to the free path, independently of the nature of the additives [78-80] (Fig. 8). The relative change in T_c is the same for such different elements as aluminum and tin. When $\xi_0/l > 1$ (l is the free path, ξ_0 is the coherence length), the change in T_c deviates from the linear law; if the dissolved impurity is electropositive (valence greater than that of the solvent) the transition temperature decreases, if it is electronegative the temperature rises.

An analogous change in T_c was observed elsewhere [81]. An analytical relationship was also found between T_c and the impurity content [7]:

$$dT_c = k_1 x + k_2 x \ln x,$$

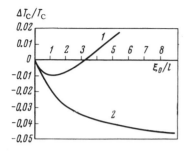

Fig. 8. Dependence of the change in critical temperature ($\Delta T_c / T_c$) on the free path (ξ_0 / l) after alloying the original component with electronegative (1) or electropositive (2) additives.

where x is the concentration of the impurity and k_1 and k_2 are constants. For high concentrations of the dissolved impurities [81] there is a rise in T_c, so that the critical temperature becomes greater than that of the solvent. The initial linear decrease in the superconducting transition temperature with increasing additive content may be explained as being due to a corresponding reduction in the free path. The subsequent increase in T_c is due to the change in the valence of the solid solution.

A substantial effect of electron concentration on T_c in contrated superconducting solid solutions was also noted by others [82]. This was attributed to the fact that, in solid solutions containing more than 5 at.% of the additive, the free path varied comparatively little, and the change in T_c was mainly due to the valence of the additive.

Among the transition metals, a similar change in T_c with changing free path was observed for tantalum [83].

As already indicated, according to the BCS theory T_c depends on the mean phonon frequency $\hbar\omega_g$, the density of states at the Fermi surface, and the interaction parameter. Since the mean phonon frequency is proportional to the Debye temperature θ_D, the following dependence of T_c on this quantity was proposed [84]:

$$T_c = c\,\theta \exp\left[-\frac{1}{N(0)V}\right].$$

At the same time, N(0) is proportional to the Sommerfeld constant γ. On this basis $N(0) = 3\gamma/2\pi^2 k^2$, where γ is the electron specific heat coefficient and k is Boltzmann's constant. At the same time $N(0) = \chi_a / 2L\mu_B^2$, where χ_a is the atomic paramagnetic susceptibility, L is Avogadro's number, and μ_B is the Bohr magneton.

In order to calculate N(0), nuclear magnetic resonance and optical and x-ray spectroscopical data may also be employed [85, 85a].

The electron specific heat, the Debye temperature, and the superconducting transition point of Ti-Zr alloys were measured by Bucher et al. [86]. In this system the number of valence electrons per atom remains constant (both metals belong to group IV A) and the curve relating the electron specific heat to composition qualitatively reproduces the composition dependence of T_c. On calculating the interaction parameter from the corresponding data, using the BCS formulas, it was found that this parameter varied continuously with composition.

In one paper [86a] the following formula was obtained for determining the temperature of the superconducting transition for superconductors with strong coupling:

$$T_c = \frac{\theta_D}{1.45} \exp\left[-\frac{1.04(1+\lambda)}{\lambda - \mu^x(1+0.62\lambda)}\right], \quad (1)$$

where $\lambda \approx N_{bs}(0)V_T$ is the dimensionless interaction constant, $N_{bs}(0)$ is the zone density of states, V_T is the parameter of electron-phonon interaction, and μ^* is the Coulomb pseudo-potential; for transition metals this approximately equals 0.13.

In deriving Eq. (1) it was considered that the superconductors had a phonon spectrum similar to that of niobium. If not, formula (1) leads to substantial errors, as in the case of mercury. For superconductors with strong coupling:

$$N_{bs}(0) = \frac{3}{2} \cdot \frac{\gamma}{\pi^2 k^2} \cdot \frac{1}{1+\lambda}. \quad (2)$$

If for a certain specific system of alloys we have the values of T_c, Θ_D, γ and use Eq. (1), we may calculate the values of λ, $N_{bs}(0)$, and V_T for this. McMillan carried out a corresponding calculation of the parameter λ and $N_{bs}(0)$ for a number of systems of alloys of the transition metals with 3d, 4d, and 5d electrons. According to these data the behavior of the parameter λ corresponds closely to that of T_c in these systems.

In another paper [86b] a method of calculating the T_c of systems of alloys for which θ_D was known was proposed. According to Hopfield's theory λ is given by the following formula

$$\lambda = \frac{\eta}{A \langle \theta^2 \rangle}, \qquad (3)$$

where A is the atomic number of the element, $\langle \theta^2 \rangle$ is the mean square phonon frequency,

$$\eta = \frac{1}{M} \left(\frac{\hbar}{k}\right)^2 \left(\frac{d\mu}{dz}\right) N_p(0),$$

M is the mass of the proton, $\hbar$ is Planck's constant, k is Boltzmann's constant, $N_p(0)$ is the density of the p states at the Fermi level, and $d\mu/dz$ is the matrix element of the atomic potential gradient.

In calculating the T_c of any system, the $\langle \theta \rangle$ for the original components is taken from published data, and then by using the data relating to θ_D is calculated for the whole system.

In order to obtain the values of η in Eq. (3) one first calculates the η for the original components from the McMillan formula for T_c Eq. (1). For intermediate alloys η is approximated linearly. Using Eq. (3), λ is then calculated for the whole system, and then Eq. (1) is used to obtain the T_c of the intermediate alloys. A calculation of this kind carried out for the T_c of the Zr–Ti, Zr–Nb, Nb–Mo, and Ta–W systems [86b] gave excellent agreement with the experimental T_c values. However, a calculation of the T_c of Nb–V alloys gave unrealistic results, so that there are clearly serious limitations to the method in question [86c].

For the Nb–Ti system the authors calculated λ, $N_{bs}(0)$ by the McMillan method and T_c (Fig. 10) by the Hopfield method. In this calculation the experimental values of θ_D and γ for alloys of this system were employed. The behavior of the parameter λ and the density of states $N_{bs}(0)$ relative to the composition agreed with that of T_c (expt.). The calculated values of T_c for this system qualitatively reproduced the form of the T_c (expt.)–composition curve (Fig. 10).

The highest superconducting transition temperature is obtained for niobium compounds; slightly lower values are obtained for compounds of vanadium. The highest known T_c values are obtained in compounds with Cr_3Si-type structure. The superconduct-

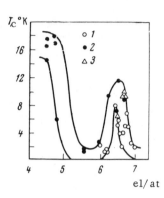

Fig. 9. Superconducting transition temperature as a function of the mean number of valence electrons per atom for the following compounds: 1) sigma phase; 2) compounds with a Cr_3Si lattice; 3) alpha manganese.

ing transition temperature of compounds with the structure of the σ-phase is appreciably lower, and lower still in the case of compounds with a Laves phase type of lattice.

The solid solutions of compounds of rare-earth and noble metals which exhibit superconductivity at low temperatures are ferromagnetic at higher temperatures.

The superconducting transition temperatures of intermetallic compounds correspond to a specific electron concentration, which for Cr_3Si-like compounds is equal to 4.7 and 5.6, in the sigma phase 6.93, and in the Laves phase 6.6 electrons/atom. These values are close to the optimum electron concentrations for pure metals (Fig. 9) [87, 88].

There is a certain relationship between the atomic volume and the superconducting transition temperature for metallic compounds. For compounds of the Cr_3Si type the optimum atomic volume is equal to 15-20 $Å^3$, for the sigma phase 15-16 $Å^3$, and for the Laves phase 15.5 $Å^3$ [87, 88].

In alloys containing bismuth, compounds with interatomic distances of 3.1 to 3.8 Å are superconducting.

A relationship between the critical transition temperature and the reduced mass of the compound was proposed by Gold [91] on the basis of the isotopic effect. If the critical temperature of several compounds is known, then Gold's rule may be used in order to determine the critical temperature for other compounds of the same type. The calculation is based on the formula

$$\frac{(T_c)_{AB}}{(T_c)_{AC}} = \frac{1 + M_A/M_B}{1 + M_A/M_C};$$

where M_A, M_B, M_C are the atomic masses. If we know the T_c of a compound AB, we may use this equation to calculate that of a superconducting compound AC. This method of calculation gives excellent results for the carbides WC, MoC, NbC, TaC, and others. By using this rule it was predicted that, if it could be made, the compound Nb_3Si would have a T_c higher than that of Nb_3Sn. This compound was in fact prepared later, but the prediction was not fulfilled [92].

A study of the relationship between the critical temperature and the minimum bismuth–metal interatomic distance in compounds of the $MeBi_2$ type (Me = alkali metal) showed that these two characteristics were linearly related to one another. The critical temperature increased with increasing bismuth–metal distance [90].

On alloying the compounds V_3Si and V_3Ga with various elements it was found that the superconducting transition point was independent of the lattice constant [93], apparently as a result of a simultaneous change in the composition of the alloys.

The influence of interatomic distance on the critical temperature should be analogous to the influence of a change in lattice

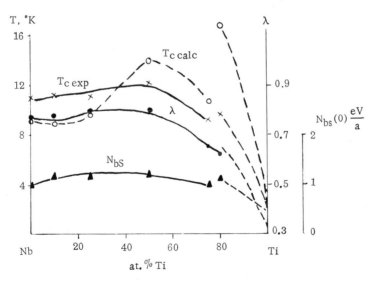

Fig. 10. Superconducting transition temperature ($T_{c\ exp}$ = experimental, $T_{c\ calc}$ = calculated), density of states $N_{bs}(0)$, and parameter λ as functions of the composition of Nb–Ti alloys.

constant. The relation between lattice constant and T_c for an alloy of any particular composition may be determined by analyzing the effect of pressure on these two characteristics. It was shown earlier [94-97] that the critical temperature might either increase or decrease with increasing pressure; for example, in the case of thallium subjected to slight pressures there was a positive value of $\partial T_c / \partial P$, whereas on applying high pressures the coefficient became negative. A relation between T_c and the lattice constant is also given in [97a].

In hard superconductors the critical current is very sensitive to a change in the structure of the alloy; it depends on the degree of deformation of the sample, on the grain size, and on the quantity, dimensions, distribution, and magnetic properties of the precipitating phases [45, 46, 98-101].

The second critical magnetic field and the temperature of the transition into the superconducting state depend chiefly on the chemical composition of the phase with the highest superconducting characteristics. However, these characteristics are also affected by the degree of deformation of the sample [51] as well as the stresses in the lattice [102]. The transition temperature of alloys is also affected by precipitating phases with appreciably differing T_c values (proximity effect) [103].

The extent of the dependence of T_c, H_c, and I_c on metallurgical factors decreases in the following order [104]: as regards microstructure I_c, H_c, T_c; as regards composition and crystal lattice structure T_c, H_c, I_c.

A great deal of work is going on at the present time on creating superconductors with a high critical temperature. It should be noted that the foregoing phonon mechanism of superconductivity will probably never yield materials with transition temperatures above 25-40°K. A possible exception is metallic hydrogen, for which the authors estimate $T_c \sim 100°K$. However, metallic hydrogen can only be obtained at very high pressures, and the possibility of its having a high transition temperature is also a matter for discussion [104b]. This point of view is confirmed indirectly by the fact that, despite the vast number of metals which have been studied, no alloys or compounds with a T_c of over 20.8°K have yet been obtained.

In view of this a great deal of attention is now being paid to the discovery of superconductors in which the attraction between the electrons might be achieved by exciton exchange, i.e., excitations of the electrons rather than the phonon type [104a]. Whereas, in fact, for superconductors with the phonon mechanism T_c is expressed by the formula $T_c = c\theta_D \exp[1/-N(0)V]$, where θ_D is of the order of 300° for the majority of metals, while $N(0)V \sim \frac{1}{5}-\frac{1}{3}$, for the exciton mechanism $T_c \sim \theta^{(e)} e^{-1/g}$, where the characteristic exciton frequency $\theta^{(e)}$ is of the order of 10^4 °K, which for $g \sim \frac{1}{5}-\frac{1}{3}$ might give $T_c = 100$°K. However, it is very difficult to create materials with superconductivity mainly due to exciton exchange, and no successful attempt of this kind has yet been made.

METHODS OF MEASURING THE CRITICAL SUPERCONDUCTING CHARACTERISTICS OF METALS AND ALLOYS

Every superconductor is characterized by three parameters: the temperature of the transition into the superconducting state (critical superconducting temperature) T_c, the critical magnetic fields H_{c1} and H_{c2}, and the critical current I_c. Each of these has its own special measuring technique.

1. Measuring the Temperature of the Transition into the Superconducting State

The critical temperature is usually determined either by directly measuring the electrical resistance as a function of temperature [105] or by an indirect method, i.e., using the change in the induction or frequency of a coil with a superconducting core at the instant of passing into the superconducting state.

In measuring T_c by the direct method one uses a sample with four contacts (Fig. 11a). This method is less accurate than one of the indirect procedures because the sample remains superconducting while even a single superconducting filament exists within it. A positive aspect of the indirect methods is the fact that by their use we may determine the relative proportions of normal and superconducting phases in the sample. A measurement of T_c by this method was described earlier [106]. In principle, the method

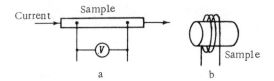

Fig. 11. Circuits for measuring T_c by reference to electrical resistance (a) and by reference to the change in the inductance or frequency of a coil (b).

relies on the fact that the self-inductance of a coil wound around the sample (Fig. 11b) varies in direct proportion to the magnetic permeability of the sample [107-109], which equals the induction B divided by the external field H. The method has the advantage that it requires no direct electrical contacts with the sample, and that it may be used for samples of irregular shapes.

Table 1 indicates methods of obtaining and measuring temperatures below 20°K.

In the inductive method of measuring the temperature of the superconducting transition, the change in the magnetic permeability of the sample is used [106, 112]. The transition is detected by

TABLE 1. Methods of Producing and Measuring Temperatures below 20°K

Temperature range, °K	Method of producing the temperature	Measuring method	Literature cited
20-13.81	Change in the vapor pressure of boiling hydrogen	Resistance thermometer Thermocouple Gas thermometer Pressure of hydrogen vapor Acoustic vibrations	[105, 110]
13.81-4.2	Cooling by gaseous helium	Resistance thermometer Thermocouple Gas thermometer	[105]
4.2-1	Change in the vapor pressure of boiling helium (He^4)	Resistance thermometer Gas thermometer Pressure of He^4 vapor	[105, 111]
Below 1	Change in the pressure of boiling He^3 Magnetic cooling	Resistance thermometer Pressure of He^3 vapor Magnetic thermometer	[105]

means of an inductance bridge (Fig. 12) [112]. A 1-Hz ac voltage is applied to the bridge, which comprises the inductances L_1 and L_2 and the R_1 and R_2. The voltage is taken from the points AA through an amplifier to a dc voltmeter, to which a signal from a phase detector is also fed. The voltmeter thus registers a steady voltage proportional to the signal taken from the bridge.

The sample is placed in one of the inductance coils (L_1). On passing into the normal state the magnetic permeability (μ) changes from 0 to 1 over a certain temperature range. The inductance $L_1 \sim \mu$, so that the signal taken from the bridge will be proportional to μ. When the sample is in the superconducting state, the bridge is compensated by means of the resistance R_1. On passing into the normal state an out-of-balance signal appears. In one particular case [112] the out-of-balance signal increased from 0 to 2 V for an ac-generator voltage of 20 V, this being very easily detected.

An apparatus of this type may be based on a gas thermometer (Fig. 13) [112]. The samples are placed inside a copper bomb, within coils set on a special copper insertion piece. The level of liquid helium falls as the helium evaporates, and the temperature accordingly changes by 0.15 deg/min. In this way the transition temperature may be measured to an accuracy of 0.2°K. Figure 14 shows the behavior of $\Delta T/T_x$ for $P_{4.2}$ = 170 mm Hg, V_b = 47 ± 8 cm^3, and V_{ds} = 6 ± 2 cm^3. The value of ΔT is calculated from the formula

$$\frac{\Delta T}{T_x} = \frac{V_b}{V_b + \frac{4,2}{300} V_{ds} \left(1 - \frac{P_x}{P_{4.2}}\right)} - 1,$$

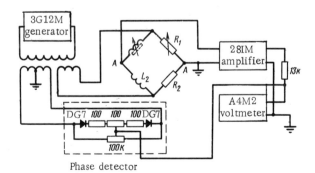

Fig. 12. Electrical circuit of an apparatus for measuring the superconducting transition temperature.

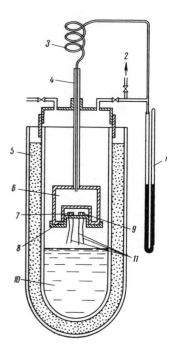

Fig. 13. Schematic diagram of the gas thermometer: 1) manometer; 2) to backing pump; 3) capillary; 4) movable piston; 5) liquid nitrogen; 6) gas thermometer; 7) sample; 8) measuring coil; 9) compensating coil; 10) liquid helium; 11) leads.

where T_x is the temperature determined, V_b is the volume of the bomb, V_{ds} is the dead space (volume of the gas in the manometer with $T > T_x$), P_x is the pressure at temperature T_x, and $P_{4.2}$ is the pressure at 4.2°K. The shaded region corresponds to the spread ΔT. The greatest spread is 0.2°K for a measuring temperature of 17.8°K. This accuracy is entirely satisfactory for studying the T_c of new alloys and establishing macroscopic laws.

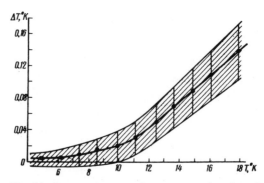

Fig. 14. Systematic measuring error as a function of measuring temperature in the gas thermometer.

The proposed measuring method has both advantages and disadvantages. One advantage is the possibility of measuring T_c on small, irregular samples and powders. The number of samples measured in a single run is quite large (ten being a typical number [111]), this being determined by the number of coils accommodated in the copper insertion piece. The disadvantages of this method include the relatively low accuracy, the worst case being 0.2°K.

The accuracy of measuring T_c may be increased by reducing the heating rate (by using a special thermostat) and increasing the accuracy of measuring T_x, by the automatic recording of the transition curve on a two-coordinate high-current automatic potentiometer. For this purpose a special thermocouple or a germanium thermometer may be employed instead of a gas thermometer.

In order to reduce the number of experiments and secure full information with only a slight expenditure of helium, a method of measuring the T_c of a superconducting system on a single sample of variable composition has been developed in the Laboratory of Rare Metals and Alloys, Institute of Metals of the Academy of Sciences of the USSR [112a].

2. Measurement of Critical Magnetic Fields

The critical magnetic fields are determined by measuring the dependence of the resistance, permeability, or magnetization of the materials on the value of the external applied field [113-116].

Measurements of magnetization are particularly valuable, as they reveal the general character of the average magnetic induction within the sample.

Inductive Method. One method of determining the critical field is based on an analysis of the magnetization curve of the sample. The ballistic method is usually employed for plotting magnetization curves [117]; it is based on the emf arising in a coil surrounding the sample as a result of a rapid change in the flux passing through the coil cross section.

The measuring coil carrying the sample is connected in series with a ballistic galvanometer. When the magnetic flux through the area embraced by the turns of the measuring coil changes, an emf is created; according to the laws of magnetic induction this will be equal to

$$E = -\omega \frac{d\Phi}{dt} 10^{-8} \text{V},$$

where $d\Phi/dt$ is the rate of change of magnetic flux. This emf will produce a current in the circuit of the measuring coil and the frame of the galvanometer, i.e., $\omega \, d\Phi/dt \, 10^{-8} = iR$, where R is the total resistance of the galvanometer circuit. On integrating the equation with respect to the time during which the magnetic flux is changing we obtain

$$\omega \Delta \Phi = 10^8 \, RQ,$$

where Q is the amount of electricity flowing through the galvanometer frame, and $\Delta \Phi$ is the change in the magnetic flux.

It follows from the theory of the ballistic galvanometer that the maximum deflection is proportional to the amount of electricity which flows through the galvanometer windings, i.e.,

$$Q = \alpha \, C_b,$$

where C_b is the constant of the ballistic galvanometer and α is its deflection.

Since the flux changes by an amount $\Delta \Phi = S \Delta B$ (ΔB is the change in the magnetic induction through the sample cross section, S is the area of the sample), the formula for the magnetic induction takes the form

$$B = \frac{C_b \, R \, \alpha}{\omega \, S} \, 10^8 \text{ G}.$$

Knowing the value of the magnetic induction in the sample, we may find the magnetization of the material in question:

$$B = \mu H + 4\pi I,$$

where μ is the magnetic permeability and H is the external magnetic field.

If we place the sample in the measuring coil, then on sharply withdrawing it from the coil we obtain an emf simply proportional to the magnetization of the sample. Remembering that $B = 4\pi I$ (I being the magnetization of the sample), the value of I may be determined from the relation

$$I = \frac{C\varphi\alpha}{4\pi\omega S},$$

where $C\varphi = C_b R 10^8$. For convenience of measurement, and also in order to increase the sensitivity of the apparatus, a system of two measuring coils in series is commonly used, these being wound in such a way that the currents flowing through them are of different signs.

In order to measure the magnetization of superconducting samples in relation to the external magnetic field, the authors used a superconducting solenoid with a fairly large region of homogeneous field. The measuring coil and sample were placed in this region. On rapidly withdrawing the sample from the measuring coil, a deflection proportional to the sample magnetization was obtained in the ballistic galvanometer.

The external field applied was varied from zero to the maximum value (20 kOe) by varying the current passing through the solenoid (0-20 A). The magnetization of the sample was recorded for preassigned values of the external field. On the basis of these results the values of $4\pi I$ were plotted in relation to H.

Typical magnetization curves of superconductors of the first and second group are presented in Fig. 2, together with those of hard superconductors.

Methods of determining the magnetic field from magnetization curves for hard superconductors are found in [118].

There are other methods of measuring the critical field. It has been suggested, in particular, that the local magnetic induction might be measured directly inside the sample by using the Faraday effect, or else by instruments capable of detecting very small fields inside the sample or on the surface [119].

<u>Measurement of the Critical Field under Pulse Conditions.</u> For superconductors with high values of the critical fields, pulse methods may conveniently be used in order to determine the critical field by reference to the appearance of an electrical resistance in the sample [120]. The experiments are carried out on samples of wire several centimeters long and several tenths of a millimeter in diameter. The external

magnetic field is created in the solenoid by passing a discharge from a condenser battery through it. The battery consists of ten electrolytic condensers ($800\,\mu$F, 300 V). Fields up to 150 kOe are obtained by using an energy of 400 J in a diameter of 0.5 cm. An additional 1.5-2 times increase in the field may be achieved by increasing the energy stored in the condenser battery. The current growth time in the solenoid is 4 msec. A voltage proportional to the current in the coil and, hence, to the magnetic field is applied to the horizontal plates of an oscillograph. In this way a "timeless" scan is obtained with respect to the magnetic field, i.e., a specific value of the magnetic field corresponds to a specific horizontal deflection of the beam.

The instant at which the superconductivity of the sample is destroyed reveals itself by the appearance of a resistance between the potential contacts of the sample. In order to detect the appearance of an electrical resistance in the sample an ac measuring current is passed through it, this having a value of a few tens of milliamps and a frequency of 35 Hz. This frequency is a convenient one for the magnetic-field pulse lengths generally employed. The signal from the potential leads is fed to the input of a resonance amplifier. In series with this, a signal compensating for interference from the sample under measurement [121] is also fed in. While the sample remains in the superconducting state, a zero line appears on the oscillograph screen. At the instant at which the field becomes equal to H_{c2} a resistance appears, the compensation is disturbed, and the oscillograph screen records the transition into the normal state. This process may be photographed. For measuring currents of 10-15 mA the critical field is independent of the current.

3. Measurement of the Critical Current

The critical current is a very important characteristic, as it governs the possible practical use of a material with high values of H_{c2} and T_c. The critical currents are usually determined by the four-probe measuring technique [122]. After the sample has been cooled to a specific temperature and its orientation relative to the applied field has been established, one of the following procedures is adopted:

1. The field is made equal to a specific value and the current is increased until a voltage drop appears at the potential contacts.
2. The current is made equal to a constant value, and the field is increased until a voltage drop appears at the contacts.
3. The field and current are increased simultaneously until a voltage drop appears at the potential contacts.

The first two measuring methods usually give matching results; the third may give lower values. When studying the critical current in this way, its value is usually determined over a wide range of external magnetic fields, since it is only in this way that the behavior of the characteristic may be determined in strong magnetic fields.

The critical current is measured in an external magnetic field created by magnets or superconducting solenoids. In the latter case the samples are placed together with the field source in a Dewar containing helium; when using ordinary magnets, however, which require no cooling, only the samples to be measured are placed in the helium (in a special Dewar with a side tube) [123].

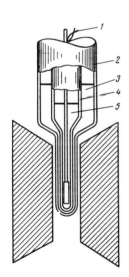

Fig. 15. Arrangement of the apparatus for measuring the critical current of the sample: 1) to supply source; 2) liquid-nitrogen Dewar; 3) liquid nitrogen; 4) liquid-helium Dewar; 5) liquid helium.

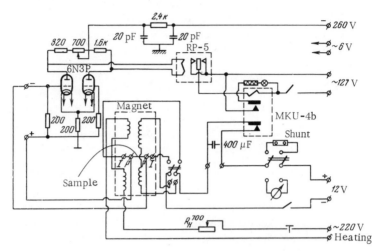

Fig. 16. Arrangement of the apparatus for measuring the critical current.

Figure 15 illustrates the measurement of the critical current in a Bitter solenoid.

In measuring the critical current, the magnetic field is usually established first, and then a current is smoothly introduced into the sample until a potential difference arises in the latter, indicating that a resistance has appeared in it. The current causing an electrical resistance to appear is regarded as critical. The sensitivity of the measurement of potential usually fluctuates between 10^{-2} and 10^{-5} V. When it is desired to determine the critical characteristics of the material, the critical current is measured in a transverse magnetic field. If the magnetic field is created by means of a solenoid in which the lines of force lie along the axis, the sample is bent into a hairpin shape so as to incorporate a section of sample perpendicular to the lines of force.

Figure 16 presents an electrical circuit for measuring critical currents [124]. The samples used here are of the wire type; they have current contacts for supplying the electric current and also potential contacts, from which the voltage is recorded as soon as it arises. The potential contacts are connected to a relay, which disconnects the current from the samples when a voltage of 20 mV has developed. The critical current is measured in an electromagnet with its lines of force perpendicular to the samples, so that there is no need in this case to give the latter a hairpin shape.

THE SUPERCONDUCTING STATE 47

The poles of the electromagnet are made of Permendur and the body of Armco iron. The electromagnet has copper windings, and rings of a niobium–zirconium alloy are fitted on to these in order to "freeze" the field. The rings are "unfrozen" by means of a heater fed from a 220 V network.

At the beginning of the experiment the magnet is placed in a glass helium Dewar and cooled to 4.2°K. Then the heater is connected and a field of specified value is created, after which the heater and the current feeding the copper windings are switched off in rapid succession. The field so created is "frozen" in the superconducting Nb–Zr rings. A steady current from an accumulator is passed through the test wire sample and its value is gradually increased by means of rheostats. On reaching the critical current (disrupting the superconductivity), a voltage drop appears at the potential terminals; this acts on an electric relay which breaks the circuit. A guard system protects the samples from overheating. Thus, by connecting all the samples in series (three were tested simultaneously in the case under consideration) and varying the magnetic field from 0 to 26 kOe the critical current may be plotted as a function of the external magnetic field. The potential and current contacts are spot-welded to nickel lobes and then to lead and copper conduits.

The quality of the contacts has a considerable influence on the results of the measurements, and they must accordingly be prepared with great care. The most reliable types are pressure contacts of a massive configuration made from superconducting material (niobium, niobium–zirconium, and so on).

The problems giving rise to these experiments frequently involve samples of different shapes, for example lamellar samples when considering the properties of rolled material. The fixing of these samples and the feeding-in of the current require a special clamping device (Fig. 17) made of copper. The surface of the copper clamps supplying the current may be covered with indium solder in order to ensure reliable contact, and then fixed to a niobium plate. Niobium is suitable for this purpose because its thermal expansion is similar to that of the material under test. The central, narrow part of the sample is placed in a groove (0.64 mm wide) in a plastic base plate and flooded with glycerin, which freezes at low temperatures and keeps the samples fixed. The potential contacts

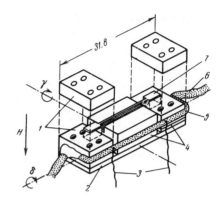

Fig. 17. Clamping device for measuring the critical current in lamellar samples at 4.2°K: 1) copper contact blocks; 2) niobium plate; 3) contacts for measuring potential difference; 4) S-8 plastic; 5) brass shunt; 6) current lead; 7) sample.

are connected to a brass shunt. The arrangement may be rotated around two axes, δ and γ (Fig. 17) making specific angles with the direction of the external magnetic field as required by the specification of the problem [125].

Another version of an apparatus for locating samples correctly in transverse and longitudinal magnetic fields is reproduced in Fig. 18 [126]. A sample 23 mm long is soldered (with pure tin) into axial apertures cut in copper holders (the ends of the sample being first copper-coated), and rests on an insulating substrate. In setting the sample across the magnetic field, the directions of the field and current are chosen so that the electromagnetic forces acting in the sample during the measurements are transmitted to the insulating base; this ensures preservation of the same constant orientation of the samples relative to the external field over a se-

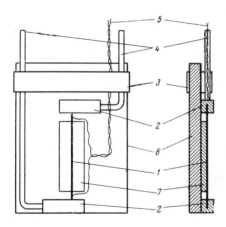

Fig. 18. Apparatus for setting up the samples: 1) samples; 2) copper holders; 3) adhesive strip; 4) current terminals; 5) potential lead; 6) organic glass holder; 7) insulating substrate.

ries of measurements. The potential leads are fixed to the sample with silver. The current is passed through copper wires 1 mm in diameter soldered to the copper holders. The whole apparatus is fixed with an adhesive strip to an organic glass holder; this ensures the absence of stresses from the sample, which might otherwise arise in the course of cooling as a result of differences in thermal expansion coefficients.

The following method of measuring the critical currents enables a large number of samples to be measured with a single Dewar-full of helium. The field is in this case created by a superconducting solenoid. After feeding current into the solenoid and reaching the maximum field, the latter is "frozen"; then it is graduated in height (the maximum field being in the center of the solenoid) so as to establish fixed values at different distances from the center to the edges.

The sample under measurement is placed in a thin-walled tube made of stainless steel, which is introduced into the working aperture of the solenoid after the "freezing" of the field. The samples are bent into the shape of a hairpin so as to make the measured part perpendicular to the magnetic lines of force. The current and potential leads, connected to an RP-4 relay (which trips on reaching a voltage of 20 mV), are also placed inside the stainless steel tube. The arrangement of the apparatus is shown in Fig. 19, which also shows the general appearance of the system after pouring helium into it. As a result of the considerable variation in magnetic field strength along the solenoid (when wound in the usual manner) and the comparatively small height−diameter ratio in the apparatus under consideration, the critical current may be measured as a function of field without recourse to repeated "unfreezing" of the field in order to obtain different values of the latter.

Thus, for example, on moving 50 mm upward from the center of the solenoid (at which point the field reached its maximum of 30 kOe in the case under consideration), the magnetic field strength fell to 7 kOe. Lower values of the field (from 7 kOe to zero) were obtained by moving 40 mm beyond the end of the solenoid.

By placing the test sample successively in different, preselected positions, the curve relating the critical current to the external magnetic field may readily be measured.

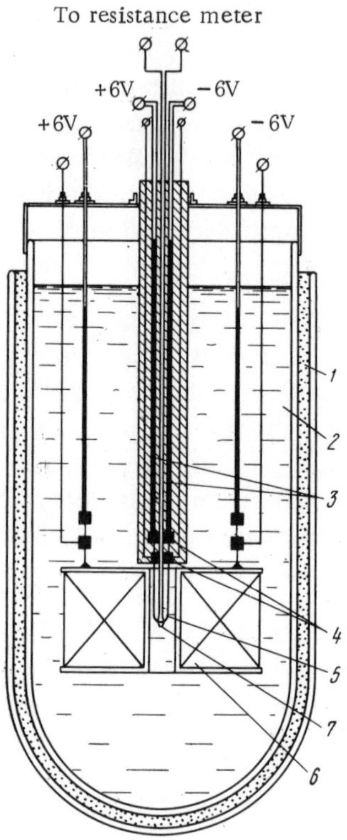

Fig. 19. Apparatus with removable samples for measuring the critical current in relation to the external magnetic field at 4.2°K: 1) liquid nitrogen; 2) liquid helium; 3) lead conduits; 4) contacts; 5) sample; 6) superconducting solenoid; 7) bismuth magnetic field meter.

A great advantage of the apparatus in question is the possibility which it offers of easily changing the test samples. These may be removed together with the central stainless steel tube, and a similar tube carrying the next sample may then be installed in the solenoid.

If necessary, an additional charge of liquid helium may be provided as the work proceeds. The consumption of helium is quite small, as the cryostat keeps the system at liquid-helium temperature. The number of samples measured in the apparatus may be as many as 12-15 for a liquid-helium consumption of about 10 liters. If a poor contact is detected between the sample and the current leads (the critical current appears to be independent of the field), the sample is replaced and a further measurement is made after repairing the contact.

A method of measuring the critical currents of superconductors in the ac mode is described in [128].

The principle of measuring the critical current in a Bitter solenoid is similar to those methods just described. In one typical case [126] the measurements were carried out in an iron-free water-cooled solenoid operating at 2 MW with a current of 10 kA, giving a magnetic field of 82.5 kOe. The opening in the solenoid was 50 mm; the axial opening of the helium Dewar placed in the solenoid was 28 mm. The axial nonuniformity of the magnetic field in the middle of the solenoid never exceeded 1% over a distance of 50 mm.

Measurement of Critical Current under Pulse Conditions. When it is required to measure critical currents in magnetic fields of great intensity, pulse methods of current measurement are often employed [120]. The field is in this case created in the same manner as when measuring critical fields in the pulse mode. Apart from the measuring current, an additional steady current of up to 10 A is passed through the sample, this being measured with a dc ammeter. In this way the breakdown of superconductivity may be studied over a range of fairly low current densities, up to 10^4 A/cm^2.

For higher current densities a special pulse system is employed for the sample. A short pulse of current increasing linearly with time is passed through the material. The screen of a double-beam oscillograph then exhibits curves representing the current passing through the sample and the voltage across the potential terminals. The arrangement of the apparatus is illustrated in Fig. 20. The generator 1 produces voltage pulses increasing linearly with time. The growth time of the pulses τ may be varied from 1 msec to 1 min, and the repetition period T from 0.5 sec to

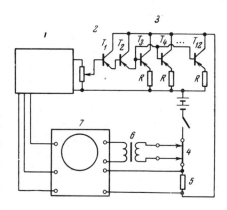

Fig. 20. Arrangement for creating linearly-increasing current pulses [120]: 1) generator; 2) potentiometer; 3) current amplifier; 4) sample; 5) resistance; 6) transformer; 7) oscillograph.

1 min. The generator is also capable of producing single pulses on triggering from an external agency (the latter is generally used for measurements in strong pulsed magnetic fields). The voltage pulses are adjusted to the right size by means of the potentiometer 2 and fed to the input of the current amplifier 3, which constitutes a "composite transistor." A P-201A transistor is used in the first stage and a P-209A in the second; in the third output stage there are ten P-209 and P210 transistors connected in parallel. The total current amplification factor is 10^5. In order to even out the currents of the output transistors T_3-T_{12}, a resistance of 0.05 Ω is connected to the emitter circuit of each. Experience showed that in this way the currents of the individual transistors differed from one another by no more than 10-15%. The amplifier is fed from a starter storage battery and produces current pulses increasing linearly with time up to an amplitude of 200 A. The sample supply circuit is connected to the input of the current amplifier, in series with a resistance of about 0.005 Ω; the voltage drop in this (proportional to the current flowing through the sample) is passed to one of the oscillograph inputs. The sweep is triggered and the oscillograph beam intensified by corresponding pulses from the generator, these being generated at the same time as the initial rise in the current through the sample.

The voltage from the potential leads of the sample is passed through a transformer to the other oscillograph input. At the instant at which the superconductivity of the sample is destroyed, a voltage pulse appears on the oscillograph screen. The position of this pulse relative to the current pulse determines the critical current of the sample in any magnetic field specified (or in the absence of a field). If the contacts deteriorate (so that local heating occurs and the superconductivity of the sample is only destroyed at its extremities), the critical current varies with the duration of the current pulse passing through the sample. This circumstance may be exploited in order to determine the rate of propagation of the normal phase in the superconductor.

In view of the shortness of the current pulses (usually 1-10 msec), this kind of apparatus may be successfully used in testing samples of superconducting wire at high current densities, since the breakdown of superconductivity in this case causes no damage to the samples.

The circuits described earlier for measuring H_{c2} at low current densities in steady fields and at high current densities may easily be combined. In this case the current situated in the pulse solenoid passes a measuring current I_1 at a frequency of 35 kHz and also a current pulse I_2 from the generator just described. The minimum pulse length ($\tau = 1$ msec) is employed, this being much shorter than the growth time of the magnetic field. A signal proportional to I_2 is passed to one input of the oscillograph, and an out-of-balance signal recording the appearance of a resistance in the sample to the other; the horizontal sweep is achieved by means of a voltage proportional to the magnetic field.

In this way, by varying the current and magnetic field (for example, by varying the instant at which the current pulse is fed into the sample relative to the magnetic-field pulse), we may obtain the relation between the critical current and the magnetic field for the material under consideration over the whole range of desired currents.

LOW-TEMPERATURE TECHNIQUE

A common characteristic of most work with superconducting devices and materials is the necessity of cooling these to temperatures lower than 20°K (methods of producing and measuring such temperatures are indicated in Table 1 on p. 38). Usually this is achieved by means of liquid helium; some of the physical properties of this and other gases are presented in Table 2.

TABLE 2. Physical Properties of Gases* [129]

Gas	Boiling point, °K (°C)	Mass of liquid, g	Volume of gas, liters	Heat of vaporization, kJ
Helium	4.2 (—269)	125.2	700	2.72
Hydrogen	20.4 (—252.8)	70.8	780	31.7
Nitrogen	77.3 (—195.9)	808	650	162
Oxygen	90.1 (—183.1)	1140	799	244
Carbon dioxide (solid)	194.6 (—78.6)	1630	829	935

*Values relate to one liter of liquid.

Work at low temperatures is carried out in cryostats and devices designed to ensure stable low-temperature conditions. A widely-used form of cryostat suitable for cooling experimental devices with minimal heat loss is illustrated in Fig. 21.

The test system is placed in the Dewar vessel. The helium Dewar consists of two concentric glass cylinders, the space between these being evacuated to a high vacuum in order to reduce heat transfer. Helium is poured into this Dewar. The helium Dewar in turn is placed in liquid nitrogen held in another double-walled glass Dewar with vacuum insulation. The inner and outer surface of the glass Dewars are usually silvered to reduce radiative heat loss. Slits are left in the silver coating to allow observation of the helium level. Thus between the medium at room temperature and the liquid helium there are four glass walls, four silvered screens, two evacuated spaces, and a screen of liquid nitrogen. Gaseous helium diffuses through the glass (special chemical glass is used for making the helium Dewars).

A study of the heat flowing to the helium showed that, in the case of well-filled Dewars, the heat transfer through the walls (i.e.,

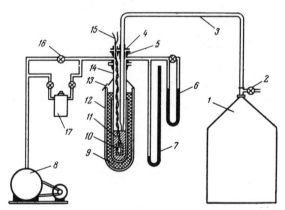

Fig. 21. Arrangement of a laboratory apparatus for working at liquid-helium temperatures: 1) Dewar for storing liquid helium; 2) gaseous helium circuit; 3) transfer line; 4) plate; 5) bushing with separation tube; 6) oil manometer; 7) mercury manometer; 8) vacuum pump; 9) liquid nitrogen; 10) sample; 11) liquid helium; 12) outer Dewar; 13) cone; 14) inner Dewar; 15) leads; 16) valve; 17) manometer.

through either the solid or the gas) was very close to zero. The main heat leakage occurs through the sample support and the electrical leads, and is associated with radiation from the upper part of the cryostat. Special measures may be taken to reduce the losses: cryostats of more complex configuration may be used, the sample support being cooled with helium vapor and the separating sleeve screened with liquid nitrogen; the losses may then be reduced to 10^{-2} liter of liquid helium per hour or even less [129]. In order to prevent the freezing of moisture from the air on the liquid-nitrogen Dewar, rubber cones (situated between the upper parts of the outer and inner Dewars) are employed.

The absolute pressure is measured with a standard mercury manometer. A differential oil manometer serves to detect any deviations from the pressure specified. A vacuum pump reduces the vapor pressure over the helium bath, in order to obtain any specified temperature. The valve and Descartes manostat provided in the installation control the flow of helium vapor through the vacuum pump in order to remove whatever amount of vapor is required to create the desired pressure from the boiling liquid helium.

The temperature in the cryostat is determined by reference to the vapor pressure. A number of other methods are employed for measuring temperature in low-temperature technology (Table 1).

The supply of helium is maintained by means of portable Dewars specifically made for this purpose.

The procedure for the filling of the cryostats requires a great deal of care owing to the extremely low heat of vaporization. For transferring the helium special transfer lines are used; these consist of two concentric stainless-steel tubes with a low thermal conductivity. The space between the tubes is evacuated. The helium is transferred into the cryostat after the outer Dewar has been filled with liquid nitrogen and the inner one with gaseous helium. One end of the siphon is let down into the cryostat and the other into the Dewar storing the liquid helium (see Fig. 21). The transfer line has to be lowered slowly, so that the cold gas evolving by virtue of the evaporation of the liquid helium may serve to cool the warm transfer line. The gaseous helium, under a slight pressure applied to the liquid-helium storage Dewar, makes the liquid helium flow into the inner Dewar of the cryostat. The level

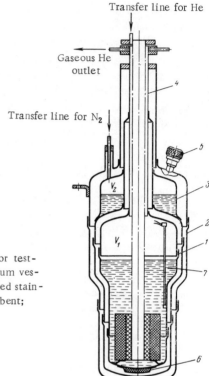

Fig. 22. Arrangement of a metal cryostat for testing superconducting solenoids [132]: 1) helium vessel; 2) body; 3) nitrogen bath; 4) thin-walled stainless steel tube; 5) evacuation line; 6) adsorbent; 7) level indicator.

of the liquid is usually observed through the slits in the coating of the Dewar. More complex helium-level indicators are frequently used, however, for this purpose [129a]. It is absolutely essential to use a helium-level indicator when working with metal cryostats. A detailed exposition of methods of working at low temperatures has been given in [130, 131].

At the present time metal Dewars are widely employed instead of the glass variety in view of their greater strength. In addition to this, metal Dewars offer greater scope when constructing complicated apparatus. The arrangement of a metal cryostat is presented in Fig. 22. The helium vessel 1 is suspended in the body 2 by a thin-walled stainless-steel tube 4, with an internal diameter of up to 100 mm. Solenoid sections are mounted in the lower part of the helium vessel. If the external diameter of the solenoid is anything up to 100 mm, it is placed in the helium vessel from the top by means of tube 4; if it is larger than 100 mm it is mounted in the helium vessel during assembly. In the latter case the diameter of the outlet tube may be reduced to a value equal to the working gap of the solenoid; this reduces the flow of heat along the tube to the helium bath. The heat insulation of the solenoid consists of high vacuum with a screen cooled in liquid nitrogen, which is transferred into bath 3. The nitrogen bath is suspended in the body by means of three thick-walled stainless steel tubes. The vacuum in the insulation space is created by periodic evacuation through the conduit 5 and maintained during operation by means of the adsorbent 6. The level of helium in the cryostat is checked during operation by reference to the indicator 7. In order to reduce the heat leak to the helium as a result of heat conduction through the exit tube 4, the latter has its upper part in contact with the nitrogen bath.

The principal components of the cryostat (helium vessel, nitrogen bath, casing) are made of sheet copper M3 or copper tubing. The surface of these is polished on the side of the vacuum

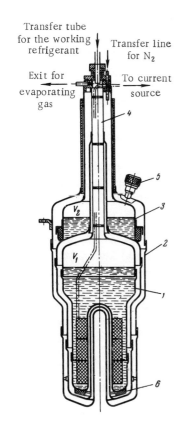

Fig. 23. Arrangement of a metal cryostat for superconducting solenoids with a working space at room temperature (notation as in Fig. 22) [132].

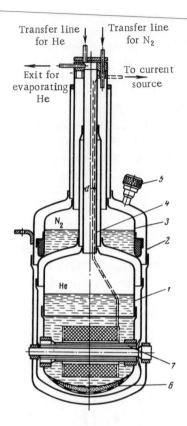

Fig. 24. Arrangement of a metal cryostat with a horizontal working space. 7) Multilayer insulation made of metallized plastic film; rest of notation as in Fig. 22.

space. The suspension tubes of the helium vessel and the nitrogen bath are made of 1Kh18N9T stainless steel and the machined parts of brass. The connecting seams are soldered with POS-50 (and, partly, with PSr-45) solder. The vacuum-tightness of the units is verified by means of a leak detector.

In many cases it is desirable to have the working space of the magnetic field at room temperature. Special cryostats exist for this purpose. Figure 23 shows the construction of a cryostat in which the working space (at room temperature) is placed vertically along the axis of the solenoid and open at the bottom (KR-22). The construction of a cryostat with a horizontal working space is given in Fig. 24 (KR-27) [132]. In the KR-27 cryostat the working space of the solenoid constitutes an open passage with its axis in the horizontal plane. For an internal solenoid diameter of 50 mm the size of the working passage at room temperature is 30-35 mm. In these cryostats the nitrogen bath is suspended on a single central tube. The nitrogen screen protecting part of the working space of the solenoid is replaced, in the KR-27 cryostat, by a multilayer insulation consisting of a metallized plastic film. The remaining parts are the same as in the cryostat illustrated in Fig. 23.

In all cases in which the sample at helium temperature is required to be placed in an external magnetic field at room temperature, special cryostats with side tubes are employed.

Apart from copper cryostats, Dewars made of stainless steel [132a] are employed for experimental investigations and in particular for technological purposes.

METALLOGRAPHY OF SUPERCONDUCTING ALLOYS

Methods of melting superconducting alloys do not differ from existing methods of producing refractory materials. The most widespread methods of melting refractory metals and determining impurities in these were set out in detail in a monograph by Savitskii and Burkhanov [133, 133a].

1. Preparation of Microsections

Samples for metallographic analysis of the structure should be cut with the least possible pressure so as to minimize possible deformation. It is essential to avoid overheating samples of alloys in which this might cause structural changes or contamination with interstitial impurities.

In preparing microsections of a superconducting wire or other parts with fine cross sections, these are placed in a mounting and covered with Wood's alloy or sulfur, or else pressed into polystyrene; in the latter case contamination of the surface of the microsection while etching is prevented. Then coarse and fine mechanical grinding and polishing follow. In processing the surface with abrasive papers it is important to check that scratches have been completely removed before passing to the next grade of paper with a smaller grain size. After final grinding on the paper

TABLE 3. Composition of Electrolytes and Conditions for the Electropolishing of Niobium, Vanadium, and Their Alloys

Polished metal	Composition of electrolyte	Voltage, V	Current density, A/dm²	Temperature, °C	Time, min	Notes
Niobium and its alloys	1) 175 ml HF (40%) 175 ml HNO₃ 650 ml H₂O	12-20 (in circuit)	20-35	50	10	As cathode a platinum crucible containing the solution is employed. Cooling is essential. On applying a potential the current density rises to 40 A/dm², then falls to the values indicated
	2) 150 ml HF (48%) 650 ml H₂SO₄ (sp. gr. 1.84)	15-25	4	25-60	5-10	
	3) 90 pts. conc. H₂SO₄ + 10 pts. conc. HF	15-20	80-130	35-40	20-30	The polishing time largely depends on the state of the surface. The electrodes are carbon or platinum
Zirconium and its alloys	1) 100 ml acetic (glacial) acid, 50 ml chloric acid (sp. gr. 1.60)	60	60-80	20	3-5	During the polishing operation the sample is to be rotated. The cathode is stainless steel
	2) 20 ml HF 10 ml HNO₃ 200 ml glycerin	9-12	20-30	24	5-10	
Vanadium and its alloys	1) 10% H₂SO₄ in water		250	22		The electrolyte is suitable for alloys in all states
	2) 20 ml H₂SO₄ (sp. gr. 1.84), 80 ml methyl alcohol, a few drops of water	10-20	20-40	20-50	A few seconds	
	3) 5-10 ml chloric acid, 95-90 ml acetic acid (glacial)	25-30	15-23	35	1-2	Concentrated chloric acid does not react with carbon-containing materials.

TABLE 4. Composition of Etchants and Etching Conditions for Microsections of Vanadium and Niobium Alloys

Metal	Composition of etchant	Notes
Niobium and its alloys	1) 1 pt. HF (48%) + 1 pt. H_2SO_4 (conc.) + 1 pt. H_2O + several drops of H_2O_2	Reveals grain sub-boundaries and boundaries
	2) 2 pts. HF + 18 pts. HNO_3 + 5 pts. glycerin	Chemical etching
	3) 10 pts. HF (conc.) + 90 pts. H_2SO_4 (conc.)	Electrolytic etching of niobium at 30-40°C. First stage: V = 1-2 V, I = 0.06 A; time 2 min; second stage: V = 5-6 V, I = 0.3 A; time $\frac{1}{4}$ to $\frac{1}{2}$ min. Reveals dislocation structure of niobium. Etchant of the same composition suitable for electrolytic etching of Nb−Zr alloys at V = 50 V, etch time 30 sec, current density 100-200 A/cm^2, room temperature
	4) 75 pts. HNO_3 + 25 pts. HF; 80 pts. HNO_3 + 10 pts. HF + 10 pts. H_2SO_4	Chemical etching of Nb−Zr alloys
	5) 50 pts. H_2O_2 + 25 pts. HNO_3 + 25 pts. C_2H_5OH + 1 pt. HF	Chemical etching of Nb−Zr alloys. Cellular structure found in Nb−50% Zr alloys
	6) 170 ml 70% HNO_3 + 50 ml 40% HF + 5 ml citric acid + 510 ml methyl alcohol	Electrolytic etching at a current density of 495-605 A/dm^2
Vanadium and its alloys	1) 5% solution of H_2SO_4 in water	Electrolytic etching for 1-5 sec. Electrodes stainless steel
	2) 5% solution of oxalic acid in water	Electrolytic etching for 5-10 sec.
	3) 1 pt. HNO_3 + 1 pt. acetic acid + 1 pt. H_2O	Chemical etching for 3-7 sec.
	4) 10 g $FeCl_3$ + 10 ml HCl + 90 ml H_2O	Chemical etching for 7-10 sec.

with the smallest grain size the sample should be washed thoroughly in running water and polished on cloth with chromium or aluminum oxide moistened in water. Polishing is carried out until grinding traces have been removed.

Mechanical polishing may be replaced by electrolytic or chemical polishing of the samples. This avoids creating a work-hardened layer on the surface and thus reveals the true structure. The quality of the polished surface depends on the nature of the metal or alloy, the composition of the electrolyte, the working parameters (temperature, current density, period of operation, agitation of the electrolyte), and the state of the surface.

The composition of the electrolytes and the conditions for the electropolishing of niobium, vanadium, zirconium, and their alloys, which include the greatest number of industrially important superconductors, are presented in Table 3 [133, 134, 137].

2. Etching of the Microsections

There are a number of methods used for revealing structure by etching: chemical dissolution, vacuum etching, electrolytic etching, highlighting of the structure by polishing, and so forth.

Table 4 presents the composition of etchants for niobium, vanadium, and Nb–V alloy microsections. In the electrolytic etching of microsections the electrolytes indicated in Table 3 for electropolishing may be used but at current densities lower than the electropolishing values. For the chemical etching of alloys of niobium with tungsten, molybdenum, and other elements a mixture of concentrated HF and HNO_3 is employed, the proportions of these depending on the composition of the alloy. For etching superconducting alloys based on other metals the corresponding etchants already mentioned in the literature may be used [133]. Thus for the chemical etching of tungsten and its alloys we may use a 3% boiling solution of H_2O_2 or the fast-acting Murakami reagent (5 cm^3 of 10% $K_3[Fe(CN)_6]$ + 0.5 g KOH); for the chemical etching of tantalum we use a mixture of hydrofluoric and nitric acids; for the etching of molybdenum, apart from a 2-10% aqueous solution of NaOH an ammoniacal solution of copper sulfate may usefully be employed.

The metallography of superconducting metals with low melting points ("soft" superconductors) and their alloys has been thoroughly described in existing treatises on aluminum [141-147, 150], beryllium [141, 142], lead [142-144, 150], indium [142], tin [142-144, 150, 151], zinc [141-143, 149, 150], the rare-earth metals and their alloys [151, 152], bismuth [153], and gallium [154].

There have recently also been some interesting investigations into the use of anodic color etching to reveal the phase composition of superconducting alloys of the Nb-Zr, V-Ga, and Nb-Sn systems [155], 156, 159-161]. A well-polished microsection forms the anode, a steady voltage being applied to this. The microsection then acquires thin oxide films on the surface; these produce interference in visible light, the coloring of the films depending on the light wavelength. The thickness of the oxide film and the coloring corresponding to this depend on the chemical composition of the sample, the electrolyte, the applied voltage, and the period of anodic oxidation. The thickness of the oxide films is also affected by the cathode material, the uniformity of the distribution, and the distance between the electrodes.

The kinetics of the decomposition of the β solid solution in alloys of the Nb-Zr system were studied earlier [156]. After mechanical polishing of the samples and electrolytic etching, the samples were subjected to anodic oxidation. For anodic oxidation an electrolyte of the following composition was employed: 4 g citric acid, 10 ml orthophosphoric acid, 20 ml lactic acid, 40 ml glycerin, 70 ml water, and 120 ml ethyl alcohol. The voltage was chosen in such a way as to ensure the best possible contrast between the base (wine red) and the precipitating phase (blue-green or yellow); the voltage was 18-22 and 55-65 V. In order to increase the contrast for microscope analysis, polarized light was employed.

A method of color etching alloys of the V-Ga system was described by Maier. Electrolytes based on acetic acid or alcohol were unsatisfactory, since these led to the preferential etching of the gallium-rich phases, and the adhesion of the oxide layers to the base was in the majority of cases inadequate. Electrolytes comprising glycerin with sodium thiosulfate ($Na_2S_2O_3$), anhydrous sodium silicate (Na_2SiO_3), sodium tetrabromate ($Na_2B_4O_7 \cdot 10H_2O$), or potassium metasulfite ($K_2S_2O_5$) are in principle suitable for a voltage of about 25 V. The best results are obtained with an electrolyte consisting of 100 ml of glycerin and 1 g of potassium metasulfite. Microsections treated in this electrolyte (for 1 min) have

TABLE 5. Colors of Phases in the V–Ga System After Anodic Oxidation

Phase	Composition, at.%	Crystal structure and lattice constants, Å	Coloring in anodic oxidation
VGa$_{\sim 4}$	< 20	Distorted type NiHg$_4$	Practically uncolored, white
V$_2$Ga$_5$	28.6	Mg$_2$Hg$_5$, tetragonal $a = 0.965$, $c = 2.70$	From ivory to brownish yellow
V$_6$Ga$_7$	46.2	Cu$_5$Zn$_8$, cubic $a = 9.17$	Brownish yellow of various tones
V$_6$Ga$_5$	54.6	Ti$_6$Sn$_5$, hexagonal $a = 8.46$, $c = 5.16$	From brown to violet brown
V$_6$Ga$_5$O$_x$	~55	NaCl, cubic, $a = 3.94$	Slightly lighter than V$_6$Ga$_5$
V$_3$Ga$_2$O$_x$	60	β-Mn, cubic, $a = 6.62$	From light violet to rose
V$_5$Ga$_3$O$_x$	62.5	Mn$_5$Si$_3$, hexagonal $a = 7.24$, $c = 4.89$	Dark violet
V$_5$(Ga, Si)$_3$O$_x$	~63	Mn$_5$Si$_3$, hexagonal $a = 7.29$, $c = 4.93$	Golden brown
(V)$_2$ V$_3$Ga	From ~62 to 75	Body-centered cubic	Violet blue
V$_3$Ga	~75	Cr$_3$Si, cubic, $a = 4.82$	From violet to light violet with increasing vanadium content
V$_3$GaO$_x$	~75	Orthorhombic, $a = 2.84$ $b = 4.59$, $c = 4.44$	From light violet to rose
V$_4$(Ga)O	~80	Martensite tetragonal $a = 3.01$, $c = 3.81$	Dark rose
V and solid solution of vanadium	~75 to 100	Body-centered cubic; a from 3.028 to 3.045	From turquoise to blue with increasing vanadium content
VO	~50	NaCl cubic, $a = 4.093$	Light blue

a satisfactory color contrast (Table 5). Double washing of the treated microsections in alcohol removes traces of electrolyte without harming or altering the oxide films.

The voltage required depends on the specified color contrast and the coloring constant of the phases required for their recognition. Alloys of different compositions require different voltages in order to achieve the best results.

The period of anodic oxidation required is determined by noting the point at which no further growth of the oxide film takes place and the current passing through the sample decreases no further.

Maier and Hoffman [155] note that the electrolytes used for alloys of the Nb–Sn system [159-161] are unsuitable for V–Ga, since the oxide films formed are dissolved in the majority of aqueous electrolytes. The method of anodic oxidation described in the paper in question [155] is suitable for pure alloys not containing oxygen.

Certain authors have obtained different colorings of the phases formed by anodic etching in the Nb–Sn system. The electrolyte employed constituted a solution of ethyl alcohol (90 cm^3), 85% lactic acid (30 cm^3), glycerin (15 cm^3), phosphoric acid (7.5 cm^3), distilled water (52 cm^3), and citric acid (3 g).

Figure 25 shows a photograph of the color microstructure of a Nb−40wt.%Sn alloy in the cast state [161a] obtained by etching in a 1% solution of orthophosphoric acid [161b] at a voltage of 24 V for 1 min. Table 6 gives the colors of the phases in the Nb–Sn system.

Fig. 25. Microstructure of a cast Nb−40 wt.%Sn alloy: 1) Nb$_3$Sn (red-brown); 2) carbide Nb$_2$C (yellow); 3) tin (bright green).

TABLE 6. Colors of the Phases in the Nb–Sn System after Anodic Etching

Phase	Nb content, at.%	Crystal structure and lattice parameters [161c]	Color of phases on anodic etching
Nb-base solid solution	100-90	bcc, $a = 3.12\text{-}3.29$ Å	Light blue
Nb_3Sn	75	A-15, $a = 5.28\text{-}5.29$ Å	Lilac to red-brown
Nb_6Sn_5	66	Orthorhombic, $a = 5.6549$, $b = 9.2307$, $c = 16.814$	Orange
$NbSn_2$	33	Orthorhombic, type $CuMg_2$ $a = 5.65$, $b = 9.85$, $c = 18.9$	Similar color to Nb_6Sn_5
Tin	0	Tetragonal, $a = 5.8197$ $c = 3.1749$, $a/c = 0.5455$	Light yellow to green

In one case [161d] the color etching of the Nb–Ga system was carried out by means of a solution consisting of ethyl alcohol, water, and phosphoric acid in the ratio 50:10:1 by volume. The voltage in etching was 80 V. The colors of the phases in the Nb–Ga system are given in Table 7.

The process of color etching is a simple and fairly reliable method of determining phases without supplementary x-ray analysis, measurements of microhardness, and so on.

3. Study of the Microstructure and Properties of Alloys

The structure of metals and alloys may be studied under the microscope by the reflection of light from the surface of the microsection. The light (so-called white) microscope is the most effec-

TABLE 7. Colors of the Phases in the Niobium–Gallium System after Anodic Oxidation

Phase	Nb	Nb_3Ga	Nb_3Ga_2	NbGa	Nb_2Ga_3	$NbGa_3$
Color	Light green	Blue-green	Violet	Orange	Yellow-orange	Light yellow

tive means of studying the microstructure of metals and alloys. In recent years, however, other methods have begun to develop; these include color and vacuum metallography, microscopy in polarized light, infrared and ultraviolet microscopy, and the use of electron, ion, neutron, and proton microscopes, which are capable of revealing even individual atoms and vacancies in the crystal lattice of refractory metals [162].

The light microscope has a resolving power of 2000 Å, and, apart from determining the phase composition, the phase distribution, and the shapes and sizes of the various phases, reveals the following substructural details: the shape and size of the subgrains, the shape and arrangement of the etch pits, and also the characteristics of twins and deformation bands. The electron microscope working in the reflection mode resolves details down to 100 Å or under.

The microstructure of alloys may be conveniently studied in the first place by considering an unetched microsection. Such microsections exhibit small regions of dark color on a light background, revealing traces of nonmetallic inclusions formed in the alloy or left in the structure during preparation. Because of their brittleness, nonmetallic inclusions may break away from the alloy completely unless very carefully polished. Polarized light is best for studying the characteristics of nonmetallic inclusions in the microscope. An examination of an unetched microsection also reveals any microporosity, a defect frequently encountered in castings of superconducting alloys, which seriously affects the possibilities of subsequent working (rolling, extrusion, etc.).

Apart from ordinary vertical illumination, oblique illumination may also be used in the metallographic microscope; this is convenient for superconducting alloys containing hard particles or phases situated in a soft base. Oblique illumination is used for "dark-field" observations [163].

The structure of metal and alloy microsections becomes much clearer after the microsection has been etched. The underlying principle is the different etching velocity of individual regions of the metal or alloy with different chemical or physical constitutions. The grains of pure metals and single-phase solid solutions have different crystallographic orientations but the same chemical composition, and on etching the degree of dissolution of

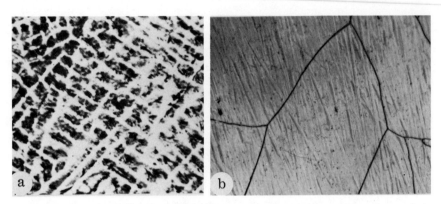

Fig. 26. Niobium—25% zirconium—25% titanium alloy in the cast (a) and annealed (b) states (magnification 300).

the different grains may differ substantially; the degree of dissolution of different phases in the alloy may also be different. Grain boundaries accommodating aggregates of impurities dissolve more readily and appear black under the microscope.

In the cast state solid solutions usually have a dendritic structure, this being characteristic of solid crystals precipitated from a liquid. In the course of crystallization the composition of the solid phase crystallizing on the surfaces of solid-solution nuclei gradually changes. Since the rate of etching in different reagents depends substantially on the concentration of one of the components in the solid solution, the structure of etched dendrites of nonuniform composition reflects light to varying extents (Fig. 26a). A dendritic liquation may frequently be distinguished at once from a true two-phase structure by analyzing the manner in which the etching intensity changes on passing from the light to the dark regions. The question as to the number of phases in the cast alloy may finally be solved by studying its structure after heat or thermomechanical treatment.

In order to make the composition of cast superconducting alloys more uniform a homogenizing anneal is usually applied (the alloys are annealed at a temperature close to the solidus). Figure 26 shows the microstructure of an alloy after a homogenizing anneal (1500°C) and subsequent quenching. We see that the alloy constitutes a single-phase solid solution — after annealing the liquation has vanished. If diffusion processes take place very slowly, the homogenizing anneal may prove ineffective in the treatment of cast alloys. In this case thermomechanical treatment is required, since the plastic deformation accelerates diffusion processes on subsequent annealing.

Figure 27a illustrates the microstructure of an Nb−50%Zr alloy containing 0.09% of oxygen; individual grains divided by a closed lattice of subgrains may readily be seen. The sub-boundaries are etched less strongly. The boundaries of the subgrains often intersect the primary grains at right angles, indicating the very low surface tension at the boundaries of the subgrains. The subgrains constitute regions differing from one another by virtue of slight disorientations. The conditions required for the formation of sub-boundaries have been fully investigated elsewhere [164]. In cast materials a substructure arises during crystallization from the melt, and the chief surfaces of separation with small disorientation angles are usually oriented in the direction of heat transfer. At large magnifications the boundaries of the subgrains appear as individual spots or etch pits, apparently representing the points at which individual dislocations come to the surface [164]. Slow cooling helps in revealing the sub-boundaries by etching, this effect evidently being associated with the segregation of interstitial impurities along the boundaries of the subgrains. The boundaries of the subgrains may remain undetected (Fig. 27b) if there are no impurities present to accelerate the rate of etching of the boundaries with small disorientation angles.

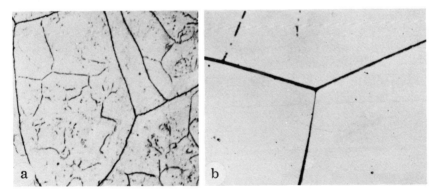

Fig. 27. Microstructure of a niobium−50% zirconium alloy containing 0.09% O_2 (a) and of a niobium −5% zirconium alloy containing 0.01% O_2 (b) after zone melting (magnification 300).

More-detailed information as to the phase and structural state of superconducting materials may be obtained by an electron-microscope examination at a magnification of some 100,000 times. Using transmission electron microscopy, the authors studied the phase transformation of the solid solution in a Nb−75at.%Ti alloy [164a]. It was discovered that in order to obtain a critical current density of the order of $1 \cdot 10^5$ A/cm^2 the optimum size of the α-phase particles had to be of the order of 10^{-5} cm, and the number of particles 10^{11} cm^{-2}. This size and density of α-phase particles was the most suitable for fixing the lines of magnetic flux.

The capacity of alloys to undergo plastic deformation is one of the most important conditions in connection with the production of a superconducting wire, strip, cable, etc. Plastic deformation is used both for the purpose of imparting a specific shape to metals and alloys (wire for winding superconducting solenoids, strip for making superconducting screens) and also for the purpose of changing the superconducting properties of alloys. Thus it is quite well known that the critical current density of superconducting alloys increases with increasing degree of deformation.

The principal mechanism of deformation in individual crystals is slip. Slip constitutes the displacement of one part of the crystal relative to another along a certain plane, called the slip plane, one or several families of possible slip planes usually being available. In addition to preferential slip planes, preferential slip directions also exist. In superconducting alloys constituting solid solutions with a body-centered lattice, the slip direction always coincides with the $\langle 111 \rangle$; however, slip occurs in any of the three slip planes (110), (123), or (112). With increasing degree of deformation a preferential orientation (texture) gradually develops, and it becomes more difficult to reveal the structure of the superconducting alloys by etching. The grains are not only drawn out in the direction of deformation but are also liable to change from an arbitrary orientation to one lying in a preferential relationship to a specific direction, which coincides with the principal direction of deformation.

Figure 28a represents the texture of 99.8% cold-worked Nb−50%Zr wire and Fig. 28b the microstructure of the same wire in the recrystallized state (after annealing at 1050°C). Recrystallization is a process in which new, stress-free crystals are created and grow in the existing system of grains, the latter having become unstable as a result of deformation.

The recrystallization temperature depends substantially on the introduction of alloying additives. The considerable effect of small traces of additives entering into the solid solution may be explained as being due to their preferential segregation along grain and subgrain boundaries, as a result of which the latter become immobile [77]. Since the recrystallization process is associated with the migration and annihilation of these interfaces, any factor constituting an obstacle to their motion will raise the recrystallization temperature.

A change in the chemical composition of the alloys produces a change in the numerical relationship between individual phases and structural constituents. Microanalysis enables us to determine the number of phases and structural constituents in the alloys if the latter are in an equilibrium state; by using the lever rule, the chemical composition of the alloy may be determined to a fairly high accuracy, i.e., we may say in advance what volume of superconducting phase is required in any specified alloy in order to ensure that a particular method of measuring its critical superconducting temperature will be effective. Thus the magnetic method of measuring the critical superconducting temperature is sensitive to the volume of superconducting phase present, the signal derived from this method being directly proportional to the relative amount of superconducting phase in a two-phase alloy. In measuring the

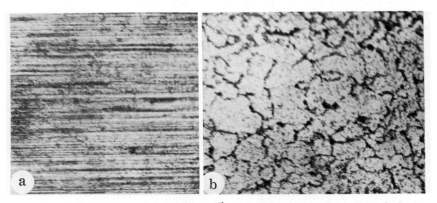

Fig. 28. Microstructure of a niobium—50% zirconium alloy in the cold-worked (a) and recrystallized (b) states (magnification 300).

critical temperature reliable results may be secured even if the amount of superconducting phase in the alloy is lower than $0.5 \cdot 10^{-4}$ cm^3 [112].

Physicochemical analysis starts from the concept that a change in the physical (including superconducting) properties of materials is primarily determined by a change in their internal structure. In view of this a study of the structure of superconducting materials is of fundamental significance. The most important information regarding the structure of the lattice is that given by x-ray diffraction down to helium temperatures. Special low-temperature apparatus is required for x-ray work in this temperature range. X-ray cryostats require attachments for the introduction of the primary beam and the extraction of the reflected beam, as well as devices for adjusting and rotating the sample. Existing low-temperature x-ray apparatus and methods of low-temperature analysis are described in [165-167]. In recent years there have been a number of investigations into the low-temperature x-ray study of compounds with the A-15 structure [168-170] and also low-temperature microstructural analysis [171]. Methods of low-temperature electron-microscope examination have also been devised [172].

Our understanding of the nature of superconducting materials is greatly aided by a study of their physical properties before and after the transition into the superconducting state, particularly such properties as the specific heat, the Debye temperature, the thermal expansion coefficient, the elastic constants, and the thermal conductivity. Measuring methods and some of the most interesting results are set out in several references [173-180].

The most reliable results are obtained by measuring a whole set of physical properties at the same time for one particular sample. At the present time a special apparatus is being developed in the Laboratory of Rare Metals and Alloys of the Institute of Metals, Academy of Sciences of the USSR; this is called the "metallographical combine" (experimental version) for simultaneously studying the structure, electrical resistance, initial melting temperature, and other properties.

At the present time we are particularly interested in the methods of mathematical statistics for studying superconducting alloys, since these enable us to use computer techniques. A method of predicting the composition, properties, and conditions of formation of various compounds based on the electron structure of the atoms and the principal physical properties of the components, developed by Savitskii and others [181], offers considerable promise. Methods of using experiment planning and computer technology in order to secure the composition–T_c diagrams of ternary superconducting systems of elements and compounds are also being developed under the direction of Savitskii [182]. These methods should substantially reduce the volume of experimental investigations required.

LITERATURE CITED

1. H. Kamerlingh-Onnes, Communs Phys. Lab., Univ. Leiden, 120b, 122b (1911).
2. W. Meissner and R. Ochsenfeld, Naturwissenschaften, 21:787 (1933).
3. B. Roberts, in: New Materials and Methods of Studying Metals and Alloys, Izd. Metallurgiya, Moscow (1966), p. 9.
3a. B. T. Matthias, T. H. Geballe, and L. D. Longinott, Science, 156:645 (1967).
4. H. Kamerlingh-Onnes, Communs Phys. Lab. Univ. Leiden, 133b, 134b (1913).
5. H. Kamerlingh-Onnes, Communs Phys. Lab. Univ. Leiden, 139b (1914).
6. W. J. de Haas and J. Voogd, Communs Phys. Lab. Univ. Leiden, 208b (1930); 214 (1931).
7. W. H. Keesom, Communs Phys. Lab. Univ. Leiden, 234b (1935); Physica, No. 35, p. 2 (1935).
8. J. E. Kunzler, Rev. Mod. Phys., 33:1 (1961).
9. E. Maxwell, Phys. Rev., 78:477 (1950).
10. C. A. Reynolds, B. Serin, W. H. Wright, and B. Nesbitt, Phys. Rev., 78:478 (1950).
11. H. Fröhlich, Phys. Rev., 79:845 (1950).
12. J. Bardeen, Phys. Rev., 80:567 (1950).
13. L. N. Cooper, Phys. Rev., 104:1189 (1956).
14. J. Bardeen, L. N. Cooper, and J. Schrieffer, Phys. Rev., 106:162; 108:1175; (1957).
15. N. N. Bogolyubov, V. V. Tolmachev, and D. V. Shirkov, New Methods in the Theory of Superconductivity, Izd. AN SSSR, Moscow (1958).
16. V. L. Ginzburg and L. D. Landau, Zh. Éksp. Teor. Fiz., 20:1064 (1950).
17. A. A. Abrikosov, Zh. Éksp. Teor. Fiz., 32:1442 (1957).
18. L. P. Gor'kov, Zh. Éksp. Teor. Fiz., 37:835 (1959).
19. B. B. Goodman, IBM J. Res. Development, 6:63 (1962).
20. B. B. Goodman, Phys. Lett., 1:215 (1962).
21. F. B. Silsbee, J. Wash. Acad. Sci., 6:597 (1916).
22. J. Bardeen, IBM Journal, 6:3 (1962).
23. E. Saur and H. Wizgall, Les Champs Magnetiques Intenses, Colloque Internat., Grenoble (1966), p. 223.
24. P. W. Anderson, Phys. Rev. Lett., 9:309 (1962).
25. K. Mendelssohn, Proc. Roy. Soc., A152:34 (1935).

26. A. A. Abrikosov, Priroda, No. 5, p. 26 (1964).
27. A. A. Abrikosov, Uspekhi Fiz. Nauk, 87(1):125 (1965).
28. L. P. Gor'kov, Zh. Éksp. Teor. Fiz., 34:735 (1958).
29. P. London and H. London, Proc. Roy. Soc., A 149:71 (1935).
30. A. A. Abrikosov, Dokl. Akad. Nauk SSSR, 86:489 (1952).
31. V. P. Ginzburg, Zh. Éksp. Teor. Fiz., 42:299 (1962).
32. W. H. Kleiner, L. M. Rotti, and S. H. Autler, Phys. Rev., 133:A1226 (1964).
33. B. B. Goodman, Phys. Rev. Lett., 6:597 (1961).
34. D. Cribier et al., Phys. Lett., 9:106 (1964).
34a. H. Trauble and U. Essmann, Phys. Status Solidi, 25(1):395 (1958).
34b. A. Selger, Metallurgical Transactions, 1:2991 (1970).
35. V. V. Shmidt, in: Metallography and Metal Physics of Superconductors, Izd. Nauka, Moscow (1965), p. 7.
35a. A. E. Bindary and M. M. Litvak, J. Appl. Phys., 34(9):2913 (1963).
36. C. J. Gorter, Phys. Lett., 1:69 (1962).
37. R. A. Camper, Phys. Lett., 5:9 (1963).
38. W. Klose, Phys. Lett., 8:12 (1964).
39. D. Saint-James and P. G. de Gennes, Phys. Lett., 7:306 (1963).
40. P. W. Anderson, Phys. Rev. Lett., 9:309 (1962).
41. J. B. Kim, C. F. Hempstead, and A. R. Strand, Phys. Rev., 131:2486 (1963).
42. D. J. Quinn and W. P. Ittner, J. Appl. Phys., 33:748 (1962).
43. Yu. F. Bychkov, I. I. Goncharov, V. I. Kuz'min, and I. S. Khukhareva, Pribory i Tekh. Éksp., 3:170 (1964).
44. Yu. F. Bychkov, I. I. Goncharov, and I. S. Khukhareva, Zh. Éksp. Teor. Fiz., 48:818 (1965).
45. M. I. Bychkova, V. V. Baron, and E. M. Savitskii, in: Metallography, Physical Chemistry, and Metal Physics of Superconductors, Izd. Nauka, Moscow (1967), p. 48.
45a. D. Kramer and G. Rhodes, Trans. AIME, 239(10):1612 (1967).
46. V. V. Baron and M. I. Bychkova, in: Metallography, Physical Chemistry, and Metal Physics of Superconductors, Izd. Nauka, Moscow (1967), p. 44.
47. N. E. Alekseevskii, O. S. Ivanov, I. I. Raevskii, and N. V. Stepanov, Fiz. Met. Metallov., 23(1):28 (1967).
48. R. S. Shmulevich, I. A. Baranov, V. R. Karasik, and G. B. Kurganov, Fiz. Met. Metallov., 21(3):379 (1966).
49. V. V. Baron, in: Metallography and Metal Physics of Superconductors, Izd. Nauka, Moscow (1965), p. 29.
50. W. V. Webb, Phys. Rev. Lett., 5:191 (1963).
51. K. J. van Gurp and D. J. van Ooijen, J. Phys., 27(7-8):51 (1966).
52. F. W. Reuter, K. M. Ralls, and J. Wulff, Trans. Metallurg. Soc. AIME, 236(8):78 (1966).
52a. D. Dew-Hughes and A. Narlicar, in: Transactions of the Tenth International Conference on Low-Temperature Physics, Izd. VINITI, Moscow (1967), p. 84.
53. C. J. Gorter, Physica, No. 2, p. 449 (1935).
54. C. P. Bean, M. V. Doyle, and A. G. Pincus, Phys. Rev. Lett., 9:93 (1962).
55. H. E. Cline, R. M. Rose, and J. Wulff, J. Appl. Phys., 34(6):1771 (1963).

56. S. V. Sudraeva, N. N. Buinov, V. A. Vozilkin, E. P. Romanov, and V. G. Rakin, Fiz. Met. Metallov., 21(3):308 (1966).
57. R. Schaw and D. E. Mapother, Phys. Rev., 118:1474 (1960).
58. J. J. Hauser and E. Helfand, Phys. Rev., 127:386 (1962).
59. J. J. Hauser and E. Buchler, Phys. Rev., 125:42 (1962).
60. J. J. Hauser and Trenting, J. Phys. Chem. Sol., 24(3):371 (1963).
61. J. J. Hauser and J. E. Kunzler, Relat. Struct. Mech. Properties Metals, Vol. 2K, HMSO, p. 766 (1963).
62. C. J. Gorter, in: Transactions of the Eighth International Conference on Low-Temperature Physics, London (1963), p. 121.
63. J. J. Gilman, Off-print, Disc. Faraday Soc., p. 38 (1964).
64. B. G. Lazarev, V. G. Khorenko, L. A. Kornienko, L. I. Krivko, A. A. Matsakova, and O. N. Ovcharenko, Zh. Éksp. Teor. Fiz., 45(6):2068 (1963).
65. V. V. Shmidt, Zh. Éksp. Teor. Fiz., 45(6):1992 (1963).
66. N. E. Alekseevskii, Uspekhi Fiz. Nauk, 95(2):2531 (1968).
66a. B. T. Matthias, E. A. Wood, and E. Corenzwit, Phys. Chem. Solids, 17:188 (1956).
67. B. T. Matthias, in: Progress in Low-Temperature Physics, Vol. 2, New York (1957), p. 138.
68. D. Pines, Phys. Rev., 109:280 (1958).
69. V. Radhakrishnan, Phys. Lett., 16(3):247 (1965).
70. E. M. Savitskii and V. V. Baron, Izv. Akad. Nauk SSSR, Metallurgiya i Gornoe Delo, No. 5, p. 3 (1963).
71. B. T. Matthias, T. H. Geballe, and V. B. Compton, Rev. Mod. Phys., 35(1):1(1963).
72. G. K. Hulm and R. D. Blaugher, Phys. Rev., 123:1969 (1961).
73. W. de Sorbo, Phys. Rev., 130:2177 (1963).
74. W. de Sorbo, Phys. Rev., 140(3A):914 (1965).
75. F. J. Blatt, Phys. Rev., 108:285 (1957).
76. L. J. Claiborne, Phys. Chem. Solids, 24:1363 (1963).
77. W. de Sorbo, P. E. Lawrence, and W. A. Healy, J. Appl. Phys., 38(2):903 (1967).
78. E. A. Lynton and B. Serin, Phys. Rev., 112:70 (1958).
79. E. A. Lynton, B. Serin, and M. Zucker, J. Chem. Phys. Sol., 3:165 (1959).
80. G. Chanin, E. A. Lynton, and B. Serin, Phys. Rev., 114:719 (1959).
81. D. P. Seraphim and D. G. Quinn, Acta Metallurgica, 9:811 (1961).
82. B. R. Coles, IBM J. Res. and Development, 6:68 (1962).
83. J. I. Budnick, Phys. Rev., 119:1578 (1960).
84. R. A. Hein, J. W. Gibson, and R. D. Blaugher, Rev. Mod. Phys., 36(1):149 (1964).
85. E. A. Linton, Superconductivity [Russian translation], Izd. Mir, Moscow (1964), p. 159.
85a. S. A. Nemnonov, Fiz. Met. Metallov., 19:64, 550 (1965).
86. E. Bucher, P. Heininger, J. Muheiv, and I. Muller, Rev. Mod. Phys., 36(1):146 (1964).
86a. W. L. McMillan, Phys. Rev., 167(2):331 (1968).
86b. J. J. Hopfield, Phys. Rev., 186(2):443 (1969).
86c. M. Ishikawa and Lois E. Toth, Phys., Rev. B., 3(6):1856 (1971).
87. R. D. Blaugher and J. K. Hulm, J. Phys. Chem. Sol., 22:134 (1961).
88. B. W. Roberts, Intermetallic Compounds, J. Wiley and Sons, Inc., New York (1967), p. 581.

89. N. E. Alekseevskii, G. S. Zhdanov, and N. N. Zhuravlev, Vestnik MGU, 3:113 (1959).
90. N. N. Zhuravlev, G. S. Zhdanov, and N. E. Alekseevskii, Vestnik MGU, 3:117 (1959).
91. L. Gold, Phys. Status Solidi, 4:261 (1964).
92. F. Galasso and J. Rule, Acta Crystallogr., 16:228 (1963).
93. Yu. V. Efimov, V. V. Baron, E. M. Savitskii, and E. I. Gladyshevskii, in: Metallography and Metal Physics of Superconductors, Izd. Nauka (1965), p. 91.
94. N. B. Brandt and N. I. Ginzburg, Zh. Éksp. Teor. Fiz., 44:1876 (1963).
95. L. S. Kan, B. P. Lazarev, and V. I. Makarov, Zh. Éksp. Teor. Fiz., 40:457 (1961).
96. M. Garfinkel and D. Mapother, Phys. Rev., 122:459 (1961).
97. I. D. Gennings and C. A. Swenson, Phys. Rev., 112:31 (1958).
97a. J. C. Phillips, Phys. Rev. Lett., 26(10):543 (1971).
98. J. Wernik, in: Superconducting Materials, Izd. Mir, Moscow (1965), p. 64.
99. R. Hake, T. Berlincourt, and D. Leslie, in: Superconducting Materials [Russian translation], Izd. Mir, Moscow (1965), p. 89.
100. J. Betterton, G. Kneip, D. Easton, and J. Scarborough, in: Superconducting Materials [Russian translation], Izd. Mir, Moscow (1965), p. 102.
101. J. D. Levingston and T. G. Alden, Transactions of the Tenth International Conference on Low-Temperature Physics, Izd. VINITI, Moscow (1967), p. 177.
102. B. G. Lazarev, E. E. Semenenko, and V. N. Kuz'menko, Fiz. Met. Metallov., 23(4):651 (1967).
103. L. N. Cooper, Phys. Rev. Lett., 6:689 (1961).
104. J. D. Levingston and H. W. Schadler, Progr. in Materials Sci., No. 12, p. 183 (1964).
104a. V. L. Ginzburg, Uspekhi Fiz. Nauk, 95(1):91 (1968).
104b. A. A. Abrikosov, Zh. Éksp. Teor. Fiz., 41(2):569 (1961).
105. H. Kamerlingh-Onnes, Communs Phys. Lab. Univ. Leiden, 133d (1913); 139f (1914).
106. S. Sh. Akhmedov and T. N. Vylegzhanina, in: Metallography and Metal Physics of Superconductors, Izd. Nauka, Moscow (1965), p. 120.
107. B. T. Matthias and J. K. Hulm, Phys. Rev., 87:779 (1952).
108. L. N. Cooper, RCA Rev., 25:405 (1964).
109. G. D. Cody, RCA Rev., 25:415 (1964).
110. U. S. Nat. Bur. Standards Circ., No. 564, p. 287 (1955).
111. J. Res. NBS, 64A:1 (1960).
112. N. D. Kozlova, Yu. V. Efimov, V. V. Baron, and E. M. Savitskii, Metallography, Physical Chemistry, and Metal Physics of Superconductors, Izd. Nauka, Moscow (1967).
112a. E. M. Savitskii, V. V. Baron, and S. D. Gindina, Dokl. Akad. Nauk SSSR, 191(2):338 (1970).
113. H. R. Hartal, in: High Magnetic Fields, John Wiley and Sons, Inc., New York (1963), p. 584.
114. T. G. Berlincourt, R. R. Hake, and A. C. Thorsen, Phys. Rev., 127:710 (1962).
115. I. L. Olsen, Helv. Phys. Acta., 26:798 (1953).
116. W. A. Cherry, RCA Rev., 25:510 (1964).

117. V. I. Chechernikov, Magnetic Measurements, Izd.MGU, Moscow (1960), p. 50.
118. C. P. Bean and M. B. Doyle, J. Appl. Phys., 33(11):3334 (1962).
119. W. de Sorbo and W. A. Healy, Cryogenics, 4:257 (1964).
120. N. E. Alekseevskii, A. N. Dubrovin, and V. S. Egorov, Dokl. Akad. Nauk SSSR, 163(5):1121 (1965).
121. N. E. Alekseevskii and V. S. Egorov, Zh. Éksp. Teor. Fiz., 45:448 (1963).
122. J. E. Kunzler, Rev. Mod. Phys., 33:501 (1961).
123. K. M. Rolls, A. L. Donnevy, R. M. Rouse, and J. M. Wulff, in: New Materials for Electronics [Russian translation], Izd. Metallurgiya, Moscow (1967), p. 37.
124. V. V. Baron, M. I. Bychkova, and E. M. Savitskii, in: Metallography and Metal Physics of Superconductors, Izd. Nauka, Moscow (1965), p. 53.
125. M. Walker and M. Fraser, in: Superconducting Materials [Russian translation], Izd. Mir, Moscow (1965), p. 158.
126. L. Rinderer and E. Saur, Z. Phys., 176:464 (1963).
127. V. L. Ginzburg, Nauka i Zhizn', No. 8, p. 37 (1967).
128. V. G. Ershov and V. R. Karasik, in: Metallography and Metal Physics of Superconductors, Izd. Nauka, Moscow (1965), p. 130.
129. J. Bremer, in: Superconducting Devices [Russian translation], Izd. Mir, Moscow (1964).
129a. Yu. F. Bychkov, O. G. Zamolodchikov, and A. N. Rozanov, Metallurgy and Metallography of Pure Metals, Atomizdat, Moscow (1969), p. 53.
130. G. K. White, Experimental Techniques of Low-Temperature Physics [Russian translation], IL, Moscow (1962).
131. R. B. Scott, Low-Temperature Techniques [Russian translation], IL, Moscow (1962).
132. A. B. Fradkov, in: Metallography and Metal Physics of Superconductors, Izd. Nauka, Moscow (1965), p. 110.
132a. V. E. Keilin, Cryogenics, Vol. 7, No. 3 (1967).
133. E. M. Savitskii and G. S. Burkhanov, Metallography of Refractory Metals and Alloys, Izd. Nauka, Moscow (1967), p. 85.
133a. E. M. Savitskii and G. S. Burkhanov, Metallography of Alloys of Refractory and Rare Metals, Izd. Nauka, Moscow (1971).
134. M. Cottin and M. Haissingsky, J. Chim. Phys., 47:731 (1950).
135. L. Ya. Popilov and L. P. Zaitseva, Electropolishing and Electro-Etching of Metallographic Microsections, Metallurgizdat, Moscow (1963), p. 262.
136. W. Rostocker, Metallurgy of Vanadium [Russian translation], IL, Moscow (1959), p. 148.
137. P. R. V. Evans, J. Less-Common Metals, 10(6):253 (1964).
138. J. Betterton, G. Kneip, et al., in Superconducting Materials [Russian translation], Izd. Mir, Moscow (1965), p. 102.
139. H. Richter, P. Wincierz, K. Anderko, and U. Zwicker, J. Less-Common Metals, 8(4):252 (1962).
140. E. J. Hughes and A. A. Johnson, J. Less-Common Metals, 12(10):408 (1966).
141. Metals Handbook, Electrolytic Polishing and Etching, ASM, Ohio (1954).
142. C. Smithells, Metals Reference Book, Vol. 1, 3rd Ed., Butterworth (1962), pp. 214-266.

143. Metals Handbook, ASM, Ohio (1948).
144. R. M. Brick and A. Phillips, Structures and Properties of Alloys, New York (1949).
145. G. Lambert, Typical Microstructures of Cast Metals, Br. Inst. Foundrymen Pub., Manchester (1957).
146. L. F. Mondolf, Metallography of Aluminum Alloys, New York (1943).
147. F. Keller and C. W. Wilcox, Metal Progr., 23:45 (1933).
148. M. C. Udy, in: The Metal Beryllium, ASM, Metals, Ohio (1955).
149. Amer. Soc. Test Mater. (part 113), pp. 803-859 (1946).
150. M. V. Mal'tsev, T. A. Barsukova, and S. A. Gorev, Metallography of Nonferrous Metals and Alloys, Metallurgizdat, Moscow (1960).
151. E. M. Savitskii, V. F. Terekhova, I. V. Burov, I. A. Markova, and O. P. Naumkin, Alloys of Rare-Earth Metals, Izd. AN SSSR, Moscow (1962).
152. K. A. Gschneidner, Rare-Earth Alloys, Vol. 4 (1961), p. 4.
153. W. Schreiter, Seltene Metalle, Leipzig (1960).
154. M. A. Filyand and E. I. Semenova, Properties of Rare Elements, Izd. Metallurgiya, Moscow (1964).
155. R. G. Maier and M. Hoffmann, Z. Metallkunde, 58(9):629 (1967).
156. G. R. Love and M. L. Picklesimer, Trans. AIME, 236:430 (April 1966).
157. D. Hauser, in: Superconducting Materials [Russian translation], Izd. Mir, Moscow (1965), p. 203.
158. W. A. Little, Phys. Rev., 134A:1416 (1964).
159. L. Wyman, J. R. Cuthill, et al., J. Res. NBS, 66A:551 (1962).
160. L. Rinderer, J. Wurm, W. Zullig, and Z. de Beer, Z. Phys., 179:407 (1964).
161. M. Hoffman and R. G. Maier, Pract. Metallogr., 3:378 (1966).
161a. E. M. Savitskii, V. V. Baron, and B. P. Mikhailov, in: Problems of Superconducting Materials, Izd. Nauka, Moscow (1970), p. 9.
161b. M. Hoffman and R. G. Marier, Prakt. Metallogr., 3(9):378 (1966).
161c. V. N. Svechnikov, V. M. Pan, and Yu. I. Beletskii, in: Metallography, Physical Chemistry, and Physical Metallurgy of Superconductors, Izd. Nauka, Moscow (1967), p. 100.
161d. L. L. Oden and R. E. Siemens, J. Less-Common Metals, 14(1):33 (1968).
162. A. H. Cottrell, J. Inst. Metals, 90:12 (1962).
163. B. I. Pogofin-Alekseev, Yu. A. Geller, and A. R. Rakhshtadt, Metallography, Oborongiz, Moscow (1950).
164. W. Rostocker and D. Dvorak, Microscope Method in Metallography [Russian translation], Izd. Metallurgiya, Moscow (1967), p. 85.
164a. M. I. Bychkova, V. V. Baron, E. M. Savitskii, S. V. Sudareva, and N. N. Buinov, in: Physical Chemistry, Metallography, and Physical Metallurgy of Superconductors, Izd. Nauka, Moscow (1969), p. 76.
165. A. A. Boiko, in: Apparatus and Methods of X-Ray Analysis, Leningrad, No. 1 (1967), p. 126.
166. A. A. Boiko and M. M. Umanskii, in: Apparatus and Methods of X-Ray Analysis, Leningrad, No. 1 (1967), p. 154.
167. D. N. Bol'shutkin, V. M. Gast, V. A. Kucheryavyi, V. S. Mironov-Kopysov, N. I. Mokrii, A. A. Prokhvatilov, and A. I. Érenburg, in: Apparatus and Methods of X-Ray Analysis, Leningrad, No. 6 (1970), p. 12.

168. B. W. Batterman and C. S. Barret, Phys. Rev., 145(1):296 (1966).
169. L. R. Testardi and T. B. Bateman, Phys. Rev., 154(2):399 (1967).
170. J. C. Phillips, Phys. Rev. Lett., 26(10):543 (1971).
171. J. Wanagel and B. W. Batterman, J. Appl. Phys., 41(9):3610 (1970).
172. U. Valdre and M. J. Goringe, J. Phys., E3(4):336 (1970).
173. A. V. Alanina, Yu. A. Dushechkin, V. A. Kucheryavyi, and B. Ya. Sukharevskii, Physics of the Condensed State, Trudy FTINT, Khar'kov, No. 4 (1969), p. 123.
174. A. Ikushima and T. Mizusaki, J. Phys. Chem. Solids, 30(4):873 (1969).
175. M. Ishikawa and L. E. Toth, Phys. Rev. B, 3(6):1856 (1971).
176. G. K. White, Cryogenics, 1:151 (1961).
177. T. A. Hahn, J. Appl. Phys., 41(13):5096 (1970).
178. L. R. Testardi and T. B. Bateman, Phys. Rev., 154(2):402 (1967).
179. E. M. Savitskii, B. Ya. Sukharevskii, V. V. Baron, É. A. Anders, V. A. Frolov, K. V. Mal'chuzhenko, and I. V. Shestakova, in: Superconducting Alloys and Compounds, Izd. Nauka, Moscow (1971).
180. E. M. Savitskii, B. Ya. Sukharevskii, V. V. Baron, A. V. Alanina, M. I. Bychkova, Yu. A. Dushechkin, I. S. Shchetkin, and M. N. Kharchenko, in: Superconducting Alloys and Compounds, Izd. Nauka, Moscow (1961).
181. E. M. Savitskii and V. B. Gribulya, Dokl. Akad. Nauk SSSR, 190(5):1147 (1970).
182. E. M. Savitskii, V. V. Baron, Yu. V. Efimov, M. I. Mychkova, and N. D. Kozlova, Dokl. Akad. Nauk SSSR, 196(5):1145 (1971).

Chapter II

Superconducting Elements

PROPERTIES OF SUPERCONDUCTING ELEMENTS

Superconductivity has been detected in 27 elements at ordinary pressures (Table 8). All these elements are metals. Thus superconductivity is a specific characteristic of the metallic type of interatomic bond. There is a point of view according to which all "true," extremely pure metals should be superconducting at 0°K.

It is important to notice the strong connection between superconductivity and the position of the elements in the periodic table. Physical theories of superconductivity designed to explain the actual phenomenon are at present quite unable to predict the existence of superconductivity in various elements or the values of the transition temperature [1a]. It is well known that the position of the elements in the periodic table is determined by the electron configuration of the isolated atom, which differs from that of the elements in real alloys and compounds. Nevertheless, the periodic law elucidates many aspects of alloy formation [1b], and in particular the laws governing changes in superconductivity. Matthias [1c] divided the elements of the periodic table into several characteristic regions on this basis (Fig. 29). In the light of subsequent experimental data we have introduced some additions to Matthias' picture.

The first main group of superconductors comprises seventeen transition metals or elements with s-d electrons. The superconducting transition temperature of these changes sharply on

TABLE 8. Superconducting Elements [1-6, 6b, 6c]

Element	T_c, °K	θ, °K	$\gamma,^*$ mJ/mole·°K^2	H_0, Oe	Type of lattice	Lattice constant, Å		
						a	b	c
W	0.012	380	0.50	1070	bcc	3.1584	—	—
Be	0.026	—	—	—	hexag.	7.01	—	10.8
Ir	0.14	410	3.10	19	fcc	5.2960	—	—
α-Hf	0.165	252	2.59	—	hcp	3.1946	—	5.0511
α-Ti	0.39	430	3.40	56	hcp	2.9504	—	4.6833
Ru	0.49	600	2.40	66	hcp	2.7058	—	4.2816
Cd	0.52	209	0.69	30	hcp	2.9788	—	5.1667
Os	0.65	500	2.35	65	hcp	2.7353	—	4.3991
α-U	0.68(?)	206	10.60	—	orthorhomb.	2.8536	5.8698	4.9555
α Zr	0.55	265	2.90	47	hcp	3.2312	—	5.1477
Zn	0.85	309	0.66	52	hcp	2.6649	—	4.9468
Mo	0.92	470	1.91	98	bcc	3.1468	—	—
Ga	1.09	317	0.60	59	orthorhomb.	4.1598	7.6602	4.5258
Al	1.19	420	1.36	99	fcc	4.0496	—	—
α-Th	1.37	168	4.65	162	fcc	5.0843	—	—
Pa	1.4	—	—	—	tetrag.	3.925	—	3.238
Re	1.7	210	2.35	193	hcp	2.7600	—	4.4580
Tl	2.4	78	2.80	171	fcc	3.4666	—	5.5248
In	3.4	109	1.81	293	tetrag.	4.5979	—	4.9467
β-Sn	3.7	195	1.75	309	tetrag.	5.8314	—	3.1814
Hg	4.15	87	1.91	412	rhombohed.	2.9863	α 70°44.6′	—
Ta	4.48	255	6.40	830	bcc	3.2980	—	—
V	5.3	338	6.70	1020	bcc	3.0282	—	—
β-La	5.9	142	6.40	1600	fcc	5.2960	—	—
Pb	7.2	96.3	3.00	803	fcc	4.9502	—	—
Tc	8.2	—	—	1410	hcp	2.7350	—	4.2816
Nb	9.2—9.4	238	7.53	1950	bcc	3.3007	—	—

*As defined on page 16.

changing the concentration of the valence electrons (Fig. 30) [1b]. In addition to this, among the other transition elements we know of two (yttrium and cerium) in which superconductivity appears at high pressures (Table 10). At the present time there are no grounds for denying the possibility that other transition metals may also prove to be superconductors under extreme conditions.

Developing his hypothesis as to the existence of several mechanisms of superconductivity [1b], Matthias called ten transition metals with s-p electrons the second group of superconducting elements. In addition to this, there are also eight nontransition elements of the B subgroup of the periodic system in which superconductivity only appears under special conditions (under pressure or in thin films).

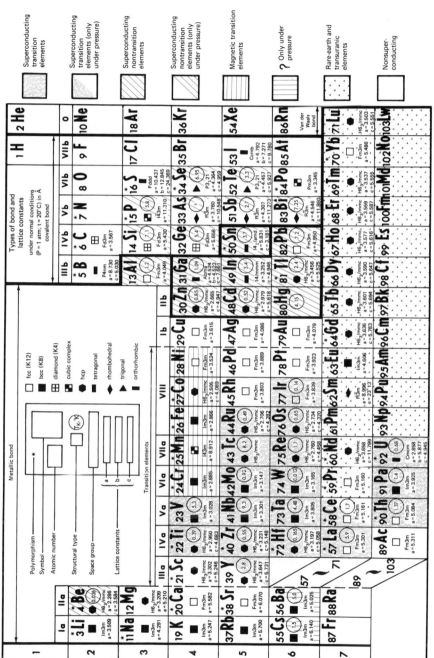

Fig. 29. Arrangement of superconducting elements in the periodic table.

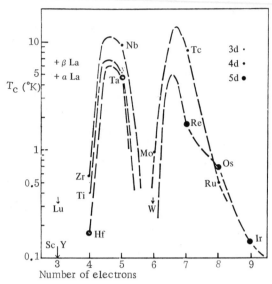

Fig. 30. Critical superconducting temperature of the elements in relation to the number of valence electrons.

Finally, it is now quite well known that certain alkali and alkaline-earth metals may exhibit superconductivity under high pressure (Table 10). Theoretical calculations show that superconductivity may occur in sodium and potassium [1d]. However, the value of T_c for these simple metals may only be very low ($< 10^{-5}$ °K).

The highest temperature of the transition into the superconducting state is found in the case of niobium. The lowest value is found in tungsten (99.999% pure single crystal) [5], being no higher than 0.1°K. It is considered that pure rhodium is a superconductor with T_c = 0.2°K [5a].

The thermodynamic critical magnetic field of the elements (extrapolated to 0°K) usually increases with increasing T_c, reaching a maximum for niobium (approximately 1950 Oe) and falling to 19 Oe for iridium (the H_0 of pure tungsten has never been determined). Information as to the Debye temperature of the elements is incomplete, and that relating to the electron specific heat may not be particularly accurate.

Superconductivity at ordinary pressure is found in ten non-transition and seventeen transition elements; these elements have a variety of crystal structures, and nine of them are polymorphic metals. It is curious to note that the high-temperature forms have a higher T_c than the low-temperature forms. The critical superconducting temperatures of various forms of polymorphic metals are indicated in Table 9.

TABLE 9. Critical Superconducting Temperature of the Polymorphic Forms of Various Elements

Element	Type of lattice	T_c, °K	Literature cited
Hg α	rhombohedral	4.15	[2]
Hg β*	tetragonal	3.9	[2]
Hg γ†		3.74	[7a]
La α	hcp	4.8	[2]
La β	fcc	5.9	[2]
Ti α	hcp	0.39	[1]
Ti β	bcc	4.0‡	[7]
Zr α	hcp	0.55	[1]
Zr β	bcc	0.5‡	[65]
Zr ω	—	0.65	—
Hf α	hcp	0.165	[1]
Hf β	bcc	—	—
Sn β	tetragonal	3.7	[1]
Th α	fcc	1.37	[1]
U α	orthorhomboid	0.68	[1]
U β	tetragonal	0.8§	[2]
U γ	bcc	1.8	—
	bcc	2.1§	[2a]
Tl α	hex	2.4	[62]
Tl β	fcc	1.75	[62]

* β-Hg is found at high pressures and is stable at low pressures after removing the applied pressure [6a].
†Deformation by tensile strain at liquid-helium temperature leads to a partial transformation of α-Hg into γ-Hg.
‡Extrapolation.
§γ form stabilized by 15% Mo, β by 2% Rh.

TABLE 10. Critical Superconducting Temperature of Several Elements Measured Either under Pressure or on Samples in the Form of Thin Films Prepared at Low Temperatures

Element	T_c, °K	Conditions of measurement or preparation	Literature cited
Vanadium	5.6	Measurement under pressure 45 kbar	[10a]
Niobium	9.7	Measurement under pressure 45 kbar	[10a]
Tungsten	4.1	Film produced at < 10°K	[8a]
Beryllium	Up to 8.4	Film produced at < 10°K	[11]
Bismuth	Up to 6.0	Film produced at < 10°K	[10]
Bismuth II	3.9	Measurement under pressure 25,000 atm	[10]
Bismuth III	7.25	Measurement under pressure 25,000 atm	[10]
Gallium	∿ 8.4	Film produced at < 10°K	[9]
Germanium	Up to 5.4	Measurement under pressure 120 kbar	[14]
Indium	Up to 4.25	Film produced at < 10°K	[2]
Silicon	7.1	Measurement under pressure 130 kbar	[2]
Tin	4.7	Film produced at < 10°K	[2]
Selenium	6.95	Measurement under pressure 130 kbar	[15]
Antimony	Up to 2.7	Measurement under pressure 85,000 atm	[12]
Tellurium	Up to 3.3	Measurement under pressure 45,000 atm	[13]
Phosphorus	4.7	Measurement under pressure > 10 kbar	[13a]
Uranium (α)	2.2	Measurement under pressure > 10 kbar	[14a]
Aluminum	3-3.6	Film obtained at T < 10°K	[61]
Tungsten (fcc)	Up to 5.5	Film obtained at T < 10°K	[8b]
Beryllium	9	Film obtained at T < 10°K	[11a]
Tellurium II	2.85	Measured under pressure, 111 kbar	[14c]
Tellurium III	3	Measured under pressure, 220 kbar	[14c]
Cesium	~1.5	Measured under pressure, 125 kbar	[14c]
Yttrium	1.2-2.8	Measured under pressure, 110-180 kbar	[14c]
Barium	5	Measured under pressure, 110-180 kbar	[14c]
Lanthanum	12	Measured under pressure, 110-180 kbar	[14c]
Cerium	1.7	Measured under pressure, 110-180 kbar	[14c]

As already indicated, the transition temperature (T_c) cannot be predicted simply on the basis of data relating to the electron specific heat and Debye temperature θ of the elements. A relationship between T_c and θ was established earlier [7]. It is worth noting that superconductors with a high T_c (Nb, La, V, Ta) also have a high electron specific heat.

As regards their properties superconducting elements may be divided into two classes: nontransition and transition superconductors, with filled and partly-filled d shells, respectively.

Pure nontransition superconducting elements constitute superconductors of the first kind (group).

At high pressures or after the condensation of films at low temperatures, the superconducting characteristics of certain elements may be considerably increased (Table 10). Thus for example the T_c of tungsten films $2 \cdot 10^{-5}$ cm thick rises to 4.1°K [8a]. The T_c of tungsten films with a metastable fcc structure is 4.6°K. In tungsten films containing the ordinary bcc phase and the metastable fcc phase the value rises from 1 to 5.5°K on reducing the size of the crystallites from 300 to 15 Å [8b]. In a number of cases superconductivity even arises in elements from which it is entirely absent under ordinary conditions of sample preparation and measurement. In gallium films condensed at temperatures below 10°K the value of T_c increases to 8.4°K instead of 1.1°K [9]. Superconductivity develops in bismuth under pressure [10] and in Be, Al, Mo, and Re films deposited on a cooled substrate [11, 11c, 11d]. According to the latest data superconductivity appears in extremely pure beryllium at ordinary pressures (T_c = 0.026°K) [11a]. After applying high pressures superconductivity was observed in antimony (T_c = 2.7°K) [12] and tellurium (T_c = 3.3°K) [13], the latter constituting a new superconductor among elements of subgroup VI B, silicon (T_c = 7.1°K), germanium (T_c = 4.85-5.4°K) [14], and selenium [15]. Evidently semiconductors acquire metallic properties under pressure at helium temperatures. It is reasonable to suppose that this is associated with the development of new "metallic" polymorphic modifications. However, these transformations are reversible: After removing the pressure these superconductors revert to semiconducting status [15a]. The development of superconductivity at helium temperatures under pressure is a characteristic feature of many semiconducting compounds (CuS, La_3S_4, La_3Se_4) as well as semiconducting elements.

However, high pressure may also exert a negative influence. For example, the T_c of lead falls under pressure (T_c = 7.24 and 7.11°K at 0 and 3345 bar) [15b]. The influence of pressure on the superconducting characteristics was studied in detail elsewhere [15c].

A considerable rise in T_c and H_c occurs on producing thin amorphous superconducting films by the condensation of metal vapor on a surface cooled by liquid helium. For example, in amorphous beryllium films ~ 100 Å thick, T_c reaches 9°K, while

$H_c \sim 100$ kOe at 4.2°K [11a]. There is also a high field in amorphous bismuth films with $T_c = 5°$K.

The internal stresses arising in such films usually raise the T_c of superconductors with noncubic crystal lattices (tin, indium, gallium). However, sometimes the distortion of the lattice leads to a slight decrease in T_c; in this case it would appear that the change in T_c is associated with the vanishing of anisotropy in the energy gap, the energy gap in these metals ordinarily being anisotropic and the value of T_c being determined by its maximum value [16, 17]. The effect of the degree of distortion of the lattice (resulting from condensation on a cold substrate) on the T_c and H_c of metallic lead and aluminum films was studied earlier [18]. For the maximum lattice distortion created by condensation on a surface cooled to liquid-helium temperature (the degree of distortion being estimated by reference to the specific electrical resistance ρ), the T_c of aluminum rose from 1.2 to 3.3°K and H_0 rose to 30 kOe. The application of high pressures reduces the transition temperature of aluminum and also that of zinc and cadmium [19]. The same applies to tin. An estimation of the change in the density of states $N(0)$ due to pressure revealed that the decrease in density in the present case was not large enough to be responsible for the observed decrease in T_c; the latter was attributed to a decrease in the value of the parameter V.

The transition temperature of tin and indium whiskers has also been measured [20]. In the case of tin there was a slight rise in T_c (from 3.7 to 4°K), while the T_c of indium remained almost exactly the same. A change from "soft" to "hard" superconductors took place in the same way as in tin films. For whiskers with a thickness less than a certain critical value, tin became an ideal superconductor of the second kind. In view of the great strength of metallic whiskers as compared with ordinary samples of the same material (the σ_b is two or three orders of magnitude higher in whiskers and tends toward the theoretical limit) an attempt was made at determining the effect of deformation on the T_c of this material; a rise of 0.3°K was obtained in this way.

Taken in the form of conventionally-prepared samples, lead has the highest superconducting parameters among all the nontransition superconducting elements. However, when studying the critical current and magnetic field of samples in the form of films, higher values of I_c and H_c were in a number of cases observed in tin [17]. Tin films 660-760 Å thick deposited on a substrate held at 4.2°K had a critical field of 4000 Oe at 1.3°K instead of the 264 Oe of large, unstressed samples, which constituted soft superconductors of the first kind. The thin tin films also exhibited a higher electrical resistance and a shorter free path of the electrons (as determined from the specific electrical resistance of the metal in the normal state). The annealing of this type of tin sample reduced the stresses arising from condensation on the cold substrate and restored the ordinary superconducting properties of the metal [20]. A similar effect was observed elsewhere [21] for tin, and also for thallium films of various thicknesses [22].

In addition to this, there are also some special methods of preparing samples of superconducting elements in porous glass, in which the superconducting properties increase sharply as a result of the formation of a very fine fibrous structure [23]. A fibrous synthetic "hard" superconductor prepared from mercury in porous aluminum by injection [24] has also been studied; the maximum critical field equalled 15,000 Oe (at 2.17°K) and the critical current density 10^3-10^4 A/cm^2. Similar results were obtained for indium in porous glass [25].

High critical fields and currents were also obtained in niobium and tantalum on preparing these in the form of thin films. The increase in the superconducting characteristics of films as compared with bulk samples may be explained by the fact that (in very thin samples) the film thickness was commensurable with the depth of penetration of the magnetic field, and also by the development of defects and large stresses in the material after condensation on a cold substrate. Niobium films 300-5000 Å thick deposited on quartz substrates retained their superconductivity in transverse magnetic fields up to 12,000 Oe, and in parallel magnetic fields up to 30,000 Oe. This was almost an order of magnitude greater than the critical magnetic field of ordinary niobium samples. The critical current density of niobium films at H = 0 equalled 10^6 A/cm^2 [26]. In tantalum films H_{c2} reached 5000 Oe (at 1°K) as opposed to 800 Oe in ordinary samples [27].

EFFECT OF DEFORMATION AND INTERSTITIAL IMPURITIES ON THE SUPERCONDUCTING PROPERTIES OF THE ELEMENTS

Pressure treatment (i.e., working by pressure, as in rolling, extrusion, etc.) has a considerable influence on the properties of superconductors. Thus after cold rolling or drawing with high degrees of deformation the critical current and magnetic properties of superconductors may undergo substantial changes. The critical current density of niobium was studied as a function of the degree of cold deformation in an earlier treatment [28]. Deformation by 99.3% increases the critical current by more than an order of magnitude in fields up to 5000 Oe. In a field of around 6000 Oe (close to the H_{c2} of the material) the critical current densities of samples with different degrees of deformation come closer together (Fig. 31). In cold-worked niobium H_{c2} equalled 6.5 kOe as against 5 kOe before rolling (for H perpendicular to the rolling plane and to the current) and increased to 11 kOe when the field was parallel to the current. In this case superconductivity was preserved for very low currents up to 16 kOe (H_{c3}) [29].

In another paper [30] devoted to the effect of the fibrous structure of cold-rolled niobium on the critical current in a magnetic field the effect of deformation on I_c and H_{c2} was also noted; it was found that the greatest values of current appeared when the field was perpendicular to the direction of the fibers and the lowest values when the field and fibers were parallel to one another.

It is interesting to note that the cold working of niobium free from impurities increased T_c to 9.67°K [22, 40].

The properties of superconducting metals may change substantially in the presence of various impurities, since the electrical resistance in the normal state then alters as well as the specific heat, the average range of the electrons, the Debye temperature, the density of states at the Fermi level, and the number and nature of crystal–structural defects arising in the presence of impurities.

In a number of cases T_c decreases for a low content (of the order of tenths of a percent) of metallic impurities in solid solution with various nontransition metals (aluminum, tin, indium, lead, zinc, cadmium, etc.) [31-35]. An exception is thallium and another is mercury [36, 37]. If the impurity is not a transition element, then a further increase in its content may lead to different results (as we shall show in Chap. IV): The transition temperature may either rise or fall, and the same may occur in the case of the critical magnetic field.

Traces of transition metals reduce the critical superconducting temperature of nontransition metals more than other elements do. Ferromagnetic impurities in many cases also severely reduce the T_c of nontransition metals. The experimental laws governing the changes in the T_c of dilute solid solutions of the substitutional type, chiefly of nontransition metals, yield a simple empirical formula for the value of T_c as a function of the molar concentration of the dissolved element and the change in the range of an electron (see Chap. I) [38]. The superconducting properties of transition metals are very substantially influenced by interstitial impurities. Superconducting transition elements generally belong to the class of refractory rare metals, which have high oxygen, nitrogen, hydrogen, and carbon affinities.

Depending on the level of impurity content, these metals may be either superconductors of the first kind or superconductors of the second kind. The effect of metallic impurities on the T_c of the transition metals is much less than that of interstitial impurities (for the same content). However, the presence of certain metallic impurities even in extremely small quantities (such as iron, for example) in a number of refractory metals (molybdenum, tungsten, iridium) reduces T_c sharply or even makes the metals nonsuper-

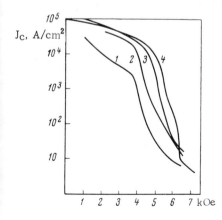

Fig. 31. Critical current density of niobium in an external magnetic field at 4.2°K as a function of the degree of deformation. Deformation, %: 1) 0; 2) 60; 3) 92.2%; 4) 99.3.

TABLE 11. Transition Temperature of Niobium of Various Degrees of Purity with Respect to Interstitial Impurities

Content of interstitial impurities, at.%	$\dfrac{R_{300°K}}{R_{4.2°K}}$	T_c, °K	Method of preparing the samples	Literature cited
Sum of impurities. < 0.3		5.1	Metalloceramic	[39]
3.83 oxygen		5.8	Quenching from 1000°	[40]
2.6 oxygen		7.0	"	[40]
1.52 oxygen		8.0	"	[40]
Not indicated	280	9.2	"	[40]
"	2.3	9.3	Film 2300 Å thick	[26]
"	500	9.46	Quenching from 1000°	[40]
"	68	9.67	Cold work	[40, 22]
0.33 nitrogen		9.12	Quenching from 1200°	[40]
1.64 nitrogen		9.24	"	[40]
3.6 hydrogen		9.22	"	[40]
	16,500	9.25		[40a]

conducting. Of the superconducting transition metals, only niobium and vanadium fail to exhibit the properties of superconductors of the first kind even after the careful removal of impurities. Such metals as tantalum, rhenium, thorium, etc., however, constitute superconductors of the first kind in the pure and unstressed form and superconductors of the second kind when they contain impurities and defects. The presence of interstitial impurities has a great influence on the level of the superconducting characteristics of the transition metals. The transition temperature of, for example, niobium (one of the most promising superconductors for practical use, particularly as a base for superconducting alloys) fluctuates from 5.1 to 9.6°K, depending on the proportion of interstitial impurities. From the data presented in Table 11 we see the great influence of gaseous impurities, especially oxygen, on the critical temperature at which niobium passes into the superconducting state.

In measuring the transition temperature of niobium samples in the form of films it was found that the temperature in question varied with film thickness. It was found in particular [21] that for the same impurity content and different thicknesses of niobium films the highest value of T_c corresponded to a film 2300 Å thick; for thinner films T_c decreased to 8°K at 400 A.

On introducing oxygen into niobium (within the solubility limit), T_c decreases from 9.46 to 5.84°K (3.83 at.% O); in the case of larger amounts of oxygen T_c increases [40]. In an alloy containing 6.43 at.% of oxygen $T_c = 9.02$°K. The decrease in the T_c of niobium was explained in the case in question [40] as being due to the negative influence of oxygen within the solubility range; the subsequent rise was attributed to the formation of oxides in the two-phase range. Later [41a], however, it was found on studying two-phase alloys of niobium with oxygen (7.2 and 12.5 at.% O) that T_c decreased with increasing oxygen content (to 8.45 and 7.85°K respectively).

The niobium–oxygen phase diagram (Fig. 32) was plotted up to 30 wt.% oxygen by Elliott [41] and subsequently refined [42, 43]. The system incorporates the compounds NbO, NbO_2, and Nb_2O_5. The first compound has a structure of the NaCl type, the second one of rutile; Nb_2O_5 has a monoclinic lattice at high temperatures and an orthorhombic one at temperatures below 830°C. The solubility limits of oxygen in niobium are different in these two papers. In the first case [42] the solubility of oxygen in niobium at 775-1100°C is 1.4-5.5 at.% and in the second [43] it is 1.1-1.4 at.% (Fig. 32b and c). For an oxygen content above the solubility limit, the monoxide NbO is formed in the alloys; the transition temperature of this phase is 1.25°K [44]. There are no data regarding the superconductivity of the other compounds. In view of the serious discrepancies in the data relating to the solubility of oxygen in niobium,

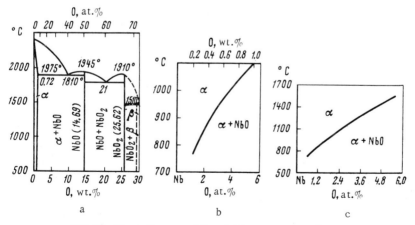

Fig. 32. Phase diagram of the niobium–oxygen system.

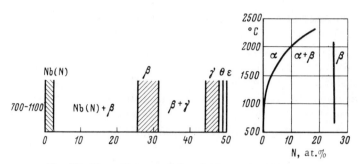

Fig. 33. Phase diagram of the niobium–nitrogen system.

any conclusions as to the relation between the effect of oxygen on the T_c of niobium and the structure of the alloys demand further investigation.

According to [40] nitrogen has less effect on the T_c of niobium. For a nitrogen content of up to 0.33 at.% $T_c = 9.12°K$, while for a higher concentration (1.64 at.%) T_c rises to 9.24°K. Niobium alloys containing 1.25-18.6 at.% N experienced a decrease in T_c (to 8.25°K) [41a] with increasing nitrogen content. The niobium–nitrogen phase diagram and the solubility limits of nitrogen in niobium have been studied repeatedly. According to the latest data [47], at 2200°C and a nitrogen pressure of $3 \cdot 10^{-1}$ mm Hg there are 9.48 at.% nitrogen in solid solution with niobium. The system contains the compounds Nb_2N and NbN, which crystallize in a hexagonal close-packed lattice and a lattice of the NaCl type, respectively. Apart from this the existence of γ, θ, and ε phases has been mentioned (Fig. 33). The solubility of nitrogen in niobium increases on raising the pressure. The change in the solutility of nitrogen in niobium with temperature has been studied several times [43, 45]. The results indicated that the maximum solubility of nitrogen in niobium at 1600, 2100, and 2300°C was 3.7, 10.0 and 18 at.%, respectively; at 1200°C it decreased to 1.25 at.%. In the Nb-rich compound Nb_2N no superconductivity appears above 1.2°K, but in NbN the T_c reaches 16.2°K [44a].

The effect of hydrogen on the T_c of niobium has been studied very little. Apart from the information presented in Table 11 it is known that for 10 and 5 at.% of hydrogen the T_c of niobium reaches 7.38 and 7.83°K respectively [48], while for 14.65 at.% of hydrogen it decreases to 6.75°K [41a].

The niobium–hydrogen system contains an interstitial Nb-base solid solution [48a, 49, 50]. The solubility of hydrogen in niobium increases with increasing pressure. Below 445°K there is a monotectoid equilibrium in this system (Fig. 34). The α phase with the orthorhombic lattice is not superconducting above 1.8°K. Niobium hydride (the β phase) is homogeneous over a wide range

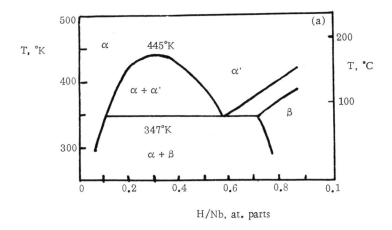

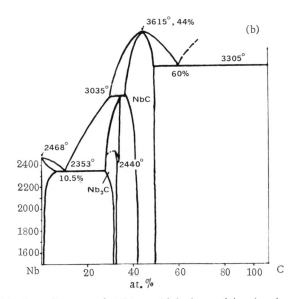

Fig. 34. Phase diagrams of niobium with hydrogen (a) and carbon (b).

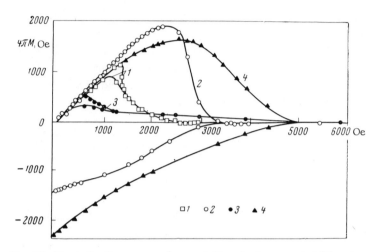

Fig. 35. Magnetization curves of pure niobium in the annealed (1) and worked (2) states and niobium containing 0.7 at.% (3) and 6.43 at.% (4) of oxygen.

of concentrations. It was concluded in one paper [50] that the changes in T_c in this system were directly related to the structure of the alloys (the value falling in the region of solid solutions but changing little in the two-phase region).

Interstitial impurities in a niobium solid solution increase its lattice constant [47]. According to the earlier-mentioned results [50], hydrogen also increases the lattice constant of niobium from 3.302 to 3.314 Å within the solubility range. Metallic impurities forming substitutional solid solutions and nonmetallic impurities forming interstitial solid solutions affect the lattice constant and critical temperature T_c of superconducting metals to different extents [40]. On adding elements of groups IV and VIA to a transition metal, in the majority of cases T_c increases with increasing lattice constant and decreases with decreasing lattice constant. In interstitial solid solutions a decrease in T_c is accompanied by an increase in the lattice parameter.

Carbon also decreases the T_c of niobium [48b]. The nearest compound to niobium itself in the Nb-C system is Nb_2C (Figs. 30 and 50b), and this exhibits no superconductivity above 1.98°K; only in the second compound with a high carbon content (over 80%) do we find $T_c \approx 11°K$ [49a, 51a].

The effect of interstitial impurities on the T_c of vanadium and tantalum has also been studied [50a, 51a, 51c].

Different levels of interstitial impurities also affect the behavior of niobium in a magnetic field. Figures 35 and 36 show the magnetization curves of niobium containing oxygen and nitrogen in a parallel field [40].

The maximum magnetization and hysteresis occur in worked niobium samples containing impurities above the solubility limit. Annealing and degassing the metal reduce its magnetization and hysteresis. For an impurity content within the solubility limit the magnetization is lower than in pure niobium, while the hysteresis is lower than in the more contaminated samples. The upper critical field rises to 6000 Oe (for 0.7 at.% oxygen at 4.2°K), while the lower critical field H_{c1} falls to 170 kOe.

We may conclude from the magnetic measurements that defects and inhomogeneities exist in niobium even for contents of 0.0006% oxygen, 0.0005% nitrogen, and under 0.0001% hydrogen. This is indicated by the small but finite deviation from precise reversibility in the magnetization curves. This effect is responsible for the appearance of a critical field H_{c3} in such materials, the maximum value of this field being determined by reference to the complete restoration of the electrical resistance ρ_n. For nio-

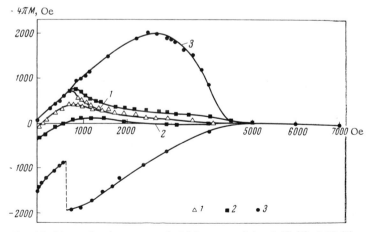

Fig. 36. Magnetization curves of niobium containing 0.23 (1), 0.33 (2), and 1.64 (3) at.% nitrogen.

bium samples containing 0.33 and 1.64 at.% of nitrogen H_{c2} equalled 5 kOe in the first case and 7 kOe in the second.

The effect of dislocation density and distribution on the magnetic properties of worked and annealed niobium and vanadium samples containing impurities constituted the subject of a special investigation [51]. The dislocation density and distribution in the material were studied in the transmission electron microscope. For this purpose samples of sheet niobium 100-200 μ thick were etched in a mixture of concentrated HNO_3 with three parts of 40% HF to a thickness of 20 μ, with subsequent electropolishing. Vanadium samples of the same dimensions were prepared by combining the etching and electropolishing in a mixture of methyl alcohol and concentrated H_2SO_4 (4:1). It was noted in this investigation that deformation in the presence of impurities (oxygen in the present case) affected the magnitude of the hysteresis and that of the frozen magnetic flux in a zero external field. The value of H_{c2} was independent of the degree of deformation, but varied with impurity content. The maximum amount of frozen flux and hence the greatest critical current corresponded to the nonuniform dislocation distribution (10^{12} cm^{-2}) created by a high degree of deformation (95%). A high dislocation density and a favorable distribution of the dislocations was obtained after brief annealing of impurity-containing metal, and also after additional deformation (bending) of samples previously subjected to a moderate degree of (rolling) reduction. If the worked samples were annealed in such a way as to give a uniform dislocation distribution and a low dislocation density (10^4 cm^{-2}), then this procedure reduced the hysteresis, the amount of magnetic flux frozen (at H = 0), and also the critical current of niobium. Qualitatively the same picture was obtained for vanadium. The critical magnetic field (H_{c2}) of the vanadium samples was much lower than that of niobium (around 1250 Oe). Figure 37 shows the

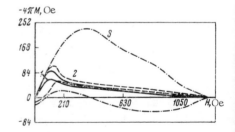

Fig. 37. Magnetization curves of worked and annealed vanadium: 1) vacuum-annealed; 2) worked (14% deformation); 3) the same (95%).

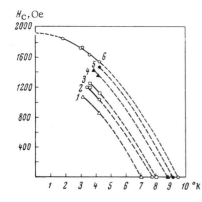

Fig. 38. Dependence of the critical magnetic field of niobium (H_{c1}) on temperature and interstitial impurity content: oxygen content, at.%: 1) 2.6; 2) 1.8; 3) 1.52; 4) 0.7; 5) 0.23 (nitrogen); 6) pure niobium.

magnetization curves of the vanadium samples after working and annealing.

When studying the magnetic characteristics of single-crystal niobium [52] it was found that hysteresis was almost entirely absent, only developing after deformation. However, even then H_{c2} never exceeded 4000 Oe. Interstitial impurities, which usually raise H_{c2}, reduce the lower critical magnetic field of niobium. Figure 38c illustrates the dependence of H_{c1} on the measuring temperature of niobium containing various amounts of impurities [40].

Niobium samples with a resistance ratio of $R_{300}/R_{4.2} \sim 10,000$ and ~ 300 were studied in [52a]; using the induction method, the temperature dependence of the fields H_{c1}, H_{c2}, and H_{c3} was determined. It was found that the experimental value of $H_{c2}(0)$ was five times greater than that calculated from the microscopic theory and depended on the purity of the sample, particularly on the content of interstitial impurities. The H_{c2} of the first sample (4000 Oe) degassed in a vacuum of $5 \cdot 10^{-11}$ mm Hg was smaller than the H_{c2} of the second sample. Both samples contained no more than 0.0005 wt.% of tantalum.

We see that an increase in the impurity content reduces both the transition temperature and H_{c1}.

An interstitial impurity content exceeding the solubility limit in niobium and the precipitation of oxides as second phases leads to a substantial rise in the critical current. The effect of oxygen content on the critical current density of niobium [40, 53] is shown in Fig. 39. For an oxygen content within the range of solubility in

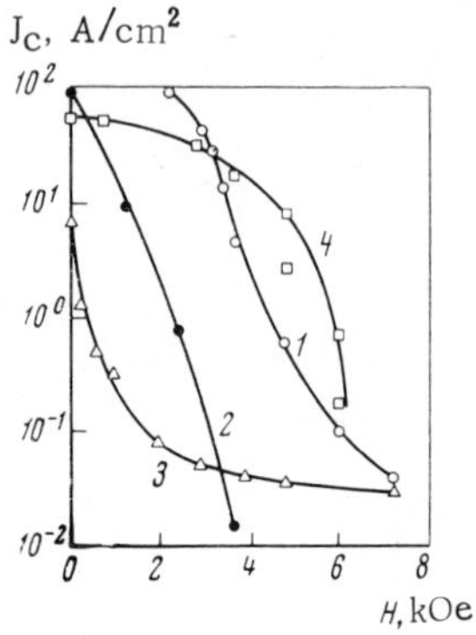

Fig. 39. Dependence of the critical current at 4.2°K on the oxygen content and deformation of niobium: 1) deformed (worked) niobium; 2) annealed niobium; 3) niobium containing 3.83 at.% oxygen; 4) 6.43 at.%.

niobium the critical current is low even in the absence of an external magnetic field. Raising the oxygen content above the solubility limit increases the critical current, particularly in high fields. The same effect was noticed by other authors [54]. In the case of niobium containing 0.7 at.% of oxygen the critical current density exceeds that corresponding to degassed niobium of a much higher purity by more than two orders of magnitude in a magnetic field of 5-6 kOe at a temperature of 4.2°K.

A number of authors have studied the anisotropy of the superconducting properties of various metals. Thus the anisotropy of the critical current density was studied for niobium single crystals [28] grown from seeds with axes in various orientations. Samples cut from these single crystals with a [100] and [110] orientation of the axes (current axes) exhibited the greatest current anisotropy, those cut in the [112] direction much less, while those with the [111] orientation showed very little anisotropy or even none at all. The greatest anisotropy of the critical current occurred in a magnetic field perpendicular to the direction of the current, having an intensity of 1-2 kOe. The effect of orientation on the superconducting properties of single-crystal niobium before and after working was considered elsewhere [55]. The critical fields H_{c1} and H_{c3} depended on the orientation, while H_{c2} did not. The effect of orientation on the critical current remained intact after the deformation (10%) of single-crystal samples.

The anisotropy of the magnetic field was also determined for rhenium single crystals [56]. Rhenium single crystals grown by electron-beam zone melting had a resistance ratio of $R_{300°K}/R_{4.2°K} = 1500$. In order to determine the anisotropy of the magnetic field the rhenium single crystals were subjected to tensile strain in such a way that the slip axis in the plane of the prism made an angle of 30° with the axis of the cylindrical sample. The critical magnetic field was greater when it was aligned parallel to the slip plane.

The foregoing experimental material shows that defects, stresses, and impurities largely determine the level and the stability of the superconducting characteristics. These characteristics are extremely sensitive to interstitial impurities in transition metals. This factor is largely responsible for the scatter in published data relating to superconducting characteristics.

We see from the information presented in this chapter regarding the transition temperature of the elements that the value never exceeds 9.5°K. Seeking superconductors with higher T_c values is one of the principal problems of superconductor research.

Considerable interest has been aroused in the suggested possibility of producing organic superconductors with high transition temperatures (over 20°C) [57]. However, it is now known that superconductivity should be unstable in a single long molecule.

Another method of securing high-temperature superconductivity has also been proposed [58]. It is suggested that superconductivity might arise in the case of an electron (exciton) mechanism of attraction between electrons in a system of the "sandwich" type, consisting of thin films of dielectric and metal, or of different chemical compounds. A superconductivity mechanism of this kind (if it could be achieved) would increase T_c by one or two orders of magnitude as compared with the phonon mechanism.

An increase in the T_c of thallium films on coating these with germanium (two or three molecular layers) was observed in one case [58b]. Factors affecting the superconductivity of thin films were discussed in detail in [61]. The question as to the production of granular films strikes us as being particularly important.

Thus there are two main models of high-temperature superconductor: bundles of parallel filaments (Little molecules) and bundles of parallel superconducting planes (Ginzburg).

An intensification of superconductivity with the aid of organic materials was observed in indium–anthraquinone films 5000 Å thick obtained by simultaneous evaporation in a vacuum of 10^{-7}

mm Hg [58a]. For relative concentrations of 10-30% anthraquinone, T_c rose to 4.6°K.

Finally, we have the proposition of Ashcroft [59] regarding the high-temperature superconductivity of metallic hydrogen. It is suggested that hydrogen of this kind may be obtained at very low temperatures and very high pressures [60], such as cannot as yet be secured on Earth (working pressure 600 kbar). However, certain research workers reject the possibility of superconductivity existing in the hypothetical metallic hydrogen. For example, Matthias [1b] asserts the latter on the basis of the empirical rule of the minimum number of valence electrons (2) in superconducting elements. However, superconductivity has now been observed under pressure in cesium, which is a univalent metal [14b].

LITERATURE CITED

1. B. T. Matthias, in: Progress in Low-Temperature Physics, Vol. 2 (1957), p. 138.
1a. B. T. Matthias, Comments Solid-State Phys., 3(4):93 (1970).
1b. N. S. Kurnakov, Selected Works, Izd. AN SSSR, Moscow, Vol. 1 (1960) and Vol. 2 (1961).
1c. B. T. Matthias, American Scientist, 58(1):80 (1970).
1d. J. P. Carbotte and R. G. Dynee, Phys. Rev., 172(2):476 (1968).
2. B. W. Roberts, Tech. Notes, 408:26 (1966).
3. B. T. Matthias, Superconductors, Interscience, New York (1962), p. 1.
3a. B. T. Matthias, T. H. Geballe, et al., Science, 151:985 (1966).
4. J. D. Levingston and H. W. Schadler, Progr. in Mater. Sci., No. 12, p. 187 (1964).
5. I. W. Gilson and R. A. Hein, Phys. Rev. Letters, 12:668 (1964).
5a. R. D. Fowler, B. T. Matthias, et al., Phys. Rev. Letters, 15(22):860 (1965).
6. E. M. Savitskii and G. S. Burkhanov, Metallography of Refractory Metals and Alloys, Izd. Nauka, Moscow (1967).
6a. M. Atoji, J. E. Schirber, and C. A. Swenson, J. Chem. Phys., 31:1628 (1959).
6b. R. W. Roberts, Superconductive Materials and Some of Their Properties, Note 408 (1967).
6c. W. Buckel, Naturwissenschaft, 58(4):177 (1971).
6d. T. Claeson et al., Phys. Stat. Sol., 42:321 (1970).
7. W. Buckel and R. Hilsch, Z. Phys., 138:109 (1954).
7a. P. R. Doidge and A. R. Eastham, Phil. Mag., 18, 153, 655 (1968).
7b. D. Köhnlein, Z. Phys., 208(2):142 (1968).
7c. M. Gey, Z. Phys., 229(1):85 (1969).
8. C. A. Reynolds, B. Serin, and W. H. Wright, Phys. Rev., 78:487 (1950).
8b. O. F. Kammerer and M. Strongin, Phys. Lett., 17(3):224 (1965).
8c. K. L. Chopra, Phys. Lett., A25(6):451 (1967).
9. W. Buckel and W. Gey, Z. Phys., 176:336 (1963).

10. N. B. Brandt and N. I. Ginzburg, Zh. Éksp. Teor. Fiz., 39:1554 (1960).
10a. D. Köhnlein, Z. Phys., 208(2):142 (1968).
11. B. G. Lazarev, E. E. Semenenko, and A. I. Sudovtsev, Zh. Éksp. Teor. Fiz., 40:105 (1961).
11a. R. L. Falge, Phys. Lett., A24(11):579 (1967).
11b. B. G. Lazarev, L. S. Lazarev, E. E. Semenenko, V. I. Tutov, and S. I. Goridov, Summaries of Contributions to the Sixteenth All-Union Conference on Low-Temperature Physics, Leningrad (1970), p. 6.
11c. P. E. Frieberthauser and H. A. Notarys, J. Vac. Sci. and Technol., 7(4):485 (1970).
11d. A. D. C. Grassie and D. B. Green, J. Phys. Chem. Solid State Phys., 3(7):1575 (1970).
12. T. R. R. McDonald, E. Gregory, G. S. Barberich, D. B. McWhan, T. H. Geballe, and C. W. Hull, Phys. Lett., 14(1):16 (1965).
13. B. T. Matthias and I. L. Olsen, Phys. Lett., 17:187 (1965).
13a. Science News, 93(24):573 (1968).
14. W. Buckel and I. W. Wittig, Phys. Lett., 17:187 (1965).
14a. J. E. Gordon, H. Montgomery, et al., Phys. Rev., 152:432 (1966).
14b. J. Witting, Phys. Rev. Lett., 24(15):812 (1970).
14c. N. B. Brandt and I. V. Berman, Summaries of Contributions to the Sixteenth All-Union Conference on Low-Temperature Physics, Leningrad (1970), p. 53.
15. New Scientist, 27:459 (1965).
15a. N. E. Alekseevskii, Uspekhi Fiz. Nauk, 95(2):253 (1968).
15b. T. M. Wu, Phys. Lett., A30(6):347 (1969).
15c. N. B. Brandt and N. I. Ginzburg, Uspekhi Fiz. Nauk, 98(1):1901 (1969).
16. E. A. Linton, Superconductivity [Russian translation], Izd. Mir, Moscow (1964), p. 12.
17. R. E. Glover and H. T. Coffey, Rev. Mod. Phys., 36(1):299 (1964).
18. B. G. Lazarev, E. E. Semenenko, and V. M. Kuzmenko, Fiz. Met. Metallov., 23:4651 (1966).
19. N. B. Brandt, N. I. Ginzburg, and I. V. Berman, Transactions of the Tenth International Conference on Low-Temperature Physics, Izd. VINITI, Moscow (1967), pp. 194-195, 269.
20. N. V. Zavaritskii, Dokl. Akad. Nauk SSSR, 86(3):501 (1952).
21. B. G. Lazarev and A. Galkin, Zh. Éksp. Teor. Fiz., 8:37 (1944).
22. G. Minnigerode, Z. Phys., 154:44 (1959).
23. C. P. Bean, M. V. Doyle, and A. C. Pincus, Phys. Rev. Letters, 9:92 (1962).
24. E. Adam and O. Smoluchowski, Transactions of the Tenth International Conference on Low-Temperature Physics, Izd. VINITI, Moscow (1967), p. 195.
25. I. H. P. Watson and Corning, Transactions of the Tenth International Conference on Low-Temperature Physics, Izd. VINITI, Moscow (1967), p. 194.
26. P. Powler, J. Appl. Phys., 34(12):3538 (1963).
27. D. Gerstenberg and P. M. Hall, J. Electrochem. Soc., 111(8):936 (1964).
28. K. S. Tedeman, R. M. Rouse, and D. Wulff, in: New Materials for Electronics, Izd. Metallurgiya, Moscow (1967), p. 84.
29. G. I. Gurp, Philips Res. Repts., 22(1):10 (1967).

30. D. Kramer and C. G. Rhodes, Trans. AIME, 233(1):192 (1965).
31. I. Ohtsuka, I. Shibuga, and T. Fukoroi, Sci. Repts. Res. Inst. Tohoku Univ., A15:67 (1963).
32. E. A. Lynton, B. Serin, and M. Zwiker, J. Chem. Phys. Solids, 3:165 (1957).
33. G. Chanin, E. A. Linton, and B. Serin, Phys. Rev., 114:719 (1959).
34. H. Gamari, Ph.D. Thesis, London Univ. (1964).
35. D. Farrel, I. G. Park, and B. R. Coles, Phys. Rev. Letters, 13:328 (1964).
36. D. Quinn, I. Park, and B. R. Coles, Phys. Rev. Letters, 13:326 (1964).
37. G. Boato, G. Gallinaro, and C. Rizzuto, Rev. Mod. Phys., 36(1):162 (1964).
38. D. P. Seraphim, C. Chion, and D. I. Quinn, Acta Metallurgica, 9:861 (1961).
39. J. K. Hulm and R. D. Blaugher, Phys. Rev., 123(5):1569 (1963).
40. W. de Sorbo, in: New Materials and Methods of Studying Metals and Alloys [Russian translation], Izd. Metallurgiya, Moscow (1966), pp. 99 and 113.
40a. G. W. Webb, Phys. Rev., 181(3):1127 (1969).
41. R. P. Elliott, Trans. AIME, 52:990 (1960).
41a. C. D. Wiseman, J. Appl. Phys., 37(9):3599 (1966).
42. A. U. Seybolt, J. Inst. Metals, 6:774 (1954).
43. E. Gebhardt, E. Fromm, and R. Rothenbacher, Metall, 18(7):691 (1964).
44. J. K. Hulm, C. K. Jones, R. Mazelsky, R. C. Miller, R. A. Hein, and J. W. Gibson, Low Temperature Physics LT9, Plenum Press, New York (1965), p. 600.
44a. K. Hechler and E. Saur, Z. Phys., 205(4):392 (1967).
45. E. Gebhardt, E. Fromm, and D. Jakob, Z. Metallkunde, 55(8):423 (1964).
46. W. Rostocker, Metallurgy of Vanadium [Russian translation], IL, Moscow (1959), p. 148.
47. A. Taylor and N. J. Doyle, J. Less-Common Metals, 13(4):399 (1967).
48. F. H. Horn and W. T. Ziegler, J. Amer. Chem. Soc., 69:2762 (1947).
48a. R. I. Walter and W. T. Chandler, Trans. Metallurg. Soc. AIME, 233(4):762 (1965).
48b. V. I. Sokolov and B. B. Kuz'min, Summaries of Contributions to the Sixteenth All-Union Conference on Low-Temperature Physics, Leningrad (1970), p. 68.
49. S. Komjathy, J. Less-Common Metals, 6(6):466 (1960).
49a. A. L. Giorgi, E. G. Szklarz, and E. K. Storms, Phys. Rev., 129(4):1524 (1963).
50. G. C. Rauch, R. M. Rose, and J. Wulff, J. Less-Common Metals, 11:99 (1965).
50a. C. Sulkowsky and J. Mazur, Trans. Tenth Internat. Conf. on Low-Temperature Physics, Izd. VINITI, Moscow (1967), p. 88.
50b. S. Rudy, St. Windisch, and C. E. Brukl. Plansee Berichte für Pulvermetallurgie, 16(1):3 (1968).
51. A. V. Narlikar and D. Dew-Hughes, Phys. Stat. Sol., 6(2):383 (1964).
51a. R. H. Willens, E. Buchler, and B. T. Matthias, Phys. Rev., 159(2):327 (1967).
51b. Yu. V. Efimov, V. V. Baron, and E. M. Savitskii, Vanadium and Its Alloys, Izd. Nauka, Moscow (1969).
52. J. A. Catterall, J. Williams, and I. E. Duke, Brit. J. Appl. Phys., 15:1369 (1964).
52a. V. R. Karasik and I. Yu. Shabalin, Zh. Éksp. Teor. Fiz., 57(6):1973 (1969).
53. I. S. Rajput and A. K. Gupta, Phys. Letts., A24(5):260 (1967).
54. W. de Sorbo, Rev. Mod. Phys., 36(1):91 (1964).
55. I. A. Gatterall, I. Williams, and D. F. Duke, Brit. J. Appl. Phys., 15:1369 (1964).

56. D. Hauser, in: Superconducting Materials [Russian translation], Izd. Mir, Moscow (1965), p. 203.
57. W. A. Little, Phys. Rev., 134A:1416 (1964).
58. V. L. Ginzburg, Uspekhi Fiz. Nauk, 95(1):91 (1968).
58a. F. Meunier, Transactions of the Bakurian Colloquium on Super-Fluidity and Superconductivity, Vol. 2, Tiflis, 1968 (1969), p 29.
58b. D. G. Naugle, Phys. Lett., A25(9):688 (1967).
59. N. W. Ashcroft, Phys. Rev. Lett., 21(26):1748 (1968).
60. A. A. Abrikosov, Astron. J., 31:112 (1954).
61b. Superconductivity Summer School of the Low-Temperature Technical Committee of the German Physical Society, Siemens AG Research Laboratories (1970).

Chapter III

Superconducting Compounds

At the present time there are more than a thousand known superconducting compounds with various crystal structures and a multitude of their alloys [1]. The critical superconducting temperature of the various metallic compounds lies between 0.012 (SnTe) and 20.98°K [$Nb_{0.79}(Al_{0.75}Ge_{0.25})_{0.21}$] [1, 2, 2a, 2b]. Certain metallic compounds have the highest critical magnetic fields and critical superconducting temperatures of all known superconductors [3].

Laue [4] showed empirically that there was a relationship between superconducting properties and the crystal structure of superconductors. The crystal lattices of all superconductors have a center of symmetry [4, 5]. Superconductivity is the more likely, the higher the symmetry of the crystal lattice. The dependence of superconductivity on crystal structure arises from the fact that the electron structure of the atoms and their spatial arrangement (crystal structure) depend on the electron density and electron energy spectrum of the metal, alloy, or compound at the Debye temperature. These determine the excitation spectrum in the immediate vicinity of the Fermi surface, which gives an indication of the possibility that the material in question may be capable of passing into the superconducting state [6, 7].

The highest superconducting properties are those of compounds with the Cr_3Si (formerly β-W) structure [3, 8]. Fairly high transition temperatures occur in sigma and Laves phases, carbides and nitrides with a cubic structure of the NaCl type, and phases of the α-Mn variety. Other crystal structures are less favorable for superconductivity. Compounds with cubic structure usually have higher superconducting transition temperatures than compounds with a hexagonal lattice [9]. Laue [4] gave a list of twenty-one classes of crystal-lattice symmetry characterizing compounds without superconductivity (C_1, D_{2d}, D_{3h}, C_s, C_{4v}, C_{6v}, C_2, D_4, D_6, C_{2v}, C_3, T, D_2, C_{3v}, T_α, S_4, D_3, O, C_4, C_6, C_{3h}). The reasons for the preferential superconductivity of certain crystal structures have never been fully elucidated, although certain hypotheses exist (these we shall discuss later).

COMPOUNDS WITH THE Cr_3Si STRUCTURE

The compound Cr_3Si discovered in 1933 [10] has a cubic structure (space group $Pm3h-O_h^3$, z = 2) [11]. This compound was originally assigned the structure of β-W. A description of the β-form of tungsten was presented in 1931 [12]. The production of the β form of metallic tungsten by reducing the lower oxide with dry hydrogen at temperatures below 520°C was described subsequently [13-15]. Later it was found that pure tungsten did not exhibit polymorphism [16]. The so-called β-tungsten contained approximately 25 at.% of oxygen. The oxygen atoms were distributed in a random manner in the structure of the oxide W_3O [17]. The existence of the compound W_3O was denied by some authors [18]. In the crystal lattice of Cr_3Si, Cr_3O, and other similar compounds the atoms were distributed in an orderly manner. It was therefore decided to refer to this group of compounds as being characterized by the "Cr_3O" type of structure [19]. However, in the opinion of Philipsborn [20] a more reasonable and appropriate title (bearing in mind the history of the corresponding discovery) was the "Cr_3Si" structural type.

A representation of the Cr_3Si type of structure is presented in Fig. 40 [11, 20]. In the unit cell of the Cr_3Si compound the atoms occupy structurally equivalent positions of two types. The chromium atoms lie in the 6(c): ±($\frac{1}{2}$, $\frac{1}{4}$, 0) positions; the silicon atoms

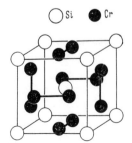

Fig. 40. Structure of the Cr_3Si type.

occupy the positions 2(a): (0, 0, 0; $\frac{1}{2}$, $\frac{1}{2}$, $\frac{1}{2}$). Equal coordinates may also be obtained for the T_d^4–$P\bar{4}3n$ and O^2–$P4_232$ space groups; however, these groups imply a lower lattice symmetry.

In A_3B compounds with Cr_3Si type of structure the B atoms occupy positions at the vertices and in the center of the cube (Fig. 40). The B atoms form a bcc substructure. Each B atom is surrounded by twelve A atoms forming a crystallographic icosahedron (combination of a pentagonal dodecahedron with an octahedron). The interatomic distance AB equals $a_0\sqrt{5/4}$ (a_0 is the lattice constant). There is no closing together of the B atoms. The eight B atoms at a distance of $a_0\sqrt{3/2}$ are the nearest such neighbors of each particular B atom. Each A atom lies in the center of an irregular tetrahedron of four B atoms. An A atom is surrounded by fourteen neighbors lying at different distances from it. The nearest neighbors of an A atom are two other A atoms at a distance $a_0\frac{1}{2}$. Eight other A atoms lie slightly further away (at a distance of $a_0\sqrt{6/4}$, 22% greater than the shortest A–A distance). If we take the AB distance as unity, then the remaining AA_1, AA_2, and BB distances respectively equal 0.89, 1.09, and 1.55 [20]. The high effective coordination numbers of the atoms (B = 12, A = 14) in the Cr_3Si type of structure make this very similar to structures of the close-packed cubic variety [21]. With the A atoms in the 6(c) position and the B atoms in the 2(a) positions the structure of compounds of the Cr_3Si type is ordered. In a disordered structure the A and B atoms are statistically distributed over the 2(a) and 6(c) positions.

In phases of composition A_3B with a structure of the Cr_3Si type the A atom and its three nearest neighbors form three chains perpendicular to one another and parallel to the cube axes (the $\langle 100\rangle$ direction). The atomic radius of the A atom calculated from the interatomic distance in these chains (the minimum for the structure under consideration is $a_0\frac{1}{2}$) is 8-14% smaller than the atomic radius generally accepted for the pure elements [22]. This indicates the presence of additional covalent bonds in the chains [17, 23]. Geller [24-26] calculated the effective radii of the elements for structures of the Cr_3Si type. These radii differ slightly from the atomic radii of Goldschmidt. Geller's calculation was criticized by Pauling [27, 28]. However, Geller's atomic radii give the lattice constants of phases with the Cr_3Si type of structure to an accuracy of ± 0.03 Å. The calculation is based on the fact that to a first approximation (to the accuracy indicated) the difference in lattice constants on changing from one type of A component to another is independent of the nature of the B component [29]. The chains of A atoms determine the unit cell dimensions of a structure of the Cr_3Si type [17]. For the same A atom, the lattice spacings of A_3B compounds of this structure are proportional to the effective radius of the B atom, which may be either greater than or less than the atomic radius of element A [29]. For the same B atom, the lattice spacings of similar compounds are proportional to the atomic radii of the A atoms.

After the discovery of the compound Cr_3Si, the compound V_3Si was also discovered (1939 [30]), then Cr_3Ge and V_3Ge (1944 [31]), Mo_3Si (1950 [32]), Mo_3Os and Mo_3Ir (1951 [33]), V_3Co and Ti_3Pt (1952 [34, 35]), V_3Ni (1954 [36]) and others. At the present time sixty-five compounds with the Cr_3Si structure are known [1, 8, 20, 37-46]. The A components in compounds with the Cr_3Si structure are transition metals belonging to groups IV-VI in the periodic table [20, 29, 47].

The following list gives the A (left) and B (right) components forming compounds with the Cr_3Si type of structure; the brackets indicate the atomic radii of the two components [38]:

Titanium (1.462) . . . . . Ir (1.357), Pt (1.387), Au (1.442), Hg (1.573), Sb (1.59)
Zirconium (1.602). . . . . Au (1.442), Sn (1.545),* Hg (1.573), Sb (1.59), Pb (1.75)
Vanadium (1.35) . . . . . Ni (1.246), Co (1.252), Si (1.319), Rh (1.345), Os (1.353), Ir (1.357), Ge (1.369), Pd (1.376), Pt (1.387), As (1.39), Ga (1.411), Al (1.432), Au (1.442), Sn (1.545), Cd (1.568), Sb (1.59), In (1.663),† Bi (1.70),† Tl (1.716),† Pb (1.75)
Niobium (1.468) . . . . . . Rh (1.345), Os (1.353), Ir (1.357), Ge (1.369), Pt (1.387), Ga (1.411), Al (1.432), Au (1.442), Sn (1.545), Sb (1.59), In (1.663), Bi (1.7), Pb (1.75)
Tantalum (1.467) . . . . . Au (1.442), Sn (1.545), Sb (1.59)
Chromium (1.282). . . . . Si (1.319), Ru (1.339), Rh (1.345), Os (1.353), Ir (1.357), Ge (1.369), Pt (1.387), Ga (1.411)
Molybdenum (1.4) . . . . Be (1.128), Si (1.319), Os (1.353), Ir (1.357), Tc (1.36),‡ Ge (1.369), Re (1.37), Ga (1.411), Al (1.432)
Tungsten (1.408) . . . . . Si (1.319), Re (1.37)

---
*The compound has the composition Zr_4Sn [41].
†The compound is obtained in a diffusion layer [37].
‡A compound $Mo_{0.46}Tc_{0.54}$ occurs with the Cr_3Si structure [39, 40].

It is characteristic that the number of compounds formed by metals belonging to the same group of the periodic table falls on passing from an element of the first long period to an element of the third long period. This selectivity is partly due to the relative size of the atoms. In order to form phases of the Cr_3Si type it is essential that the difference in the atomic radii of the components corresponding to the coordination number 12 should be smaller than 15% [29]. However, there is another factor which influences the formation of such phases. Vanadium forms the greatest number of phases of this kind. The absence of any such phases formed between niobium and silicon, nickel, or cobalt may be attributed to the unfavorable relationship of the atomic radii; the absence of compounds such as Nb_3As and Nb_3Pd, however, cannot be explained by the size factor, as there is a favorable relationship between the sizes of the niobium and other atoms in these cases. Tantalum has an atomic radius almost equal to that of niobium, yet forms far fewer phases of the Cr_3Si type. Alloys of hafnium (a group IV element) exhibit no such phases at all (although the possible existence of hafnium phases with a Cr_3Si structure has been affirmed in view of the presence of outer d electrons in this case [48]). Among tungsten alloys only two such compounds are known, although it has been suggested that phases of the Cr_3Si type might occur in the W–Ir and W–Os systems [45]. All this shows how limited the effect of the size factor really is. The electron characteristics clearly exert a major influence. The effect of the electron relationships of the components on the stability of phases of the Cr_3Si is confirmed by the changes in the compositions of the phases in the Cr–Pt, Cr–Ir, and Cr–Os systems [42]. However, this question has never been finally resolved [47].

New types of Cr_3Si compounds such as Nb_3Si and Nb_3In formed only at high temperatures and pressures have been mentioned [43, 46, 49, 50], as well as others obtained by diffusion in thin films (V_3In, V_3Tl, V_3Bi [37]).

It is considered that the production of a compound V_3Al will enrich this group of compounds by providing a superconductor with a high value of T_c (10-17°K) [2a, 50a, 50b]. The existence of a V_3Al phase with the Cr_3Si type of structure (a = 4.926 Å) in alloys was first observed in [50c]. However, no success was achieved in producing this phase in a homogeneous, single-phase state in subsequent investigations [50a, 50b, 50d, 50e, 50f]. The formation of this phase is probably due to the influence of impurities. It is also possible that it may be obtained under extreme conditions.

In all known compounds of the Cr_3Si type the B components are elements (metals or nonmetals) lying to the right of the Mn–Tc–Re line in the unfolded Mendeleev Table [17, 29, 45, 51]. Whereas the number of elements capable of constituting the A components is limited, the B components may be elements varying substantially electrochemically.

The ratio of the atomic radii of the elements constituting the components of compounds with the Cr_3Si structure lies in the range $0.81 \leq r_B/r_A \leq 1.08$ [17, 47]; for the majority of compounds $r_B/r_A \approx 1$. It was shown earlier [52] that compounds of vanadium with the Cr_3Si type of structure were characterized by a wider range of atomic size ratios of the components (0.925-1.3).

The majority of compounds with structures of the Cr_3Si type have narrow ranges of homogeneity. However, deviations from stoichiometric composition may in fact occur [20]. In binary vanadium systems, for example, in the case of a large difference in the atomic radii of the components, phases of the Cr_3Si type have an exact stoichiometric composition or a very narrow range of homogeneity (1-3 at.% of one of the components), and they are also formed by peritectic reactions [52]. For a small difference in the atomic size ratio ($0.97 \leq r_B/r_A \leq 1.065$), however, these phases are formed in vanadium systems by way of the ordering of a solid solution with a bcc lattice, and they then have quite a wide range of homogeneity [52]. The ranges of homogeneity of Cr_3Si-type compounds extend in the direction of a higher content of the transition metal [20]. The chains of A atoms remain unbroken in this case, while the excess atoms occupy the positions of the B component. If the chains of A atoms are not interrupted by B atoms, then the B atoms never occupy the A positions. Only weak bond forces therefore exist between the B atoms, while the A atoms are strongly coupled. Chains of A atoms are a characteristic feature of the Cr_3Si type of structure [20]. The existence of chains of A atoms clearly explains the formation of the filamental structure of the compound V_3Si noted earlier [53]. An extension of the range of homogeneity in the sense of an increase in the content of the B component only appears at high temperatures (for example, in the case of the V–Ga system [54]), and is evidently associated with at least a partial disruption of the chains of A atoms. There are two phases, VOs [55] and Tc_3Mo_2 [56], with the Cr_3Si type of structure with a deviation from stoichiometric composition very unusual for this type of phase; the compound Tc_3Mo_2 has a high critical temperature (15°K [56] or 13.40°K [40]).

The structure of the Cr_3Si type is particularly favorable for superconductivity [3, 8, 9, 57-61]. Of the sixty-five known binary compounds with this structure, superconductivity is found in thirty-four (Table 12), in nine of these the T_c is greater than 10°K.

The superconductivity of certain compounds has never yet been studied; it is quite possible that the remaining Cr_3Si-type compounds also possess superconductivity, but only at temperatures close to absolute zero. According to Matthias [59, 64c], the maximum T_c values occur for compounds with an electron concentration of about 4.7 and 6.5 electrons/atom.

It is interesting to note that in certain investigations the superconducting properties of these compounds were associated not with a cubic lattice of the Cr_3Si type but with a tetragonal structure formed at low temperatures as a result of a martensitic transformation. Thus a martensite-type transformation (detected by x-ray diffraction) was observed in V_3Si single crystals at temperatures above the critical point T_c [65]. Below the temperature of the martensite transformation the degree of tetragonality of the lattice (the axial ratio c/a) became steadily greater, reaching a final value of 1.0025. The structure of Nb_3Sn in the superconducting state is also tetragonal with $c/a = 1.0042$ [65a]. The martensitic transformation in Nb_3Sn and V_3Si is a phase transformation of the first kind [65b]. On studying thin films of V_3Si under the electron microscope at 11-300°K it was found that down to 20°K the structure of these samples remained almost constant. Below 20°K a complex banded structure developed, apparently as a result of stresses. The distance between the bands was 100-250 Å, while their thickness was 100 Å. An analogous banded structure was observed elsewhere [65d] at temperatures of under 50°K in the compound V_3Ga. It was considered that these regions might accommodate fixed filaments of magnetic flux in a superconducting state. Several authors [67-69, 69a] mentioned a sudden change in the specific heat and electrical resistance of the compounds V_3Si, V_3Ga, V_3Ge, V_3Sn, and Nb_3Sn at low temperatures. For V_3X com-

TABLE 12. Critical Superconducting Temperature of Compounds with the Cr_3Si Structure [1, 40, 44, 49, 56, 61-64b]

Compound	T_c, °K	$T_{expt.}$, °K	a, Å	r_A/r_B	Electron concentration, electrons/atom
Ti_3Sn	5.80	—	5.217	0.920	4.25
Ti_3Ir	5.40	—	5.009	1.077	5.25
Ti_3Pt	0.58	—	5.032	1.054	5.5
Ti_3Au	—	1.20	5.096	1.014	3.25
Ti_3Hg	—	1.20	5.188	0.929	3.5
Zr_3Au	0.92	—	5.482	1.111	5.75(11)* or 3.25(1)
Zr_3Pb	0.76	—	—	0.916	4.0
Zr_4Sn	0.85	—	5.65	1.037	4.0
Zr_3Sb	—	1.20	5.634	1.006	4.25
Zr_3Hg	—	1.20	5.558	1.018	3.5
V_3Si	17.0; 17.1	—	4.722—4.728	1.020	4.75
V_3Ga	16.8	—	4.816 (4.83)	0.954	4.5
V_3In	13.9	—	5.28—5.56	0.811	4.5
V_3Sn	7.0; 3.8	—	4.94—4.96	0.871	4.75
V_3Ge	6.01	—	4.769	0.933	4.75
V_3Pt	2.83	—	4.814 (4.808)	0.970	6.25
V_3Sb	0.80	—	4.941	0.846	5.0
V_3Au	0.74	—	4.883	0.933	6.5(11)* or 4(1)
V_3Rh	0.98	—	4.784 (4.767)	1.001	6.0
V_3As	—	1.00	4.74 (4.75)	0.968	5.0
V_3Co	—	0.35	4.675	1.075	6.0
V_3Ir	—	0.35	4.786 (4.785)	0.992	6.0
$V_{2.67}Ir_{0.33}$	1.39	—	—	—	—
V_3Cd	—	4.20	4.92—4.95 (4.806)	0.861	4.25
V_3Tl	—	4.20	5.21—5.25	0.787	4.5
V_3Bi	—	4.20	4.72	0.794	5.0
V_3Ni	—	4.20	4.70 (4.710)	1.082	6.25
V_3Pb	—	4.2	4.937	0.771	4.75
V_3Al	—	4.2	4.926	0.942	4.5
V_3Pd	—	4.2	4.816	0.978	6.25
$Nb_{0.8}Sn_{0.2}$	18.5	—	5.290	0.952	4.8
Nb_3Sn	18.05	—	5.289	0.952	4.75
Nb_3Al (cast)	17.5; 18.0	—	5.187	1.025	4.5
Nb_3Al (annealed)	17.1	—	5.186	1.025	4.5
Nb_3Ga (annealed)	14.9; 16	—	5.171	1.040	4.5
Nb_3Ga (sintered)	12.5—13.2	—	—	1.040	—
Nb_3Au	11.5	—	5.2027	1.018	6.5(11)*
Nb_3In	9.2	—	5.303	0.883	4.5
Nb_3Pt	9.20—10.9	—	5.153	1.058	6.25
Nb_3Ge	6.9	—	5.166 (5.174)	1.072	4.75
Nb_3Ge	go 17	—	—	—	—
Nb_3Rh	2.5—2.64	—	5.115	1.091	6.0

*On the assumption that the valence of the gold equals 11.
†At high pressure and high temperature.

TABLE 12 (Continued)

Compound	T_C, °K	$T_{expt.}$, °K	a, Å	r_A/r_B	Electron concentration, electrons/atom
Nb_3Bi †	2.25	—	5.320	0.864	5.0
Nb_3Ir	1.7	—	5.131	1.082	6.0
Nb_3Os	1.05	—	5.121	1.085	5.75
Nb_3Sb	1.02	—	5.262	0.923	5.0
Nb_3Pb	—	—	5.270	0.839	4.75
Ta_3Au	13—16	—	5.222	1.017	4.0
$Ta_{4.3}Au$	0.55	—	—	—	—
Ta_3Sn	8.35	—	5.276—5.278	0.950	4.75
Ta_3Sb	—	—	5.259—(5.260)	0.923	5.0
Cr_3Os	4.03	—	4.677	0.948	6.50
Cr_3Ru (cast)	2.15	—	4.683	0.957	6.50
Cr_3Ru (sintered)	3.3	—	—	0.957	—
Cr_3Rh	0.06	—	—	—	—
Cr_3Ir	0.45	—	4.668	0.945	6.75
Cr_3Ga	—	1.02	4.645	0.909	5.25
Cr_3Ge	—	1.20	4.623	0.936	5.5
Cr_3Pt	—	0.30	4.706	0.924	7.0
Cr_3Si	—	1.20	4.645	0.972	5.5
Cr_3Rh	0.06	—	4.656	0.953	6.75
Mo_2Tc_3	15—13.4	—	4.934	1.029	5.4
Mo_3Ir	8.8; 9.6	—	4.972	1.031	6.75
Mo_3 $(Mo_{0.4}Pt_{0.6})$	8.8	—	—	1.010	6.6
Mo_3Os	7.20	—	4.973	1.034	6.5
Mo_3Ge	1.43	—	4.933	1.021	5.5
Mo_3Si	1.30	—	4.893	1.061	5.5
Mo_3Ga	0.76	—	4.943	0.992	5.25
Mo_3Al	0.58	—	4.950	0.977	5.25
Mo_3Sn *	—	0.35	—	—	—
Mo_3Be	—	1.20	4.89	1.242	5.0
W_3Si	—	1.20	4.910	1.069	5.5
Cr_3O	—	1.02	—	1.441	6.0
V, Mn_3Si	9.6-12.5	—	—	—	—

pounds with a structure of the Cr_3Si type there is a specific interrelationship between the temperature of the martensite transformation and the temperature corresponding to the transition into the superconducting state ($T_m/T_c \approx 1.5$), as may clearly be seen from the following data [68-70]:

Compound	V_3Si	V_3Ga	V_3Ge	V_3Sn	Nb_3Sn
T_c, °K	16.8	14.1	6.9	3.7	18.0
T_m, °K	23.5—28.6	20.6	10.0	5.4	36.0
T_m/T_c.	1.4—1.7	1.5	1.45	1.46	2.0

The relation between the phase transformation and superconductivity is supported by the absence of such a transformation in V_3Ir ($T_c < 0.3°K$) down to 4.6°K [71]. However, in certain superconducting samples of V_3Si also no martensite transformations appear [67, 72]. In tin-enriched samples of Nb_3Sn also, no phase transformation is observed [65b]. Thus the question as to the relation between the superconductivity of Cr_3Si-type compounds and the structural transformation requires a more detailed study.

Recent x-ray work established the existence of a martensitic transformation in tin-enriched superconducting $Nb_3(Sn, Sb)$ alloys [72a].

The crystal structure and physical properties (including superconductivity) of various materials are determined by the electron structure. Attempts have been made to determine the electron structure of compounds of the Cr_3Si type and establish their relationship to the crystal structure and critical temperatures. Weger [73] noted considerable anisotropy in the electron properties of compounds of the Cr_3Si type due to their structural characteristics. The presence of chains of A atoms along the (100) type of axis (Fig. 41a) determines the quasiunidimensional character of electron motion in compounds of this type. The Fermi surface is extremely anisotropic and is formed by planes of the (100) type (Fig. 41b). The severe temperature dependence (in the normal state) of the nuclear magnetic-resonance frequency displacement (the Knight shift) [74], the electron magnetic susceptibility [74, 75], and also the quantity $1/T_1 T$ (T_1 is the longitudinal nuclear magnetic-resonance relaxation time) [73] indicates a considerable temperature dependence of the density of states at the Fermi sur-

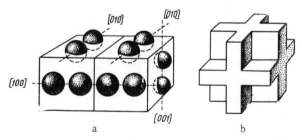

Fig. 41. Crystal structure of compounds of the V_3X type (a) and Fermi surface of the 3d electrons of vanadium in the strong-coupling approximation (b).

face. There is also a substantial change in the band structure of Nb_3Sn on varying the temperature below 100°K [76]. The dependence of the Knight shift, the magnetic susceptibility, and the quantity $1/T_1T$ on temperature is the more considerable, the higher the superconducting transition temperature. This relationship is quite natural, since the effective attraction between the electrons and the temperature dependence of the density of states on the Fermi surface are due to the same cause: electron–phonon interaction [71]. Since the strong electron–phonon interaction in superconducting compounds of the Cr_3Si type is due to the quasiunidimensional motion of the electrons, it is reasonable to suppose that in compounds of this type which exhibit no superconductivity down to extremely low temperatures (for example, V_3Ir) the "unidimensionality" in the motion of the electrons is largely obliterated, i.e., the wave functions of the electrons lying along different axes of the crystal lattice of these compounds overlap each other to a considerable extent [71]. The existence of such relationships facilitates the motion of the electrons in at least one direction without intersecting planes and conical atomic nodal surfaces, and is probably a cause of superconductivity [76a]. However, existing data [73-76] are insufficient to provide quantitative relationships indicating the critical temperatures of the compounds.

Only empirical and semiempirical relationships in fact exist for determining the T_c values of superconducting compounds as functions of their other parameters (to a certain degree of approximation). Thus the critical temperature of compounds with the Cr_3Si type of structure formed by niobium and vanadium with elements of subgroups III B and IV B increases with diminishing atomic and mass number of the B components: Nb_3In (9.2°K [50]) → Nb_3Ga (14.5°K [77]) → Nb_3Al (17.1-18.0°K [77-80]); V_3Sn (3.8-6.0°K [72, 80]) → V_3Ge (6.01°K [58]) → V_3Si (17.0°K [58, 81].

Figure 42a shows the mean atomic volumes and structural data for compounds of the Cr_3Si type with known critical temperatures [62].* The majority of compounds fall into the wide band of atomic volumes indicated by the broken lines. The highest T_c corresponds to the greatest mean atomic volume. Only the compounds

---

*In this and subsequent figures the arrows indicate the temperature at which the material has still not passed into the superconducting state.

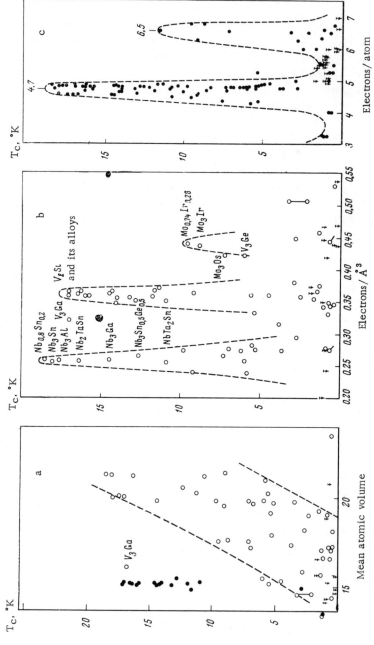

Fig. 42. Dependence of the critical temperature T_C of compounds with the Cr_3Si structure on their mean atomic volume (a), the mean number of valence electrons per unit volume (b), and the mean electron concentration (c).

V_3Si, V_3Ga, and some of their alloys (black circles in Fig. 42a) have a small atomic volume and a high T_c [1, 29]. This graph takes no account of the degree of ordering in the structure of the compounds. In the curve relating T_c to the mean number of valence electrons per unit volume (Fig. 42b) three maxima occur.

For the majority of materials with high transition temperatures the number of valence electrons per unit volume is 0.26 electron/$Å^3$. The maximum lies rather higher for the silicides; this corresponds to the small atomic volume occupied by the silicon atoms. Certain compounds of vanadium and molybdenum lie higher still. There is a sharper correlation between T_c and the mean number of valence electrons per atom (Fig. 42c). This clearly shows the extent to which the density of states affects T_c and confirms the validity of the empirical relationship of Matthias. The central points of the maxima correspond to the lower boundary of the peaks derived from the Matthias rule (4.7 and 6.5 electrons/atom). A similar relationship is found on studying other compounds and metallic solid solutions [82]. There are only five compounds with the Cr_3Si structure in which the mean electron concentration equals four electrons/atom or under; the critical temperatures of four of these compounds agree closely with the Matthias rule on assuming that the valence of gold has risen to 11 [83].

In addition to their high superconducting transition temperatures, compounds with the Cr_3Si structure have very high critical magnetic fields [1, 3, 84-89]. The greatest critical magnetic fields among all known superconductors are those of the compound $Nb_3(Al, Ge)$ (420 kOe at 4.2°K) [2c, 88a], Nb_3Al (250 kOe at 0°K) [89a], and Nb_3Sn (245 kOe at 0°K [89] and around 200 kOe at 4.2°K [90, 91]). It was considered until recently that V_3Ga had the highest critical magnetic field (350-500 kOe [92-94]), i.e., dH_c/dT exceeded 40 kOe/deg. These values evoked doubts, since the upper critical field appeared to be twice the maximum value calculated from the paramagnetic energy of the normal state [95]. On refining the data in question it was found that V_3Ga was inferior to the aforementioned compounds not only with respect to T_c but also with respect to the critical magnetic field (210 kOe at 0°K [89] and 196 kOe at 4.2°K [96]). The lower values of the critical field may be explained as being due to the effect of Pauli paramagnetism [96]. The critical field of V_3Si at 0°K equals 235 kOe; however, at 5-11°K the critical fields of V_3Si and Nb_3Sn are equal to one another [89].

A study of the tunnel effect in a multilayered film sample of $V_3Si/Al/AlO_x/Pb$ at 1.3-1.7°K revealed an energy gap of $\Delta = 2.8 \pm 0.2$ mV in V_3Si [97, 98]. On superimposing a longitudinal magnetic field (H = 50 kOe) at 4.2°K the gap remained unaffected. The gap only completely vanished from V_3Si at 17°K, i.e., at the critical temperature of this superconductor. Other authors [99] studied the I–U tunnel characteristics of a point contact between an electrolytically-oxidized niobium or tantalum needle and macroscopic samples of superconducting compounds; they obtained the 1.2°K energy gaps of Nb_3Sn ($\Delta = 2.8$ mV, $2\Delta/kT_c = 3.6$), V_3Ge ($\Delta = 0.85$ mV, $2\Delta/kT_c = 3.2$), and V_3Si ($\Delta = 1.3$ mV, $2\Delta/kT_c = 1.8$). In some cases [99] a value of $\Delta = 2.8$ mV ($2\Delta/kT_c = 3.8$) was also obtained for the compound V_3Si. The reason for this effect was not fully established. It may have been due to the existence of several discrete values for the expanded energy gap. Other authors [100] indicated the possible existence of two energy gaps in Nb_3Al and V_3Sn; the values of these were estimated from the nuclear relaxation of Nb^{93} and Al^{27} in Nb_3Al and V^{51} and Si^{119} in V_3Sn: 1.0 kT_c, 2.3 kT_c, 1.2 kT_c, and 3.5 kT_c respectively. The anisotropy of the gap in the energy spectrum of superconductors is discussed elsewhere [100a].

INTERSTITIAL PHASES AND CERTAIN OTHER COMPOUNDS OF METALS WITH NONMETALS

In the foregoing we considered the properties of silicides with the Cr_3Si structure. However, there is a very large number of metallic phases with structures of other types; these are formed by the interaction of metals with nonmetals: carbon, nitrogen, oxygen, hydrogen, boron, silicon, and others. Table 13 gives the T_c of compounds formed between transition metals and nonmetals.

The highest superconducting transition temperatures are those of niobium, molybdenum, vanadium, tantalum, and tungsten with nonmetals. The critical temperatures of the compounds formed by any particular transition metal with various nonmetals decreases in the following order: nitride, carbide, boride, silicide [101, 106]. The lowest T_c occur for germanides. In the compounds formed by various transition metals with one particular nonmetal, the probability that superconducting properties will be observed increases with diminishing acceptor capacity of the transition metal

SUPERCONDUCTING COMPOUNDS   121

TABLE 13. Critical Temperatures of Carbides, Nitrides, Borides, Silicides,* and Germanides [1,101-105a]

Nitride	T_C, °K	Carbide	T_C, °K	Boride	T_C, °K	Silicide	T_C, °K	Germanide	T_C, °K
ScN	<1.38 †	ScC	<1.33	ScB_2	<1.30	$ScSi_2$	<1.0	$ScGe_2$	1.30—1.31
				ScB_4	<1.34				
—	—	YC	<1.33	YB_6	<1.30	YSi_2	<0.30	YGe_2	3.8
LaN	1—1.35	YC_2	3.88	LaB_6	<1.30	$LaSi_2$	2.5	$LaGe_2$	1.49
		CeC_2	<0.2	CeB_6	Paramagnetic	$CeSi_2$	<1.00	$CeGe_2$	Ferromagnetic
PrN	<1.33	PrC_2	<0.2	PrB_6	»	$PrSi_2$	Ferromagnetic	$PrGe_2$	»
—	—	NdC_2	<0.2	NdB_6 ‡	Superconductor Paramagnetic	—	—	$NdGe_2$	»
—	—	SmC_2	<0.2	SmB_6	<1.28	—	—	—	—
—	—	GdC_2	<0.2	EuB_6	Paramagnetic	—	—	—	—
—	—	TbC_2	<0.2	GdB_6	»	—	—	—	—
—	—	YbC_2	<0.2	TbB_6	<1.28	—	—	—	—
				YbB_6	<1.28				
TiN	4.86—5.8	TiC	<1.20	TiB	<1.20	Ti_5Si_3	<1.20	Ti_5Ge_3	<1.20
				TiB_2	<1.28	TiSi	<1.20	$TiGe_2$	<1.20
						$TiSi_2$	<1.20		
ZrN	8.9—10.7	ZrC	<1.20	ZrB	2.8—3.4	Zr_4Si_3	1.1	Zr_5Ge_3	<0.30
				ZrB_2	<1.80	Zr_5Si_3	<1.20		
						Zr_6Si_5	<1.20	ZrGe some composition	<1.20
						$ZrSi$	<1.20		
						Zr_4Si	<1.20	$ZrGe_2$	0.30
						Zr_2Si	<1.20		
						Zr_3Si_2	<1.20		
						$ZrSi_2$	<1.20		

* The table includes no silicides with the Cr_3Si structure, the superconducting properties of which were considered earlier.
² The sign < means that no superconductivity was found down to the temperature indicated.
‡ At around 3°K certain samples exhibit superconductivity, while others are obvious paramagnetics.

TABLE 13 (Continued)

Nitride	T_c, °K	Carbide	T_c, °K	Boride	T_c, °K	Silicide	T_c, °K	Germanide	T_c, °K
HfN	6.2–6.6	HfC	<1.23	HfB	3.1	HfSi$_2$	<1.02	—	—
VN	7.5–8.2	V$_2$C	<1.20	VB	<1.20	V$_{0.6}$Si$_{0.4}$	<1.00	—	—
V$_5$N$_2$	<1.20	VC	<1.20	VB$_2$	<1.20	V$_5$Si$_3$	<0.30	VGe some composition	<1.20
NbN*									
NbN	15.6	NbC	6.0–11.1	NbB	8.25	VSi$_2$	<1.20	Nb$_2$Ge	1.90
Nb$_2$N	<1.20	Nb$_2$C	<1.98	Nb$_2$B	<1.28	Nb$_2$Si	<1.20	Nb$_5$Ge$_3$	<1.02
				Nb$_3$B$_1$	<1.28	Nb$_5$Si$_3$	<1.20	NbGe$_2$	<1.20
Ta$_2$N	<1.20	Ta$_2$C	<1.93	Ta$_2$B	3.12	NbSi$_2$	<1.20		
TaN	<1.20	TaC	3.3–10.2	TaB	4.0	TaSi	4.28–4.33		
TaN†				Ta$_3$B$_4$	<1.28	Ta$_2$Si$_2$	<1.20	TaGe$_2$	<1.20
				TaB$_2$	<1.20	TaSi$_2$	<1.20		
						Ta$_3$Si$_2$	<1.20		
CrN	<1.28	Cr$_4$C	<1.20	Cr$_2$B	<1.20	Ta$_5$Si	<1.20		
		Cr$_7$C$_3$	<1.20	CrB	<1.28	CrB	<1.20	CrGe	<1.20
		Cr$_3$C$_2$	<1.20	CrB$_2$	<1.28	CrB$_2$	<1.20	Cr$_3$Ge$_2$	<1.20
Mo$_2$N	5.0	Mo$_2$C	2.78–12.2‡	Mo$_2$B	4.74	Mo$_3$Si$_2$	<1.20	MoGe$_{0.7}$	4.20
MoN	12.0	MoC	7.7–13	MoB	0.5	MoSi$_{0.7}$	1.34	MoGe$_2$ (α)	<1.20
				Mo$_2$B$_5$	<1.28	MoSi$_2$	<1.20	Mo$_3$Ge$_2$	<1.20
								MoGe$_2$ (β)	<1.20
W$_2$N	<1.28	W$_2$C	2.74–5.2	W$_2$B	3.10	W$_3$Si$_2$	2.84	—	—
WN	<1.38	WC	<1.28	WB	<1.28	WSi$_2$	<1.20		
				W$_2$B$_5$	<1.28		—		
ReN$_{0.34}$	4–5	—	—	Re$_2$B	2.80	Mn$_3$Si	<4.2	—	—
	—	—	—	—	—		—		
	—	TcC§	3.85	—	—		—		

SUPERCONDUCTING COMPOUNDS

Nitride	T_c	Carbide	T_c	Carbide	T_c	Boride	T_c	Silicide	T_c	Germanide	T_c
—	—	—	—	Fe$_3$C	1.30	—	—	FeSi	<1.28	FeGe$_2$	<1.02
—	—	—	—	—	—	—	—	CoSi$_2$	1.22	CoGe$_3$	<1.02
—	—	—	—	—	—	—	—	CoSi	Not superconducting. (T_{exp} not exactly known)	—	—
—	—	—	—	—	—	Ru$_2$B	<1.20	NiSi	<1.90	—	—
—	—	—	—	RuC	2.00	Ru$_7$B$_3$	2.58	NiSi$_2$	<1.00	—	—
—	—	—	—	RhC	<1.28	Rh$_2$B	<1.28	—	—	Rh$_5$Ge$_2$	2.12
—	—	—	—	—	—	RhB	<0.30	RhSi	<0.30	RhGe	<1.0
—	—	—	—	—	—	Rh$_7$B$_3$	<0.20	—	—	—	—
—	—	—	—	—	—	—	—	PdSi	0.93	PbGe	<0.30
—	—	—	—	—	—	Os$_2$B	<1.02	OsSi	<0.30	—	—
—	—	—	—	—	—	IrB	<1.28	IrSi	<1.02	IrGe	4.70
—	—	—	—	—	—	—	—	IrSi$_3$	<1.02	—	—
—	—	—	—	—	—	PtB	<1.28	PtSi	0.88	—	—
Th$_3$N$_4$	<1.20	—	—	ThC	<1.20	ThB	1.77	Th$_3$Si$_2$	<1.20	ThGe	<1.20
—	—	—	—	—	—	ThB$_2$	<1.20	ThSi$_2$ (α)	3.16	—	—
—	—	—	—	—	—	ThB$_6$	<1.28	ThSi$_2$ (β)	2.41	—	—
UN	<1.20	—	—	UC	<1.20	—	—	USi	<1.10	UGe$_3$	<0.30
—	—	—	—	—	—	—	—	USi$_2$	<0.30	—	—

* Obtained by diffusion saturation for 70 h at a pressure of 10 atm and a temperature of 1300°C [105d].
† Structure of the NaCl type; obtained at 1800°C and pressures up to 100 kbar [105b].
‡ The T_c of Mo$_2$C with the orthorhombic structure equals 12.2°K, with the cubic 13.5°K; a deficit of carbon in the compound reduces T_c.
§ Excess carbon occurs in the sample.

TABLE 14. Critical Magnetic Fields of Compounds with the NaCl Structure [89, 107, 108]

Compound	T_C, °K	H_{C_2}, Oe	H_{C_1}, Oe	$T_{meas.}$, °K
NbN	14.6—16	153000		0 *
MoC	13	107500		1.2
$Mo_{0.6}C_{0.4} + 2\%$ VC	12.2	100000		1.2
$Mo_{0.56}C_{0.44}$	13	93500	87	1.2
NbC	9	16900	120	4.2
$Nb_{0.4}Ta_{0.6}C$	11.8	14100	190	1.2
TaC	10.2	4600		1.2

*Extrapolation.

atom [101]. For compounds formed between one particular transition metal and one particular nonmetal a decrease in the relative content of the latter leads to a decrease in critical temperature; the T_C of MeC and MeN are usually higher than the T_C of Me_2C and Me_2N (see Table 13).

Among the various compounds of transition metals with nonmetals, of particular interest are compounds with a comparatively simple crystal structure: the fcc lattice of the NaCl type. Compounds with this type of structure have high superconducting transition temperatures and high critical magnetic fields (Table 14). A characteristic of these compounds is the existence of wide ranges of homogeneity, the upper boundary of which usually corresponds to the stoichiometric composition (Berthollides) [109]. In working with

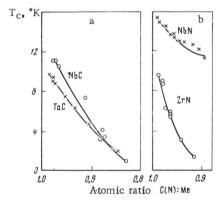

Fig. 43. Dependence of the T_C of niobium and tantalum carbides (a) and niobium and zirconium nitrides (b) on the relative proportion of the nonmetal within the range of homogeneity of the compound.

these compounds it is difficult to obtain alloys of identical composition and hence identical properties. Existing differences in published data regarding the properties of the same nominal compounds are largely attributable to differences in the actual compositions of the samples studied.

The maximum critical temperature of certain nitrides and carbides corresponds to stoichiometric composition. Within the homogeneous range of these compounds, a reduction in the proportion of the nonmetal is accompanied by a monotonic fall in T_c, for example, the carbides TaC and NbC (Fig. 43) [110], the nitrides NbN, ZrN [101, 111], and HfN [112], in which the superconductivity practically vanishes for small proportions of the nonmetal (T_c = 1.0°K for $NbC_{0.75}$ [113]). The forbidden band width increases at the same time [101]. According to one reference [102] the maximum critical temperature of MoC corresponds to a composition differing from stoichiometric; it decreases on increasing or decreasing the nonmetal content (Fig. 44).

Very often monocarbides of exact stoichiometric composition are practically impossible to produce. It has been suggested [114], for example, that $VC_{0.88}$ is the compound with the highest carbon content capable of being produced by the ordinary techniques of sample preparation. It may well be that the stoichiometric carbide VC has superconducting properties. It was found [113] by extrapolating the T_c values of NbN samples of various compositions (15.6°K for $NbN_{0.91}$ and 17.51°K for $NbN_{0.94}$) that the hypothetical stoichiometric compound NbN should have $T_c \approx 18°K$. The linear extrapolation of T_c to the stoichiometric composition NbC gives a value of 14°K [114].

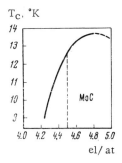

Fig. 44. Variation of T_c within the range of homogeneity of MoC.

The rise in T_c on approaching the stoichiometric composition of the carbides and nitrides indicated cannot be explained by the change in the concentration of valence electrons. For these cases it would appear more useful to consider the electron band structure [110]. The unusually low values of the electron specific heat coefficient and the magnetic susceptibility of niobium nitride and carbide indicate that the carriers are mainly situated in the sp bands [113]. It was suggested by Rajput and Gupta [112] that NbC and NbN occur at the peak of the curve relating the energy to the density of states $N(E)$. This is related to the fact that in the present case the density of states at the Fermi surface $N(0)$ changes very little as a result of the slight transfer of 5s electrons into the 4d band. Measurements of the specific heat, magnetic susceptibility, and Knight shift also indicate that the conduction band of NbN either has no d character at all or only a very poor one [115]. The behavior of this superconductor is similar to that of other superconductors with strong phonon coupling [115]. Zirconium carbide, which remains "normal" down to 1.2°K, behaves in a different manner, since both its components lie in the same group of the periodic table. Carbon has little effect on the energy-band width of zirconium; it increases $N(0)$ very slightly.

Gold [116] studied the isotropic effect and established a relation between the critical temperature of AB-type compounds and their reduced mass. Knowing the critical temperatures of several such compounds, we may determine the unknown critical temperature of the compound AC from the formula

$$\frac{(T_c)_{AB}}{(T_c)_{AC}} = \frac{1 + \dfrac{M_A}{M_B}}{1 + \dfrac{M_A}{M_C}},$$

where M_A, M_B, M_C are the atomic masses. This rule is valid for the carbides of the transition metals; calculations applied to other compounds prove less successful.

High-temperature and -pressure technology have led to the production of a new group of superconductors (sesquicarbides) with high transition temperatures: La_2C_3 5.9-11°K, Y_2C_3 6-11.5°K, Th_2C_3 3.8-4.1°K, and others [117a, 117b, 117c, 117d]. The alloying of these carbides with transition metals leads to a rise in their T_c ($La_{0.8}Th_{0.2}C_{1.6}$ up to 14.3°K, $Y_{0.65}Th_{0.35}C_{1.35}$ up to 16.8°K) [117e, 117f].

Recent investigations into the phases formed between transition metals and nonmetals have led to the discovery of several new types of quite complicated structures [117]. Compounds of the transition metals with carbon, hydrogen, nitrogen, and other nonmetals are often united into a single class of interstitial structures in which the nonmetallic atoms with small atomic radii are situated in the accessible spaces between the relatively large metal atoms [118]. In binary carbides and nitrides the small nonmetallic atoms have six neighbors. The ratio of the atomic radius of the nonmetal to that of the metal lies between 0.41 and 0.59. For complex carbides the ratio of the atomic radii may be greater [117, 119]. However, more complicated structures are then formed: carbides of the perovskite type, η carbides, H phases, and phases of the β manganese (composition T_3M_2X), and Mn_5Si_3 types (composition T_5M_3X).* A characteristic structural element of these structures is an octahedral group of six atoms of the transition element centered by a nonmetallic atom. The different arrangements and combinations of these octahedral groups create complex interstitial structures (Fig. 45) [117]. These phases are formed in ternary systems of transition metals and their alloys with metals from the main groups of the periodic system in the presence of nonmetallic atoms with small atomic radii.

In recent years some two hundred carbides and nitrides with complex structures of this kind have been discovered. Nowotny and colleagues [117, 120] made a detailed study of the compositions and crystal structures of these compounds and established their characteristic components. However, little is yet known as to the superconducting properties of these compounds.

On the basis of the crystal-chemical similarity between these complex phases and the binary carbides of transition metals, as well as the high degree of order in their crystal structure, we may well expect them to be superconducting materials [105]. Of the various crystal structures of complex carbides and nitrides the most favorable in this respect is the structure of β manganese distorted by interstitial atoms of nonmetals [105, 121]. The octahedral group of atoms is arranged in a nonuniform manner in the matrix lattice of the T_3M_2X or $T_3T_2'X$ compounds [117]. The matrix

---

*T is the transition metal atom, M is the atom of the B subgroup, X is the nonmetallic atom (interstitial element) with a small atomic radius.

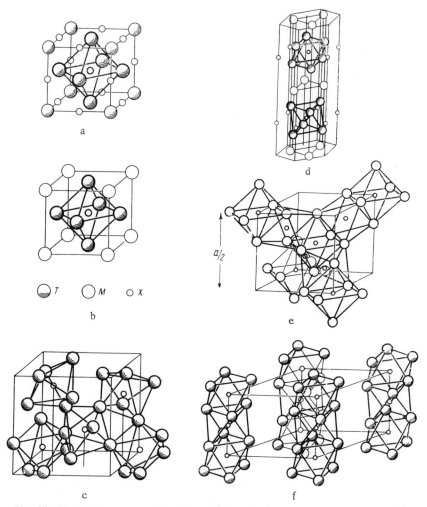

Fig. 45. Crystal structure of interstitial phases with separated octahedral groups: a) NaCl type; b) perovskite type; c) β-Mn type with filled octahedral spaces; d) Cr$_2$AlC type; e) η-carbide type; f) Mn$_5$Si$_3$ type with filled octahedral spaces.

lattice constitutes an ordered lattice of the β manganese type. Phases with such structures are stabilized by carbon and nitrogen. No phases stabilized by boron have yet been observed. Only two phases with such structures have yet been examined for superconductivity. Both of these are in fact superconductors: Nb$_3$Al$_2$N (T_c = 1.3°K [121]) and Mo$_3$Al$_2$C (T_c = 10°K, H_{c2} = 15,600 Oe, H_{c1} = 91 Oe at 1.2°K [107]).

Phases of composition T_5M_3X or $T_3T_2'MX$ have a structure of the Mn_5Si_3 type [117]. This structure is stabilized by carbon and nitrogen, more rarely by oxygen and boron. The nonmetal atoms lie in the octahedral spaces of the transition metal atoms. The octahedral groups form, as it were, columns along the c axes in a crystal lattice of the Mn_5Si_3 type. The M components are chiefly silicon, germanium, tin, and lead. Recently a superconductor with a similar structure and a high transition temperature was found: $Mo_{4.8}Si_3C_{0.6}$ with $T_c = 7.6°K$ [105]. However, other phases of this structure studied ($V_5Si_3C_{0.05}$, $Nb_5Si_3C_{0.05}$, $V_5Ge_3C_{0.05}$, $Nb_5Ge_3C_x$, $Nb_{2.5}Zr_{2.5}Ge_3C$, and $Ti_3Mo_3Si_2C$) failed to exhibit superconductivity above 1.1-1.2°K.

The extensive class of H phases is unfavorable toward superconductivity [120]. These H phases have the composition T_2MC and a structure of the Cr_2AlC or Ti_2SC type [117]. The M components of these phases are elements of the main II-VI subgroups of the periodic table. The structure of the H phases is lamellar in a direction normal to the c axis. The x axis of the unit cell is three times larger than the period of the simple subcell. However, the ratio $c/3a$ always differs from the ideal value, which indicates a considerable proportion of nonmetallic bond [117]. This is evidently responsible for the absence of superconducting phases among such compounds. Of course not all H phases have yet been examined for superconductivity. The electron concentration of the H phases, equal to 3.5-4.75 electrons/atom, is also unfavorable toward high critical temperatures [121]. According to existing data [62] regarding the electron concentration of compounds favorable toward high T_c (4.7 and 6.6 electrons/atom), superconductivity could apparently only exist in H phases with high concentrations of the valence electrons per atom.

The T_c of the compounds MeC_8 and MeC_{16} in the K–C, Rb–C, and Cs–C systems varies with the composition of the alloys, reaching a maximum (0.55, 0.151, and 0.135°K respectively) for an excess of the alkali metal [121a, 121b].

Individual superconducting compounds are also found among compounds of the transition metals with other nonmetals (Table 15), particularly among semiconducting compounds. However, sulfides, phosphides, selenides, and tellurides of transition metals usually exhibit no superconductivity, and certainly not at the higher temperatures [122, 123].

TABLE 15. Superconductivity of the Sulfides, Phosphides, Selenides, and Tellurides of Transition Metals

Compound	T_c, °K	Compound	T_c, °K	Compound	T_c, °K
La_3S_4	8.25	Rh_2P	<1.03	$NiTe_2$	<1.20
CeS	<1.06	Rh_5P_4	1.22	$RuTe_2$	<0.32
Ce_3S_4	<1.28	$Ni_{0,6-0,48}Fe_{0,19-0,31}P_{0,21}$	<0.99	RhTe	<1.06
NbS	<1.28	$Ni_{0,16-0,64}Co_{0,64-0,16}P_{0,2}$	<0.99	$RhTe_2$[5*]	1.51
NbS_2[2*]	6.1—6.3	La_3Se_4	8.6	$ThTe_2$[6*]	<1.06
WS	<1.30	$NbSe_2$[2*]	6.2—7	Pd_4Te	<0.32
FeS	<1.02	$MoSe_2$	<1.26	Pd_3Te	0.76
FeS_2[1*]	(×)	$RuSe_2$	<0.32	$Pd_{1,1}Te$	4.07
NiS	<1.28	Rh_xSe_y	6.0 (max)	PdTe	2.3—3,05
RuS_2	<0.32	$Rh_{0,53}Se_{0,47}$	6.0	$PdTe_{1,02}$	2.56
$Rh_{17}S_{15}$	5.8	$RhSe_{1,75}$	6.0	$PdTe_{1,04}$	2,11
Pd_4S	<0.32	Rh_2Se_5	1.04	$PdTe_{1,06}$	2,11
IrS	<0.32	Pd_xSe_y	2.5 (max)	$PdTe_{1,08}$	1.88
$IrS_{2,6}$	<0.32	$Pd_{6-7}Se$	0.66	$Pd_{1,75}Te_2$[7*]	2.25—1.93
LaP	<1.68	Pd_4Se	0.42	$Pd_{1,5}Te_2$[7*]	2.21—1.87
V_3P	<1.00	Pd_5Se_2	2.30	$Pd_{1,25}Te_2$[7*]	2.20—1,90
VP	<1.01	Pd_2Se	2.20	$Pd_{1,05}Te_2$[7*]	1.77—1.74
Cr_3P	<1.01	$Pd_{17}Se_{15}$	<0.32	$PdTe_2$	1.69
CrP	<1.01	PdSe	<0.32	$PdTe^2$[*]	1.53—1.46
Mo_3P	7.0—5,31	$PdSe_2$	<1.50	$PdTe_{2,1}$	1.89
MoP	<1,01	$IrSe_2$	<0.32	$PdTe_{2,3}$	1,85
W_3P	2.26	$IrSe_{2,9}$	<0.32	IrTe	<0.30
WP	<1.01	$NbSe_{0,25}Te_{0,75}$	4.4	$IrTe_2$	<0.32
MnP	<1.01	$NbSe_{0,2}Te_{0,8}$	<2.0	$IrTe_3$	1.18
Fe_2P	<0.97	PdSeTe	1.20	PtTe	0.52
FeP	<0.97	YTe	1.02	$PtTe_2$	<1.20
CoP	<0.97	La_3Te_4[3*]	2.45	$Ni_{0,05}Pd_{0,95}Te_2$	1.40
Ni_3P	<1.01	La_3Te_4[4*]	3.75	$Ni_{0,1}Pd_{0,9}Te_2$	1.30
Ni_2P	<1.01	$NbTe_2$	<2,00	$Pd_{0,9}Pt_{0,1}Te_2$	1.65
Ru_2P	<1.1	CoTe	<1.20	$Pd_{0,95}Pt_{0,05}Te_2$	1.71—1.65
		NiTe	<1.60		

[1*] Synthesized at high pressures
[2*] Single crystal
[3*] Metalloceramic sample annealed at 1450°C for 1.5 h
[4*] Sample melted in a molybdenum crucible
[5*] Low-temperature form
[6*] High-temperature form
[7*] Unannealed alloys
× signifies semiconductor

The highest superconducting transition temperatures among all these compounds are found in the γ-lanthanum sulfide La_3S_4 [124] and La_3Se_4 [127] with a defective structure of the Th_3P_4 type. The T_c of the phases depends on the composition; it increases with increasing lanthanum content, and at the stoichiometric composition equals $8.25 \pm 0.2°K$ for La_3S_4 and about $8.6°K$ for La_3Se_4. With increasing lanthanum content the electron concentration of the phase (determined from the Hall constant) also increases. Samples with high sulfur contents exhibit semiconducting properties below 10°K. A reduction in the lanthanum content in the La_3Se_4 compounds (within the limits of homogeneity) causes a corresponding decrease in T_c [129]. Alloys with lower lanthanum contents become insulators. Such a sharp change in the electrical properties of this compound is evidently associated with a change in the character of the bond in the lattice of the compound with changing composition. This is of fundamental significance and requires urgent investigation. In the compound La_3Se_4 the conductivity is proportional to the number of conduction electrons, as derived from the difference in valences, while T_c is fairly well described by the BCS equation [130].

The critical temperature of 5.8°K in $Rn_{17}S_{15}$ (fairly high for this group of compounds) is evidently due to the favorable cubic structure of this compound [1]. However, another sulfide, $Pd_{17}S_{15}$ with a similar structure has no superconducting properties [122].

The phosphides Mo_3P (7-5.31°K) and W_3P (2.26°K) have a body-centered tetragonal structure. The compounds Cr_3P are not superconductors although they have a similar crystal structure. The nonsuperconducting vanadium phosphide V_3P (structure of the Fe_3P type) is not isomorphic with the foregoing compounds. No superconductors are found among phosphides with other crystal structures [123].

In the palladium–tellurium system there are two superconducting phases: a nickel–arsenide type of phase PdTe, the critical temperature of this depending considerably on the composition of the alloy, and $PdTe_2$, with the trigonal structure of CdI_2. A fairly high superconducting transition temperature (6°K) occurs in the case of $RhSe_2$ with a cubic structure of the FeS_2 type. For a palladium–selenium compound containing 85-87 at.% Pd, T_c equals 0.66°K. However, this quantity is far higher than that given by the

BCS theory for such concentrations of the valence electrons [122, 130]. Superconducting phases with similar structures occur among the compounds of transition metals with simple metals; these will be discussed later.

Data relating to the superconductivity of compounds of transition metals with arsenic are presented in Table 16.

Superconducting compounds with nonmetals are also formed by nontransition metals. At the present time three such compounds are known: CuS (1.62°K), GaN (5.85°K), and PbP$_{\sim 1}$ (7.8°K) [62]. However, the majority of such compounds are not superconductors at temperatures above 1.0°K (Table 17). Covalent compounds of a nonmetallic type such as the carbides and nitrides of boron and silicon are also not superconducting.

Among the compounds of nontransition metals with nonmetals superconductors with low critical temperatures are quite possible. The T_c of $Ge_{0.976-0.987}Te$ equals 0.1-0.3°K [135]. The addition of silver to the compound GeTe yields samples with T_c equal to 0.41 and 0.21°K for hole concentrations of $64 \cdot 10^{20}$ and $27 \cdot 10^{20}$ cm^{-3} respectively [138].

TABLE 16. Superconductivity of Compounds of Metals with Arsenic [1, 62, 131]

Compound	T_c, °K	Compound	T_c, °K	Compound	T_c, °K
AsCo	<1.1*	AsNi$_{0.25}$Pd$_{0.75}$‡	1.34	As$_3$Pd$_5$	1.95¶
As$_2$Co	<1.1	AsNi$_{0.12}$Pd$_{0.88}$	1.39	As$_3$Rh$_5$	<1.1
AsCu$_3$	<1.28†	As$_{0.5}$Ni$_x$Pd$_y$	≯1.60	AsRh$_2$	<1.1
As$_2$Cu	<1.57	AsNi$_{0.25}$Pd$_{0.75}$§	1.6	AsRu	<1.1
As$_4$Cu$_{18}$Sb$_3$	<0.30	AsPd	<1.02	AsRu$_2$	<1.1
As$_2$Fe$_3$	<1.30†	AsPd$_2$	1.7	As$_{0.26}$Sb$_{0.74}$	<1.32†
As$_2$Mo	<1.30†	AsPd$_3$	<1.1	As$_x$Sn$_y$	<4.2†
AsNi	<1.28	As$_2$Pd$_5$	<1.1	As$_4$Mo$_5$	<4.2

*The sign < indicates the lowest test temperature at which the superconductivity of the material has been studied.
†Measurement of electrical resistance
‡Structure of the C2 type
§Structure of the B8$_1$ type ($a = 3.66$, $c = 5.03$ Å).
¶It is considered that the superconductivity of the arsenic-palladium alloy is due to the formation of As$_3$Pd$_5$.

TABLE 17. Compounds of Nontransition Metals and Nonmetals with Nonmetals Which Have Been Examined for Superconductivity [1, 2, 102, 132-138]

Compound	T_c, °K	Compound	T_c, °K	Compound	T_c, °K
CuSi	<1.28 *	Al_2Pb_4Si	<1.02	SnTe	0.012
Cu_3Si	<1.28	CaB_6	<1.28	PbP	7.8†
CuP	<1.28	CaCe	<1.70	PbS	<1.28
CuS	1.62	HgS	<1.30	PbSe	<1.26
$Cu_{1.8}S$	<1.30	B_4C	<1.28	PbTe	<1.28
CuS_2‡	1.48—1.53	BN	<1.28	SbS_3	<1.28
CuSe	<1.28	GaN	5.85	Bi_2Ge	<1.28
CuTe	<1.26	In_2Te	<1.37	Bi_2S	<1.90
$CuTe_2$‡	?—1.30	InTe	<1.37	Bi_2S_3	<0.10
CuSSe‡	1.5—2.0	$In_{1-x}Te$ (0 ≤ x ≤ 0.18)	1—4	Bi_2Se_3	<1.26
$CuSe_2$‡	2.3—2.43	In_4SbTe_3	1.4—1.5	Bi_2Te_3	<1.26
CuSeTe‡	1.6—2.0	CSi	<1.28	STl_2	<1.30
AgS	<1.90	GeTe	0.172	SeN	<1.30
Ag_2S	<1.28	$Ge_{0.976—0.987}Te$	0.1—0.3		
Ag_2Se	<1.28	GeTe(Ag)	0.21—0.41		
Ag_2Te	<1.28	SnS	<1.28		

* The symbol < means that on measuring the compound down to this temperature no superconducting transition occurred.
† Mixture of phases
‡ Synthesized under high pressure

Known compounds of nontransition metals with nonmetals also have low critical fields. For example, the H_c of GeTe equals 95 Oe at 0°K [134], that of SnTe 1.38 Oe at its critical temperature (0.012°K) [2]. The critical fields of the majority of the sulfides, phosphides, tellurides, and selenides of transition metals are also fairly small. For example, the electrical resistance of a metalloceramic sample of lanthanum telluride La_3Te_4 is restored at 1.4°K in a field of 8 kOe, and that of a sample melted in a molybdenum crucible in a field of 12 kOe [126].

Recently a superconducting compound of a metal with a halogen was obtained (Ag_2F with a hexagonal structure of the anti-CdI_2 type, T_c = 0.066°K, H_c = 2.5 kOe at 0°K [139]), possibly opening a new group of superconducting compounds. Earlier [1] only the compound $AgCl_2$ had been studied; this showed no superconductivity above 1.9°K. On considering the general properties of the hal-

TABLE 18. Superconductivity of Oxides [1, 2, 62, 140-143]

Oxide	T_c, °K	Oxide	T_c, °K	Oxide	T_c, °K
Ag_2O	<1.28 *	Sn_2O_3	<1.30	$Ti_{0.573}Rh_{0.287}O_{0.14}$	3.37
CdO	<1.30	Ti_2O_3	<1.20	$Zr_{0.61}Rh_{0.285}O_{0.105}$	11.8
CoO	<1.28	TiO	0.68	Zr_3V_3O	7.5
CuO	<1.28	$Ti_{1.1}O$	<0.21	$Tl_{0.3}WO_3$	2.00—2.14
Cu_2O	<1.28	Tl_2O_3	<1.28	$In_{0.3}WO_3$	<1.25
Mn_2O_3	<1.28	UO_2	<1.20	$Ba_{0.3}WO_3$	<1.25
MoO_2	<1.30	VO	<1.20	$SrTiO_3$	0.2—0.39
Mo_2O_5	<1.28	V_2O_3	<1.28	$SrTiO_3(Nb)$	0.38
Mo_3O_5	7.20	WO_2	<0.3	$(Ba_{0.1}Sr_{0.9})TiO_3$	0.1—0.6
NbO	~1.25†	WO_3	<0.3	$(Ca_{0.31}Sr_{0.7})TiO_3$	0.1—0.6
NiO	<1.28	$LaNiO_3$	<1.30	$Na_{0.3-0.8}WO_3$	<0.3
PbO_2	<1.02	$Cu_{0.287}Ti_{0.573}O_{0.14}$	<1.02	$Ba_{0.13}WP_3$	1.9
Rh_2O_3	<1.28	$Pd_{0.285}Zr_{0.61}O_{0.105}$	2.09	$BaTiO_3$	<0.1
SnO	3.81 (?)	TiReO	5.74	$CaTiO_3$	<0.1

* The symbol < means that on measuring down to this temperature the sample failed to pass into the superconducting state.
† Blurred (diffuse) transition.

ides and the type of chemical bond in their lattices we can hardly expect superconducting halides to have high superconducting properties.

Certain oxides of metals possess superconductivity, but the majority are nonsuperconducting at temperatures over 1°K (Table 18). Latest investigations have shown that certain oxides pass into the superconducting state at fairly low temperatures. For example a blurred transition into the superconducting state occurs at about 1.25°K in NbO [138]. In the same investigation it was found that titanium oxide of exactly stoichiometric composition became superconducting at 0.68°K. However, even a slight deviation from stoichiometric composition ($Ti_{1.1}O$) led to the vanishing of the superconductivity. The reason for such a sharp change in the properties of this oxide is uncertain, since a slight deviation in the composition produced only minute changes in the structure of the compound. No such effect is found in other oxides. A nonstoichiometric composition of the oxides Me_xWO_3 and Me_xTiO_3 (where Me = thallium, strontium, barium, sodium, etc.) leads to a broadening of the transition from the superconducting to the normal state and slight fall in T_c, although not so large as in TiO.

The compounds Me_xWO_3 and Me_xTiO_3 are only semiconducting at very low temperatures (see Table 18). The greatest critical temperature is that of $Tl_{0.3}WO_3$ (2-2.14°K), which has a hexagonal structure [142]. The compound $SrTiO_3$ is also a superconductor of the second kind at low temperatures (0.4°K) with a parameter $\varkappa$ depending on the mobility of the electrons in the normal state [141]. In a polycrystalline sample of this compound T_c equals 0.2-0.3°K. Alloying with niobium increases T_c to 0.38°K.

The critical temperature of the semiconducting compounds $(Ba_x, Sr_{1-x})TiO_3$ and $(Ca_x, Sr_{1-x})TiO_3$ varies with chemical composition and electron concentration n. For $n = 6 \cdot 10^{19}$ cm^{-3} the critical temperatures of these compounds vary from 0.1 to 0.6°K, depending on the composition [140]. The maximum critical temperature is slightly higher than in the pure reduced compound $SrTiO_3$. Samples without any strontium, $CaTiO_3$ and $BaTiO_3$, show no superconducting properties at all. The superconductivity of these compounds containing strontium is evidently associated with their ferroelectric properties [140].

The question as to the superconductivity of semiconducting compounds is discussed in full detail elsewhere [143b].

Complex carbides with the perovskite structure have not yet been examined for superconductivity. According to some authors [14] who have studied the superconducting properties of $Ba_{0.13}WO_3$ and $Na_{0.3-0.8}WO_3$, a structure of the perovskite type is unfavorable toward the development of superconductivity.

Hydrides ($LaH_{2.45}$, $NaNH_3$, $NbH_{0.88}$, $NbH_{0.99}$, Ta_xH_y, TiH_2, and ZrH_2) are not superconductors above 1-2°K [1]. The compounds $Ag_7O_8NO_3$, Ag_7O_8F, and $Ag_7O_8BF_4$ are superconductors at 1.04, 0.15, and 0.3°K respectively [143a].

SIGMA AND LAVES PHASES AND SIMILAR COMPOUNDS

Apart from compounds with the Cr_3Si and NaCl structures, sigma and Laves phases often have high superconducting transition temperatures; so do compounds with structures of the α-manganese type. Certainly the T_c of known compounds with such structures are lower than those of the phases considered earlier.

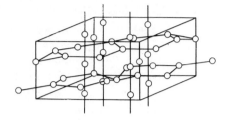

Fig. 46. Structure of the σ phases.

The crystal structure of the σ phases (Fig. 46) is isomorphic with that of β uranium and constitutes a complex tetragonal lattice with thirty atoms per unit cell (space group $D_{4h}^{14}-P_2^4/mnm$ [11, 144, 145]). The atoms of β uranium belonging to different crystallographically-equivalent groups have different electron configurations (no less than four valence states) [146]. The great similarity between the structures of β uranium and the sigma phases accordingly implies that the latter are only formed in systems involving transition metals which, as a result of their partially-filled inner electron orbits, are capable of forming the electron coordination required [119]. The d electrons play a specific part in stabilizing the sigma phases [147]. It has been suggested that sigma phases are formed for a particular electron concentration of the corresponding alloys [148]. The electron concentration characteristic of the sigma phases equals 6.2-7.0 electrons/atom [149], or, preferentially, 6.4-6.9 electrons/atom. The number of electron vacancies in the d [150] explains the experimental data far better. For the formation of the sigma phases the dimensional factor is also important ($0.93 \le r_A/r_B \le 1.5$) [47, 151]. However, such concepts as the differences in the atomic radii, the electron concentration, and so on are insufficient to solve the problem of the conditions underlying the formation of sigma phases [147]. These phases tend to be formed mostly in alloys of group VA-VIA metals with metals belonging to groups VIIA-VIIIA of the periodic table.

There is a state of order in the arrangement of the atoms in the sigma-phase structure [47, 147]. Of the five nonequivalent positions of the atoms in the structure of the sigma phases, three are occupied by atoms of one specific type only. The remaining two positions may be occupied by atoms of any of the components. The latter fact explains the existence of wide ranges of homogeneity in the majority of the σ phases [62, 147]. The maximum superconducting transition temperature of the σ phases equals 12°K (Table 19).

TABLE 19. Critical Superconducting Temperature of the σ Phases [62, 149, 152-154]

Compound	T_c, °K	Lattice constants, Å		Electron concentration, electrons/atom
		a	c	
$Al_{>0.25}Nb_{<0.75}$	7—12	—	—	4.5
$Mo_{0.3}Tc_{0.7}$	12.0	9.5091	4.9448	6.6
Nb_3Ir_2	9.8	9.834	5.052	6.7
$Nb_{0.63}Ir_{0.37}$	2.4	9.86	5.06	6.48
$Mo_{0.42}Re_{0.58}$	8.4	9.59	4.97	6.68
$Mo_{0.5}Re_{0.5}$	4.7;* 7.3	9.61	4.98	6.5
$W_{0.4}Tc_{0.6}$	7.9	—	—	—
$Mo_{0.6}Ru_{0.4}$	7.0	9.55	4.95	6.8
Mo_3Ir	6.8	9.631	4.956	6.75
$Mo_{0.74}Ir_{0.26}$	6.7	9.63	4.96	—
Fe_3Re_2	6.55*	—	—	7.56
Re_3V	6.26*	—	—	6.44
$Re_{0.76}V_{0.24}$	4.52	9.45	4.88	6.52
$Mo_{0.62}Os_{0.38}$	5.65	9.60	4.93	6.76
$Re_{0.52}W_{0.48}$	5.2; 4.43*	—	—	6.52
$Re_{0.7}W_{0.3}$	4.9	—	—	6.7
$Re_{0.6}W_{0.4}$	4.9	—	—	6.6
$Ru_{0.5}W_{0.5}$	5.12	9.63	5.01	6.5
Ru_xW_y	5.2	—	—	—
$Ru_{0.4}W_{0.6}$	4.67	9.57	4.96	6.8
$W_{0.72}Ir_{0.28}$	4.46	—	—	9.84
Nb_3Rh_2	4.04	9.80	5.07	6.6
W_xOs_y	4.4	—	—	—
$W_{0.66}Os_{0.34}$	3.81	9.63	4.93	6.68
$W_yOs_{0.23-0.33}$	2.5—3.6	—	—	—
$Nb_{0.62}Pt_{0.38}$	4.01	9.91	5.13	6.90
$Nb_{0.625}Pt_{0.375}$	3.73	9.91	5.12	6.88
$NbRe_3$	5.27*	—	—	—
$Nb_{0.4}Re_{0.6}$ †	2.5	9.77	5.14	6.2
$Nb_{0.5}Re_{0.5}$ ‡	2.0—3.8	9.79	5.10	6.0
$Cr_{0.42-0.33}Re_y$	2.50 (max.)	9.26—9.32	4.805—4.845	—
$Cr_{0.45-0.3}Re_y$	1.8—2.4	—	—	—
$Cr_{0.4}Re_{0.6}$	2.15	—	—	6.60
Ta_3Rh_2	2.35	9.80	5.09	6.6
$Cr_{0.6}Ru_{0.4}$	2.10	—	—	6.8
Cr_2Ru	2.02	—	—	6.67
$Cr_{0.5}Ru_{0.5}$	1.30	—	—	7.0
Nb_xPd_y	2.0	—	—	—
$Nb_{0.6}Os_{0.4}$	1.78; 1.85	9.844	5.056	6.2

*Results of Savitskii and Khamidov [154a].
†Plus a phase of the α-Mn type with a = 9.773 Å.
‡Plus a cubic phase with a = 3.184 Å.
§Plus a phase of the α-Mn type with a = 9.783 Å.

TABLE 19 (Continued)

Compound	$T_c°$, K	Lattice constants, Å		Electron concentration, electrons/atom
		a	c	
$Pt_{0.3}Ta_{0.7}$	1.0	9.93	5.15	6.5
$Re_{0.6}Ta_{0.4}$	1.4	9.77	5.09	6.2
$Ta_{0.65}Ir_{0.35}$	1.2	—	—	4.2
$AlTa_3$†	<1.02	9.97	5.19	4.5
$AuTa_2$	<1.2	—	—	3.7
$AuZr_3$	<1.02	10.58	5.53	4.25
CrRe	<1.02	9.20	4.78	6.5
MnRe	<1.0	9.23	4.79	7
$Os_{0.3}Ta_{0.7}$	<1.2	9.88	5.14	5.9
$Pt_{0.2}Ta_{0.8}$	<1.2	10.02	5.20	6
Zr_2Re,* $HfRe_2$*	<4.2	—	—	—
TaRe,*Cr_2Re_3,*MnRe*	<4.2	—	—	—

*See p. 137.
†See p. 137.

The change in the T_c of the sigma phases in relation to their mean electron concentration exhibits a sharp maximum at 6.7 electrons/atom (Fig. 47, [62]). Any deviation from this optimum electron concentration is accompanied by a decrease in the critical temperature. The only exception is the sigma phase of the Nb–Al system, which has high critical temperatures at 4-5 electrons/atom.

The graph of T_c plotted against the mean atomic volume has a wide maximum at 15-16 Å³/atom (Fig. 47) [62].

The correlation between T_c and the electron concentration was demonstrated [62] with due allowance for the average molecular weight of the compounds. The electron concentration corresponding to the maximum critical temperatures (6.7 electrons/atom) is the most favorable one for the formation of the sigma phase [150]. On simultaneously considering the valence and unit-cell volume we find that the range ~ 0.37-0.47 electrons/Å³ corresponds to high critical temperatures of the sigma phases [62].

The sigma phases have fairly high critical magnetic fields. For example, when studying the W–Tc system it was found that H_c rose sharply on formation of the sigma phase [153].

The Laves phases of composition AB_2 have a close-packed structure of three types: the cubic structure of the $MgCu_2$ type and the hexagonal $MgZn_2$ and $MgNi_2$ structures (Fig. 48) [155, 156].

In the cubic structure of the MgCu$_2$ type (space group Fd3m–O_h^7, z = 24) eight magnesium atoms lie in the (a) position and sixteen copper atoms in the (d) position. A characteristic geometrical sign of the MgCu$_2$ type of structure is the absence of conjugation between atoms of different types [17]. The interatomic distances between the pairs of AA or BB atoms are in fact equal to twice the atomic radii of the components. The maximum occupation of space corresponding to the ideal state of close packing for the composi-

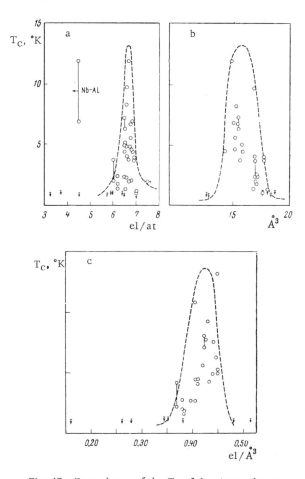

Fig. 47. Dependence of the T_c of the sigma phases on the electron concentration (a), the mean unit-cell volume (b), and the number of valence electrons per unit volume (c).

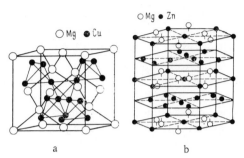

Fig. 48. Crystal structures of Laves phases of the MgCu$_2$ (a) and MgZn$_2$ (b) types.

tion AB$_2$ occurs for a ratio of $r_A/r_B = \sqrt{3}/\sqrt{2} = 1.225$ [17, 19]. In the MgCu$_2$-type structure the lattices formed by the A and B atoms are set into each other in such a way that each A atom is surrounded by twelve B atoms, while each B atom is surrounded by six A atoms. The coordination numbers 12 and 6 are clearly the maximum possible for the composition AB$_2$ [17].

The MgZn$_2$ type of structure differs from the foregoing in the same way as a close-packed hexagonal from a close-packed cubic lattice. In the hexagonal MgZn$_2$ lattice (space group P6$_3$/mmc–D$_{6h}^4$, z = 12) four nickel atoms lie in the (f) positions with $z = \frac{1}{16}$, two zinc atoms in (a), and six zinc atoms in (h) with $z = \frac{5}{6}$.

The MgNi$_2$ type of structure is intermediate between MgCu$_2$ and MgZn$_2$.

The atoms of the A components of the Laves phases are larger than those of the B components. The large A atoms occupy the space inside tetrahedra with B atoms at the corners. The three structures of the Laves phases are obtained by different sequences of alternation of layers of A and B atoms.

A variety of elements may constitute the components of Laves phases. The same element may be both an A and a B component. Only Laves phases containing N, B, C, P, As, S, O, Se, Te, F, Cl, and Br are unknown. This is in conformity with the typically metallic character of the Laves phases.

Laves phases are formed if the ratio of the atomic radii of the components is approximately equal to 1.225 [17]. In practice Laves phases are formed for ratios of the atomic radii equal to

1.1-1.6 [119]. The chemical composition of the phases is determined not by the valence relationships but by the geometrical properties of the space capable of accommodating the atoms in such a way as to minimize the energy of the lattice as a whole. Thus it is the spatial factor which determines the possible formation and stability of Laves phases. The type of crystal structure of the Laves phases is determined by the electron structure of the components and hence by the electron concentration of the compounds. The effect of electron concentration on the crystal structure was clearly demonstrated in certain ternary magnesium systems [157]. The boundaries of the ranges of homogeneity for each type of Laves-phase structure occur at approximately the same values of electron concentration, i.e., under conditions in which almost all the quantum states in the corresponding Brillouin zones are occupied [158].

At the present time more than 100 different Laves phases have been discovered [17, 47]. Superconducting properties have only been studied in some of these compounds. Data relating to the critical temperatures of Laves phases of the $MgCu_2$ and $MgZn_2$ type are presented in Table 20. The compound $HfMo_2$ is the only Laves phase with the $MgNi_2$ structure in which the superconducting properties have been studied. In this compound no superconductivity appeared down to 1.02°K [62]. Many Laves phases with the $MgCu_2$ or $MgZn_2$ structure are superconductors, particularly if one of the components is a transition metal. Laves phases formed by elements with a low valence (1-2) do not exhibit superconductivity [62].

The critical superconducting temperature of any Laves phase lies below 10°K. The mean unit-cell volume of compounds of the $MgZn_2$ type corresponding to the maximum T_c equals 15 Å^3, the electron concentration 7 electrons/atom, and the number of electrons per unit volume 0.36 electron/Å^3 (Fig. 49) [62]. The optimum volume for compounds with the $MgCu_2$ type of structure is 17-20 Å^3 (Fig. 50) [62]. The electron concentration of the majority of phases of the $MgCu_2$ type equals 6.67 electrons/atom. It is interesting to note that the compound ZrV_2, which has the highest critical temperature of all Laves phases (8.8°K), also has the lowest electron concentration (4.67 electrons/atom). The lowest electron concentration occurs in the Laves phases formed by bismuth. However, the mean unit-cell volumes of the compounds Bi_2K, Bi_2Rb,

TABLE 20. Critical Superconducting Temperature of Laves Phases [1, 62, 159-161]

Compound	T_C, °K	Lattice constants, Å		Electron concentration, electrons/atom
		a	c	
Structure of the MgCu$_2$ type				
ZrV$_2$	8.8	7.439	—	4.67
LaOs$_2$	6.5	7.737	—	6.33
ThIr$_2$	6.50	7.664	—	7.33
CaRh$_2$	6.40	7.525	—	6.67
SrRh$_2$	6.2	7.706	—	6.67
CaIr$_2$	4—6.15	7.545	—	6.67
BaRh$_2$	6.0	7.852	—	6.67
SrIr$_2$	5.7	7.700	—	6.67
CeRu$_2$	4.90	7.535	—	6.33
CsBi$_2$	4.75	9.746	—	3.67
RbBi$_2$	4.25	9.609	—	3.67
ZrIr$_2$	4.10	7.359	—	7.33
KBi$_2$	3.58	9.501	—	3.67
ThRu$_2$	3.56	7.651	—	6.66
LaAl$_2$	3.24	8.13	—	3.00
YIrOs	2.60	—	—	6.67
YIr$_2$	2.18	7.50—7.52	—	7.0
ZrW$_2$	2.16	7.621	—	5.33
RhHf$_2$	1.98	—	—	—
BiAu$_2$	1.84	7.958	—	2.33 (I) 9.0 (II)
LaRu$_2$	1.63	7.702	—	6.67
YPt$_2$	1.1	7.590	—	7.67
CeCo$_2$	1.5	7.165	—	—
PbAu$_2$	1.18	7.94	—	—
ScIr$_2$	1.03	7.348	—	7.0
SrPt$_2$	0.7	—	—	—
ZrNi$_{1.5}$V$_{0.5}$	0.43	7.068	—	7.17
Ce$_x$Gd$_y$Ru$_{0.66}$	3.20—5.20	—	—	—
(Ce$_x$Pr$_{1-x}$)Ru$_2$	1.4—5.3	—	—	—
GdIr$_2$	*	7.550	—	7.00
GdRh$_2$	*	7.514	—	7.00
Gd$_x$Th$_{1-x}$Ru$_2$	3.60 (max.)	—	—	—
NdIr$_2$	*	7.605	—	7.00
NdIrOs	*	—	—	6.66
PrIr$_2$	*	7.621	—	7.00
NdPt$_2$	*	7.694	—	7.33
NdRh$_2$	*	7.564	—	7.00
NdRu$_2$	*	7.614	—	6.33
PrOs$_2$	*	7.663	—	6.33
PrPt$_2$	*	7.709	—	7.66
PrRh$_2$	*	7.575	—	7.00
PrRu$_2$	*	7.624	—	6.33

TABLE 20 (Continued)

Compound	T_C, °K	Lattice constants, Å		Electron concentration, electrons/atom
		a	c	
VTa_2	< 4.2	7.17	—	5.00
$NaAu_2$	<1.02 †	7.814	—	1.00
$MgCu_2$	<1.02	7.034	—	1.33
$MgLa_2$	<1.02	8.763	—	2.66
$CaAl_2$	<1.02	8.035	—	2.66
$CaPd_2$	<1.02	7.665	—	7.33
$CaPt_2$	<1.02	7.625	—	7.33
$SrPt_2$	<1.02	7.777	—	7.33
$BaPd_2$	<1.02	7.953	—	7.33
$BaPt_2$	<1.02	7.920	—	7.33
AlCuZr	<1.02	7.322	—	2.7
$AlLu_2$	<1.02	7.744	—	3
YAl_2	<1.00	7.860	—	3
YRh_2	<0.35	7.459	—	7
$ScAl_2$	<1.02	7.579	—	3
$LaPt_2$	<1.00	7.774—7.763	—	7.7
$LaRh_2$	<1.00	7.646	—	7
$LaIr_2$	<1.00	7.686	—	7
$CeRh_2$	<0.35	7.538	—	7
$CeIr_2$	<0.37	7.571	—	7
$GdPt_2$	<1.00	7.637	—	7.7
$GdRu_2$	<1,2	—	—	6.3
$TiBe_2$	<1.02	6.451	—	3.7
ZrW_2	< 1.2	—	—	—
$ZrMo_2$	< 1.2	—	—	—
$ZrCo_2$	<1.02	6.935	—	7.3
$HfMo_2$	< 0.05	—	—	—
HfW_2	<1.02	7.594	—	5.3
$ThOs_2$	<1.02	7.704	—	6.7
UAl_2	<1.12	7.811	—	4
UFe_2	<1.06	—	—	7.3
UIr_2	<0.35	7.496	—	8
UOs_2	<0.37	7.509	—	7.3
MnNiZr	<0.35	7.070	—	7
Structure of the $MgZn_2$ type				
VTa_2	—	—	—	—
$ZrRe_2$	6.8; 5.9	5.262	8.593	6.0
$ZrRe_{2-x}$	5.2 ‡	—	—	5.86
$HfRe_2$	4.8; 5.61	5.239	9.584	6.0
$HfRe_{2+x}$	(5.8—5.9)‡	—	—	(6.58)
YOs_2	4.7	5.303	8.786	6.33
$ScOs_2$	4.60	5.179	8.484	6.33
$V_{45}Nb_{0.5}Zr$	4.3	—	—	—
$ScRe_2$	4.2‡	—	—	—

TABLE 20 (Continued)

Compound	T_C, °K	Lattice constants, Å		Electron concentration, electrons/atom
		a	c	
$LuOs_2$	3.49	5.254	8.661	6.33
$ZrOs_2$	3.0	5.219	8.538	6.67
$HfOs_2$	2.69	5.184	8.468	6.67
$YIr_{1.5}Os_{0.5}$	2.40	—	—	6.83
$YOsRe$	2.00	—	—	6.0
$ZrRu_2$	1.84	5.144	8.504	6.67
YRe_2	1.83	5.396	8.819	5.67
$ScRu_2$	2.24	—	—	—
$Sc_{1.2}Ru_2$	1.67	5.119	8.542	6.12
YRu_2	2.42	5.256	8.792	6.33
$LuRu_2$	0.86	5.204	8.725	6.33
$ZrAl_2$	<0.35	5.282	8.748	3.33
$ErRu_2, GdOs_2$	*			6.33
$GdRu_2$	*	5.271	8.904	6.33
$NdIn_{1.5}Os_{0.5}$	*	—	—	6.83
$NdOs_2$	*	5.368	8.926	6.33
$NdReOs$	*	—	—	6.00
$PrOs_2$	*	5.368	8.945	6.33
$SmOs_2$	*	5.336	8.879	6.33
$BeGe_2$	<1.75	4.27	6.92	4.66
$CaMg_2$	<1.02	6.23	10.12	2.00
$ScRe_2$	<4.2‡	5.27	8.59	5.66
YRe_2	<4.2‡	—	—	5.66
$GdRe_2$	<4.2‡	5.412	8.827	5.66
$TbRe_2$	<4.2‡	5.397	8.797	6.00
$HoRe_2$	<4.2‡	5.378	8.785	5.66
$ErRe_2$	<4.2‡	5.363	8.758	5.66
$TiZn_2$	<1.02	5.064	8.210	2.66
$ZrZn_2$	<0.1	—	—	2.66
$ZrCr_2$	<0.35	5.102	8.239	5.33
$MoBe_2$	<1.68	—	—	3.33
WBe_2	<1.68	4.446	7.289	3.33
$ReBe_2$	<1.68	4.354	7.101	3.66
$ZrV_{0.5}Co_{1.5}$	<0.35	4.982	8.114	6.66
$NbCoV$	<1.02	4.928	8.08	6.33
Structure of the $MgNi_2$ type				
$HfMo_2$	0.05	—	—	5.3

*Different (not constant) values of T_C were measured.
†The symbol < means that no superconductivity was obtained down to this temperature.
‡Data of Savitskii and Khamidov.

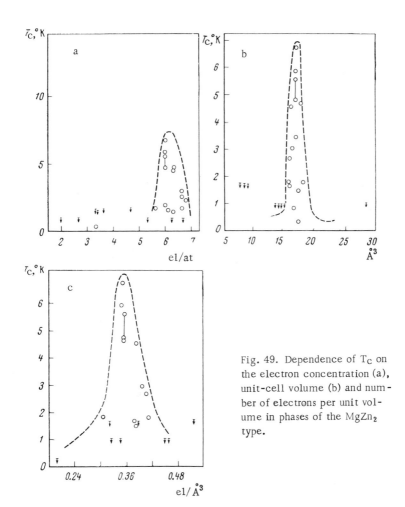

Fig. 49. Dependence of T_C on the electron concentration (a), unit-cell volume (b) and number of electrons per unit volume in phases of the $MgZn_2$ type.

and Bi_2Cs are only half those of the majority of superconducting Laves phases.

Many Laves phases are characterized by polymorphic transformations [147]. In the majority of cases the high-temperature form of these compounds has a structure of the $MgZn_2$ type and the low-temperature form one of the $MgCu_2$ type. The effect of polymorphism on the superconducting properties of the Laves phases has never been studied. However, since the T_c of $MgZn_2$-type compounds is usually lower than that of the $MgCu_2$ type, we

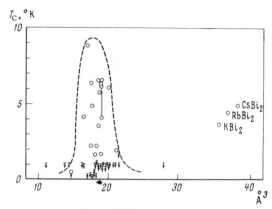

Fig. 50. Dependence of the T_c of MgCu$_2$-type phases on the mean unit-cell volume.

can hardly expect higher critical temperatures in the high-temperature forms of the compounds.

The critical magnetic fields of Laves phases are comparatively low. For example, the H_c of the Laves phase ZrV$_2$, which has a high critical temperature, equals 30 kOe at 4.2°K [162].

The χ phase has a body-centered cubic structure with 58 atoms in the unit cell and is isomorphic with α manganese (Fig. 51 [163]. In this complex ordered structure there are four subgroups of atoms with coordination numbers 16, 13, and 12. Different positions of the atoms correspond to different atomic states. In the binary χ phases the valence of the different atoms and the occupa-

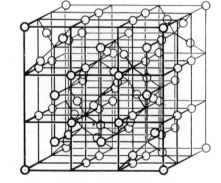

Fig. 51. Structure of α-manganese.

TABLE 21. Critical Temperatures of the χ Phases
(Structure of the α-Mn Type) [1, 62, 165]

Compound	T_C, °K	Lattice constant a, Å	Electron concentration, el/at	Compound	T_C, °K	Lattice constant a, Å	Electron concentration, el/at
$NbTc_3$	10.5	9.625	6.5	$Re_{0.64}Ta_{0.36}$	1.46	9.765	6.28
$MoRe_3$	9.89; 9.26	—	6.75	$Re_{24}Ti_5$	6.6	9.587	6.48
$MoRe_4$	7.85*	—	6.8	$Re_{0.83}Ti_{0.17}$	5.1	9.595	6.49
Tc_6Zr	9.7	9.636	6.57	$Hf_{0.14}Re_{0.86}$	5.86	—	6.57
$Nb_{0.18}Re_{0.82}$	9.7; 8.89	9.641	6.64	Hf_5Re_{24}	4.30*	—	6.48
$Nb_{0.38}Re_{0.62}$	2.45	9.770	6.24	Zr_5Re_{24}	4.75*	—	6.48
$Nb_{0.32}Re_{0.68}$	4.5	9.730	6.36	Al_5Re_{24}	3.35	9.60	6.32
$NbRe_3$	5.27*	—	—	NbOs	2.86	9.760	6.5
$Nb_{0.26}Re_{0.74}$	7.2	9.688	6.48	$NbOs_2$	2.52	9.655	7.0
$Nb_{0.2}Re_{0.8}$	9.1	9.648	6.6	$Nb_{0.6}Pd_{0.4}$	2.47–2.04	9.77	7.0
$Nb_{0.14}Re_{0.86}$	8.5	9.610	6.72				
$Nb_{0.4}Re_{0.6}$	2.36	9.781	6.2	Sc_5Re_{24}	2.2*	—	6.3
Re_3W	9.0; 6.45*	—	6.75	TaOs	1.95	9.773	6.5
Re_6Zr	7.40	9.698	6.57	Al_2Mg_3	<0.35†	10.55	—
Re_3Ta	6.78	—	6.5	$Hf_{0.5}Re_{0.5}$	<1.02	—	—
$TaRe_3$	4.75*	—	—	$V_3Ni_5Ge_2$	<0.35	8.928	—
$Re_{0.65}Ta_{0.35}$	1.58	9.762	6.3	$V_{2.8}Si_2Co_{5.2}$	<1.02	8.774	—
				$Fe_5Si_2V_3$	<0.37	8.851	—

*Results of Savitskii and Khamidov [154a].
†The symbol < means that no superconducting transition occurred down to the temperature in question.

tion density of different positions in the crystal lattice have as yet been little studied [62]. In the opinion of Kasper [163] the χ phase is not an electron compound, and its formation is mainly governed by the spatial (steric) factor. However, the electron structure of the components and the electron concentration of the alloys have an undoubted influence on the development of the χ phases [164].

Almost all the binary χ phases are formed by transition metals. The measured critical temperatures of the χ phases are presented in Table 21. The dependence of T_c on the electron concentration, the number of electrons per unit volume, and the unit-cell volume of the χ phases is illustrated in Fig. 52 [62]. The maximum T_c corresponds to 6.6 electrons/atom, 16 Å^3, or 0.43 electron/Å^3. The unit-cell volume of all the superconducting χ phases differs very little (by ±4%) from the optimum value for superconductivity, corresponding to the maximum value of T_c.

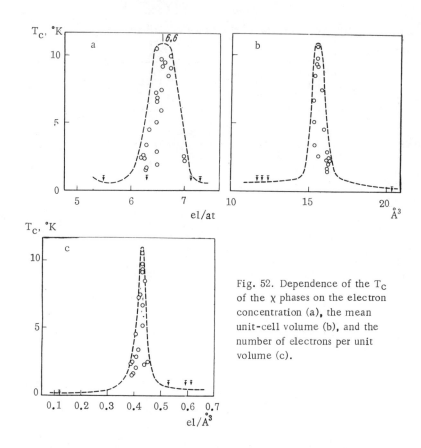

Fig. 52. Dependence of the T_c of the χ phases on the electron concentration (a), the mean unit-cell volume (b), and the number of electrons per unit volume (c).

It is generally considered that the σ-, δ-, P-, χ-, R-, and μ- phases and also compounds with the structure of α-manganese form a class of similar structures characterized by the existence of hexagonal or pseudohexagonal networks in one or more planes of the reciprocal lattice and a specific shape of the Brillouin zone [147]. The differences between these structures are due to the influence of the space factor and the dimensions of the Brillouin zone itself. Not all the characteristics of these phases have yet been established.

The superconducting properties of the μ phases (hexagonal lattice of the W_6Fe_7 type, space group $R3m-D_{3h}^5$, z = 39), R phases (hexagonal lattice, space group $R3-C_{3i}^2$, z = 159), and δ phases (tetragonal lattice with 56 atoms per unit cell) have not yet been studied [62]. A relationship between these structures and the structure of the sigma phases suggests that superconductivity may well be expected among these compounds.

The binary systems of rhenium offer a favorable opportunity of comparing the critical temperatures of the σ-, λ-, and χ-phases. The compounds of rhenium with transition metals are mainly divided into three types: σ phases with the β-U structure, λ-phases with the structure of $MnZn_2$, and χ phases with the structure of α-Mn [166]. Savitskii and colleagues [166, 167] considered various laws relating the T_c of rhenium compounds to the type of crystal structure, the electron concentration, the electron structure of the components, and a number of other factors.

The dependence of the T_c of compounds with various types of structure on the position of their components in the periodic table relative to rhenium and on the concentration of the valence electrons is illustrated in Fig. 53 [166, 167]. As the alloying element approaches rhenium, the electron concentration increases, and hence the electron structure of the atoms becomes more complicated. Likewise the critical temperature of the compounds increases. This law is particularly clearly evident for compounds with the same type of structure formed by metals belonging to the same period; it is characteristic of the λ and χ phases formed in binary rhenium systems with transition elements of the three long periods, and for the σ phases of metals belonging to the second and third long periods of the Mendeleev Table [166]. Exceptions are

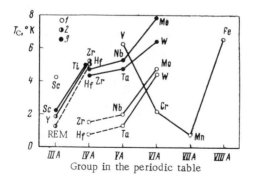

Fig. 53. Dependence of the T_c of rhenium compounds on the position of the second component in the periodic table 1) σ-phase; 2) λ-phase; 3) χ-phase. REM — rare-earth metal.

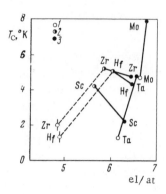

Fig. 54. Change in the T_c of rhenium compounds in relation to the type of crystal structure and the electron concentration. Notation as in Fig. 53.

the σ phases of metals belonging to the first long period. This is evidently associated with the presence of ferromagnetism in some of these compounds.

Another relationship associated with the structure of the electron shells of the second element lies in the fact that for rhenium compounds with the same type of crystal structure the T_c is greater for those compounds which are formed by metals with a less complicated electron shell, i.e., by a metal of the second period rather than the third long period (for example, the T_c of the σ-, χ-, and λ-phases in the systems with zirconium, niobium, and molybdenum are higher than those in the systems with hafnium, tantalum, and tungsten) [166].

A comparison of the T_c values of compounds of various structural types carried out by Savitskii and colleagues [166, 167] shows that the highest value of T_c among all the phases of one particular system incorporating rhenium occurs in the case of the Laves phases with the $MgZn_2$ type of structure. In order of diminishing T_c the crystal structures of the rhenium compounds may be placed in the following sequence: $\lambda \rightarrow \chi \rightarrow \sigma$ (Fig. 54) [166, 167].

SUPERCONDUCTING COMPOUNDS WITH OTHER TYPES OF STRUCTURES

So far we have considered phases with the highest critical temperatures. The crystal structures of other types are less-favorably disposed toward superconductivity. However, individual

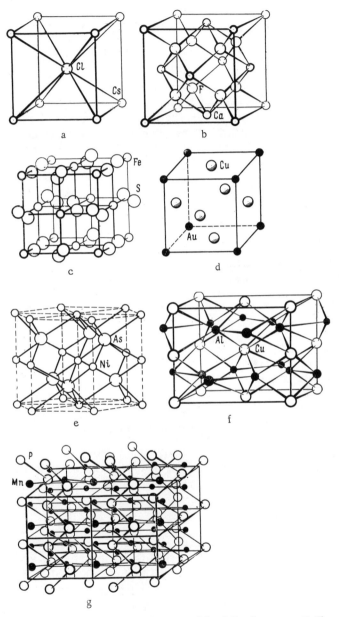

Fig. 55. Crystal structures of phases of the following types: CsCl (a), CaF_2 (b), FeS_2 (c), Cu_2Au (d), NiAs (e), $CuAl_2$ (f), and MnP (g).

superconductors do occur among compounds of other structures. The crystal lattices of some of these are illustrated in Fig. 55.

Cubic Structure of the CsCl Type. Among the compounds possessing this type of structure there are only three superconductors: $Mg_{0.47}$, $Tl_{0.53}$ (2.75°K, a = 3.628 Å), OsTi (0.46°K, a = 3.077 Å), and RuTi (1.07°K, a = 3.067 Å) [62]. According to Savitskii and colleagues [168], the high-temperature form of V_3Ga has a lattice of the CsCl type. When the crystal structure of V_3Ga changes from the Cr_3Si to the CsCl form the critical temperature falls sharply. The high-temperature form of this compound exhibits no superconductivity above 4.2°K.

The superconducting compound CoU (1.70°K) has a distorted structure of the CsCl type.

Cubic Structure of the FeSi Type. Among compounds with this structure there are two superconductors with low critical temperatures: AuBe (2.64°K) and GaPt (1.74°K) [62].

Cubic Structure of the CaF_2 Type. Only two superconductors are known in this case: $CoSi_2$ (1.22°K) and Ga_7Pt_3 (2.9°K) [62].

Cubic Structure of the FeS_2 Type. Several compounds with the pyrite structure have fairly high critical temperatures (Table 22). There are also some ternary alloys exhibiting superconducting properties with a slightly different (disoriented) structure constituting a distorted version of the FeS_2 type.

The crystal structure in question provides yet another example of the strong influence of ordering on the superconducting properties of compounds. The critical temperatures of disordered compounds are much lower than those of compounds with an ordered structure of the FeS_2 type.

Cubic Structure of the Cu_3Au Type. Among the large number of compounds with this structure only six superconductors have been found [62]. The highest critical temperature occurs in the case of $InLa_3$ (T_c = 10.4°K, a = 5.07 Å). Another interesting compound is $Bi_{26}Tl_{74}$, which in the ordered state has a T_c of 4.15°K and in the disordered state 4.4°K. This is the only compound in which T_c is higher in the disordered than in the ordered state. The reason for this effect is as yet unknown. The

TABLE 22. Critical Temperatures of Phases with the FeS$_2$, Fe$_3$Th$_7$, NiAs, CuAl$_2$, and MnP Structures [62, 161]

Compound	T_C, °K	Lattice constants, Å		Electron concentration, electrons/atom
		a	c	
Structure of the FeS$_2$ type (pyrite)				
Rh$_{0.36}$Se$_{0.64}$	6.0	6.015	—	7.09
Rh$_{0.53}$Se$_{0.47}$	6.0	—	—	7.59
RhTe$_2$*	1.51	6.441	—	7.0
AsNi$_{0.12}$Pd$_{0.88}$	1.39†	—	—	7.5
AsNi$_{0.25}$Pd$_{0.75}$	1.34†	—	—	7.5
PdSb$_2$	1.25	6.459	—	6.33
AuSb$_2$	0.58	6.658	—	3.67 (I) or 7.0 (II)
CoAs$_2$	<1.1	—	—	6.0
IrTe$_{2.67}$	<0.35	6.410	—	7.0
PiBi$_{2(\alpha)}$	<1.45	6.683	—	6.67
AsPd	<1.02‡	—	—	7.5
AsPdSe	<1.2	6.092	—	7.0
Distorted structure of the FeS$_2$ type				
BiPtSe	1.45	6.42	—	7.0
BiPdTe	1.20	6.656	—	7.0
PdSeTe	1.20	6.533	—	7.0
BiPtTe	1.15	6.59	—	7.0
BiPdSe	1.0	6.432	—	7.0
PdSbSe	1.0	6.323	—	7.0
PtSbSe	<1.2	6.33	—	7.0
PtSbTe	<1.2	6.48	—	7.0
Structure of the NiAs type				
NiBi	4.25	4.070	5.35	7.5
Bi$_2$Pd$_3$	3.7—4.0	—	—	—
BiPt	1.21—2.4	4.315	5.490	7.5
PdTe	2.30	—	—	8.0
BiRh	2.02—2.2	4.094	5.663	7.0
Bi$_{0.1-1}$PtSb$_{0.9-0}$	1.21—2.05	4.13—4.32	5.47—5.48	7.5
PtSb	2.10	—	—	7.5
AsNi$_{0.25}$Pd$_{0.75}$	1.6†	3.66	5.03	7.5
PbSb	1.50	—	—	7.5
AuSn	1.25	4.32	—	5.52
PtSn	0.37	4.11	—	5.44
BiNi$_x$Rh$_y$	§	4.08—4.09	5.36—5.67	—
Bi$_x$NiSb$_y$	<1.4	3.94—4.07	5.13—5.34	—
BiMn	<1.28	4.30	6.10	6.0
NiAs	<1.28	—	—	7.5
MnSb	<1.00	—	—	6.0
CoTe	<1.00	—	—	7.5

TABLE 22. (Continued)

Compound	T_c, °K	Lattice constants, Å		Electron concentration, electrons/atom
		a	c	
RhTe	<1.06	3.99	5.66	7.5
IrSb	¶	—	—	7.0
IrTe	<0.35	3.930	5.386	7.5
NiSb	<0.35	3.942	5.155	7.5
NiTe	<1.00	—	—	8.00
Structure of the Fe$_3$Th$_7$ type				
Rh$_3$Th$_7$	2.15	10.031	6.287	5.5
Ni$_3$Th$_7$	1.98	9.885	6.225	5.8
Fe$_3$Th$_7$	1.86	9.823	6.211	5.2
Co$_3$Th$_7$	1.83	9.833	6.200	5.5
Ir$_3$Th$_7$	1.52	10.06	6.290	5.5
Os$_3$Th$_7$	1.51	10.02	6.285	5.2
B$_3$Rh$_7$	<0.35	7.47	4.78	6.5
B$_3$Ru$_7$	<2.58	7.465	4.715	6.5
Pt$_3$Th$_7$	<1.02	10.126	6.346	5.8
Structure of the CuAl$_2$ type				
CoZr$_2$	6.30	6.367	5.513	5.67
AuPb$_2$	4.42	7.325	5.655	3.0 (I) or 6.33 (II)
CuTh$_2$	3.49	7.28	5.75	3.0
AuTh$_2$	3.08	7.42	5.95	3 (I) or 6.33 (II)
PdPb$_2$	2.95	6.849	5.833	6.0
RhPb$_2$	2.66	6.664	5.865	5.67
AgIn$_2$	2.30—2.46	6.883	5.615	2.34
AgTh$_2$	2.26	7.56	5.84	3.0
CuAl$_2$	1.02	6.060	4.874	2.33
BTa$_2$	3.12	—	—	4.33
BCr$_2$	<1.20	—	—	5.0
BMo$_2$	<4.74	—	—	5.0
BW$_2$	<1.20	—	—	5.0
AlTh$_2$	<0.35	7.614	5.857	3.67
MnSn$_2$	<1.12	—	—	4.67
FeCu$_2$	<1.02	5.911	4.951	5.33
CoSn$_2$	<1.02	6.361	5.452	5.67

Structure of the MnP type

Compound	T_c, °K	a	b	c	Electron concentration, electrons/atom
GeIr	4.70				6.5
AuGa	1.2	6.40	6.27	3.42	—
GeRh	0.96	5.70	6.48	3.25	6.5
PdSi	0.93	6.133	5.599	3.381	7.0
PtSi	0.88	5.932	5.595	3.603	7.0
GePt	0.40	6.088	5.733	3.701	7.0

TABLE 22 (Continued)

Compound	T_c, °K	Lattice constants, Å			Electron concentration, electrons/atom
		a	b	c	
AsCo¶	<1.1	5.9	5.1	3.5	7.0
GePd†	<0.35	6.259	5.782	3.481	7.0
CrP	<1.02	—	—	—	5.5
IrSi*	<1.02	5.558	3.211	6.273	6.5
RhSb	<0.35	6.340	5.955	3.876	7.0

---

*Low-temperature form.
†The superconductivity is attributed to the formation of $AsPd_2$.
‡The symbol < means that no superconductivity occurs down to the temperature in question.
§ Different values obtained in the measurement.
¶ Measuring temperature not indicated.

compound Bi_3Zr also has a fairly high T_c (5.62°K). The critical temperatures of the remaining compounds with this structure are low, for example, in $AlZr_3$ T_c = 0.72°K.

Complex Cubic Structure. Recently a number of superconducting compounds $Be_{22}Me$ with a complex cubic structure obtained by introducing small quantities of transition-metal atoms into the simple metallic matrix have been discovered [169]. The structures of these compounds are of the same type as that of the first-discovered compound $Be_{22}Re$ (a = 11.56 Å) [170]. There are four such compounds: $Be_{22}Re$ (T_c = 9.65°K), $Be_{22}Tc$ (5.25°K), $Be_{22}Mo$ (2.25°K), and $Be_{22}W$ (4.14°K) [169, 170]. In the isoelectron series $Be_{22}Mo_{1-x}W_x$ the critical temperature depends almost linearly on the mass number. The introduction of small quantities of iron or manganese into the compound $Be_{22}Re$ leads to a sharp decrease in T_c.

Hexagonal Structure of the NiAs Type. Many NiAs phases exhibit superconductivity at temperatures below 5°K Table 22). The T_c of certain NiAs phases vary over a fairly wide temperature range with varying composition (for example, in BiPt 1.2-2.4°K). The substantial variation in the T_c of these phases may be associated with the different numbers of vacancies in the crystal lattice; thus, the compound BiPt may have up to 0.8% vacancies in the quenched state.

TABLE 23. Superconducting Compounds with Various Crystal Structures [26, 62, 121b]

Compound	$T_c, °K$	Compound	$T_c, °K$	Compound	$T_c, °K$	Compound	$T_c, °K$
Cubic structure							
$Rh_{17}S_{15}$	5.80	$IrZn_2$	0.78—0.65	$AuAl_2$	<0.34[1*]	$NbRh_3$	<1.43
$CoTi_2$	3.44	Au_4Al	0.7—0.4	$AuGa_2$	<0.32	$NbIr_3$	<1.6
$RhHf_2$	2.02	$PtAl_2$	0.55—0.48	$AuZn_2$	<0.32	$IrHf_2$	<1.6
$Nb_{0.26}U_{0.74}$	1.85	$CoHf_2$	0.50	$CaPt$	<0.34	$PtSn_2$	<0.34
$AuZn_3$	1.28	Au_5Ca	0.38—0.34	$AlPt$	<0.34	$Nb(Ir_{0.92}Pt_{0.08})_3$	<1.6
Al_3Mg_2	0.84	$Ag_{3.3}Al$	0.34	VIr_3	<4.20	$(Mn_{0.97}V_{0.03})_3Si$	<4.2
Tetragonal structure							
Nb_3Sn_2[6*]	16.6	FeU_6	3.86	α-BiLi	2.47	$CdHg$	<1.77
$BaBi_3$	5.69	YGe_2	3.80	MnU_6	2.32	$Nb_{0.6}Ru_{0.4}$	<1.2
$Nb_{0.85}Ir_{1.15}$	4.75	$Nb_{0.96}Rh_{1.04}$	3.76	CoU_6	2.29	β-VIr	<1.36
$PbBi_2(\beta)$	4.25	V_2Ga_5	3.55	$BiNa$	2.25		
WBe_{13}	4.10	$PtPb_4$	2.80	$InSb$	2.1—1.6		
Hexagonal structure							
$BiIn_2$	5.6	Hg_2Na	1.62	Ag_5Sr	<0.34	$(Nb_{0.15}\cdot Ir_{0.85})\cdot Pt_3$	<1.6
$TaSi$	4.25—4.38	Au_5Sn	1.1—0.7	Ag_5Sn	<0.34	$Nb(Ru_{0.4}, Pt_{0.6})_3$	<1.6
$RhBi_4(\gamma)$	2.70	$TlAl_3$	~0.75	Ag_5K	<0.34	$Nb(Ru_{0.15}, Pt_{0.85})_3$	<1.6
$La_{0.75}Y_{0.25}$	2.50	Au_5Ba	0.4—0.7	Ag_5Rb	<0.34	$Nb(Rh_{0.9}, Nb_{0.1})_3$	<1.6
β-ThSi	2.41	Au_5B	0.35—0.7	$CaCu_5$	<0.34	$Nb(Ir_{0.85}, Pt_{0.15})_3$	<1.6
$La_{0.55}Lu_{0.45}$	2.20	$PtBi_2(\beta)$	0.155	$CaZn_5$	<0.34	$Nb(Ir_{0.15}, Pt_{0.85})_3$	<1.6
$MoIr$	1.85	Ag_5Ba	<0.34	$(Nb_{0.65}\cdot Zr_{0.35})Pt_3$	<1.6	$Nb(Os_{0.14}, Pt_{0.86})_3$	<1.6
Hg_3Li	1.7					$TiVSi$	<4.2
Orthorhombic structure							
$MoIr$	8.8	$RhBi_3(\beta)$	3.2	$NbSn_2$	2.60	α-VIr	<1.6
$Nb_{0.85}Ir_{1.15}$	4.6	$Nb_{0.9}Rh_{1.1}$	3.07	$PtSn_4$	2.38	$NbPt$	<1.39
$NiBi_3$	4.06	$Nb_{0.85}Rh_{1.15}$	3.00	Rh_5Ge_3	2.12	$NbPt_2$	<1.46
$PdBi$	3.7	Nb_6Sn_5	2.8	$LaGe_2$	1.49	$TaRh_2$	<1.39
Monoclinic structure							
$Nb_{0.75}Rh_{1.25}$	2.7	$PdBi_2(\alpha)$	1.70				

[1*] The symbol < means that no superconductivity occurred down to the temperature indicated.
[2*] Compounds with unknown structures or structures not accurately established are included.
[3*] Synthesized at a pressure of about 40,000 atm at high temperature.
[4*] At 1500°C.
[5*] At 1000°C.
[6*] The high T_c of Nb_3Sn_2 (or Nb_6Sn_5) is due to small quantities of Nb_3Sn in the alloy. The T_c of Nb_3Sn_2 itself is lower than 2.8°K [177a].

TABLE 23 (Continued)

Compound	T_c, °K	Compound	T_c, °K	Compound	T_c, °K	Compound	T_c, °K
			Unknown structure 2*				
ReTiO$_5$	5.71	Bi$_3$Sn³*	3.67—3.63⁵*	Bi$_2$Ir³*	2.2—1.7	Bi$_3$Zn³*	0.87—0.8
BiRu³*	5.74*	Bi$_3$Mo³*	3.7—3.0	ReYOs	2.0	CoBi³*	0.5
HgSn$_6$	5.1—4.8					BiCo³*	0.49—0.42
BiRu	4.12—3.31⁵*	Mg$_2$In	3.2	BiCu³*	1.40—1.33	InBi	<0.5
In$_2$Bi	5.1	Bi$_2$Ag³*	3.0—2.78	Bi$_4$Mg—BiMg³*	1.0—0.7		
In$_5$Bi	4.1	BiZr$_3$³*	2.84—2.35	Bi$_3$Fe³*	1.0—0.75		
Bi$_3$Sn³*	3.77—3.72⁴*	AgBi³*	2.78	LaAg	0.9		
		BiRe$_2$³*	2.2—1.9				

See p. 156 for 2*, 3*, 4*, 5*.

<u>Hexagonal Structure of the Fe$_3$Th$_7$ Type</u>. All these compounds are encountered solely in thorium alloys; their critical temperatures are low (Table 22). The mean unit-cell volume of these compounds is greater than the mean volume of the majority of other compounds.

<u>Tetragonal Structure of the CuAl$_2$ Type</u>. There are eight superconductors of this type with T_c values between 2.2 and 6.3°K (Table 21). Phases of the CuAl$_2$ type may be formed by elements with large differences in atomic radii (r_A/r_B between 1.080 abd 1.499) [62]. However, the effect of the atomic dimensions on the superconducting properties of compounds of the CuAl$_2$ type has not been studied.

<u>Tetragonal Structure of the Ti$_3$P Type</u>. According to recent data [171], the compound Nb$_3$Si has a tetragonal structure of the Ti$_3$P type (space group $P4_2/n-C_{4h}^4$) with a = 10,230 Å, c = 5,189 Å. The critical temperature of this compound equals 1.5°K [62].

<u>Orthorhombic Structure of the MnP Type</u>. The critical temperature of known superconducting compounds of this structural type are presented in Table 22.

Apart from those just described, there are a number of other superconducting compounds the crystal structure of which has never been established. Data relating to these compounds are presented in Table 23.

Certain compounds not constituting superconductors when formed under ordinary conditions acquire superconductivity on subjection to high pressures or when produced in the form of a thin film [62]. Bismuth compounds of this type exhibit a specific relationship between the T_c value and the shortest interatomic distance in the crystal lattice; these compounds are only superconductors for interatomic distances of 3.1-3.8 Å [62].

Thus, judging from available experimental data, the crystal structure is a necessary but not always sufficient factor determining the presence or absence of superconductivity in particular compounds. On the other hand, there is no doubt whatsoever that the highest critical superconducting temperatures are obtained for compounds in which the spatial distribution of the atomic components resembles that of the Cr_3Si structure.

EFFECT OF ALLOYING ELEMENTS AND IMPURITIES ON THE STRUCTURE AND PROPERTIES OF COMPOUNDS

A study of the superconductivity of compounds and solid solutions based on the latter may lead to the discovery of new materials with high superconducting characteristics, promote the development of the metallography of superconducting materials, and create the theoretical foundations for the discovery of new superconductors. Such investigations are also important in order to establish the laws governing the changes taking place in the critical temperatures and other superconducting characteristics with changing composition, structure, and state of processing of the alloys as well as the electron structure and other characteristics of the material in question. Such an analysis should enable us to predict what elements and what proportions of these are required in order to create superconducting alloys with specified critical parameters, and how to produce superconducting materials for hydrogen, neon, and nitrogen temperatures [172, 173]. Finally, such investigations should help us to develop the best possible technology for producing superconducting compounds with the maximum superconducting characteristics [172].

It is of particular interest to study alloys based on compounds with the Cr_3Si structure, since, as we have already seen, these have the highest superconducting parameters of all known superconductors.

The effect of alloying on the critical superconducting temperature of compounds of the Cr_3Si type has been considered by a num-

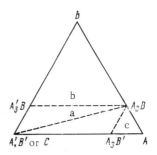

Fig. 56. Diagram illustrating the probable behavior of solid solutions based on compounds of the A_3B type with the Cr_3Si type of structure in relation to the nature of the alloying element. A = transition metal; B = element from the main subgroups of the periodic table; C = interstitial element.

ber of authors. A systematic investigation into the solubility of a third component in V_3Si and its effect on the T_c of the compound was carried out [81, 174] along three sections of the V–Si–element system, where the "element" was O, C, B, Ce, La, Ti, Zr, Nb, Cr, Mo, Mn, Re, Pd, Ge, or Sn (Fig. 56). Recently we also studied the effect of alloying elements on the T_c of V_3Ga [175]. The critical temperature of solid solutions based on Nb_3Sn was studied elsewhere [176, 177].

Data relating to the critical temperatures of solid solutions based on compounds with the Cr_3Si type of structure are presented in Table 24.

When interstitial elements are dissolved in A_3B compounds with the Cr_3Si structure, limited solid solutions may be formed, the maximum extent of these being reached for an A:B ratio of 3:1 (section a in Fig. 56). When transition metals are dissolved, solid solutions of the substitutional type may occur, the range of homogeneity of these being directed along the $A_3B-A_3^1B$ section (section b in Fig. 56). Germanium, tin, and other atoms of the main subgroups of the periodic table may replace the atoms of the B components in compounds with the Cr_3Si type of structure. The isomorphic substitution of these elements occurs along the $A_3B^1-A_3B$ section (section c in Fig. 56).

1. Effect of Transition Metals on the Properties of Cr_3Si-Type Compounds

The laws governing the formation of solid solutions on alloying A_3B compounds of the Cr_3Si type with transition metals were studied in detail for V_3Si-base alloys [81, 174].

When transition metals dissolve in V_3Si, the existence of isostructural compounds (Mo_3Si and Cr_3Si) [179] and favorable dimen-

TABLE 24. Superconducting Transition Temperature of Solid Solutions with the Cr_3Si Structure [1, 62, 174-178]

Composition	T_C, °K	Composition	T_C, °K	Composition	T_C, °K
\multicolumn{6}{c}{1. Based on vanadium compounds}
$(V_{0.998}. La_{0.002})_3Si$	16.48	$V_3Si (0.4\% Fe, Mn)$	16.3	$(V_{0.926}. La_{0.074})_3Ga$	7.8
$(V_{0.98}. La_{0.02})_3Si$	15.92	$V_3Si (1.3\% Fe, Mn)$	14.4	$(V_{0.93}. Ce_{0.07})_3Ga$	7.5
$(V_{0.96}. Ce_{0.04})_3Si$	15.32	$V_3Si_{0.99}Be_{0.01}$	16.6	$(V_{0.918}. Ti_{0.082})_3Ga$	13.4
$(V_{0.9}. Ti_{0.1})_3Si$	10.9	$V_3Si_{0.76}Be_{0.24}$	15.6	$(V_{0.914}. Zr_{0.086})_3Ga$	9.9
$(V_{0.9}. Zr_{0.1})_3Si$	13.2	$V_3Si_{0.9}Al_{0.1}$	14.0	$(V_{0.913}. Hf_{0.087})_3Ga$	6.4
$(V_{0.86}. Zr_{0.14})_3Si$	8.8	$V_3Si_{0.98}Al_{0.02}$	16.12	$(V_{0.938}. Nb_{0.062})_3Ga$	13.5
$(V_{0.99}. Nb_{0.01})_3Si$	15.8	$V_3Si_{0.8}Al_{0.2}$	14	$(V_{0.934}. Ta_{0.066})_3Ga$	12.3
$(V_{0.9}. Nb_{0.1})_3Si$	12.8	$V_3Si_{0.76}Al_{0.24}$	11.2	$(V_{0.933}. Cr_{0.067})_3Ga$	12.6
$(V_{0.87}. Nb_{0.13})_3Si$	<14.0	$V_3Si_{0.6}Al_{0.4}$	10.1	$(V_{0.933}. Mo_{0.067})_3Ga$	11.7
$(V_{0.9}. Cr_{0.1})_3Si$	11.3	$V_3Si_{0.87}Ga_{0.13}$	13.3	$(V_{0.961}. W_{0.039})_3Ga$	11.5
$(V_{0.93}. Cr_{0.07})_3Si$	<14.0	$V_3Si_{0.79}Ga_{0.21}$	10.7	$(V_{0.966}. Mn_{0.034})_3Ga$	9
$(V_{0.99}. Mo_{0.01})_3Si$	16.0	$V_3Si_{0.72}Ga_{0.28}$	10.5	$(V_{0.976}. Re_{0.024})_3Ga$	10.8
$(V_{0.988}. Mo_{0.012})_3Si$	14.0	$V_3Si_{0.68}Ga_{0.32}$	9.4	$(V_{0.955}. Fe_{0.045})_3Ga$	10.1
$(V_{0.97}. Mo_{0.03})_3Si$	10.4	$V_3Si_{0.52}Ga_{0.48}$	7.8	$(V_{0.927}. Co_{0.073})_3 Ga$	8.4
$(V_{0.93}. Mo_{0.07})_3Si$	5.59	$V_3Si_{0.5}Ga_{0.5}$	8.6—11.9	$(V_{0.931}. Ni_{0.069})_3Ga$	11.9
$(V_{0.84}. Mo_{0.16})_3Si$	5.1	$V_3Si_{0.46}Ga_{0.54}$	7.4	$V_3 (Ga_{0.999}. Zn_{0.001})$	10.2
$(V_{0.8}. Mo_{0.2})_3Si$	4.54	$V_3Si_{0.25}Ga_{0.75}$	8.1	$V_3 (Ga_{0.994}. Cd_{0.006})$	11.2
$(V_{0.67}. Mo_{0.33})_3Si$	<1.9	$V_3Si_{0.1}Ga_{0.9}$	10.6	$V_3 (Ga_{0.937}. In_{0.063})$	11.5
$(V_{0.99}. Mn_{0.01})_3Si$	16.25	$V_3Si_{0.9}Ge_{0.1}$	14.0	$V_3 (Ga_{0.957}. Ge_{0.043})$	13.1
$(V_{0.93}. Mn_{0.07})_3Si$	15.5	$V_3Si_{0.8}Ge_{0.2}$	11.8	$V_3 (Ga_{0.78}. Sn_{0.22})$	6.9
$(V_{0.9}. Mn_{0.1})_3Si$	11.9	$V_3Si_{0.6}Ge_{0.4}$	8.7	$V_3 (Ga_{0.974}. Pb_{0.026})$	13.2
$(V_{0.83}. Mn_{0.17})_3Si$	8.1	$V_3Si_{0.4}Ge_{0.6}$	7.2	$V_3 (Ga_{0.793}. Sb_{0.207})$	6.9
$(V_{0.94}. Re_{0.06})_3Si$	<14.0	$V_3Si_{0.2}Ge_{0.8}$	6.1	$V_3 (Ga_{0.975}. Bi_{0.025})$	14.7
$(V_{0.936}. Re_{0.064})_3Si$	4.7	$V_3Si_{0.8}Sn_{0.2}$	12.85	V_2NbSn	5.5
$(V_{0.98}. Fe_{0.02})_3Si$	13.4	$V_3Si_{0.6}Sn_{0.4}$	9.7	$V_{1.5}Nb_{1.5}Sn$	7.4
$(V_{0.9}. Ru_{0.1})_3Si$	2.90	$V_3Si_{0.4}Sn_{0.6}$	9.8	$V Nb_2Sn$	9.8
$(V_{0.993}. Co_{0.007})_3Si$	16.4	$V_3Si_{0.2}Sn_{0.8}$	6.4	$V_{0.5}Nb_{2.5}Sn$	14.2
$(V_{0.89}. Pd_{0.11})_3Si$	7.5	$V_3Si_{0.986}Sb_{0.014}$	10.5	V_2TaSn	2.8
$V_3Si (0.25\% Fe. Mn)$	17.0	$(V_{2.67}. Ir_{0.33}) Jr$	11.39	$V Ta_2Sn$	3.7
\multicolumn{6}{c}{2. Based on niobium compounds}
$(Nb_{0.95}. V_{0.05})_3Au$	8.4	$Nb_3Sn_{0.9}Ga_{0.1}$	18.1—15.3 *	$Nb_3Rh_{0.7}Au_{0.3}$	4.6
$(Nb_{0.9}. V_{0.1})_3Au$	6.5	$Nb_3Sn_{0.8}Ga_{0.2}$	17.8—17.4 *	$Nb_3Rh_{0.5}Au_{0.5}$	6.6
$(Nb_{0.85}V_{0.15})_3Au$	5.1	$Nb_3Sn_{0.6}Ga_{0.4}$	16.0—13.5 *	$Nb_3Rh_{0.3}Au_{0.7}$	9.5
$(Nb_{0.83}V_{0.17})_3Au$	4.6	$Nb_3Sn_{0.4}Ga_{0.6}$	14.6—14.0 *	$Nb_3Rh_{0.1}Au_{0.9}$	10.8
$(Nb_{0.75}. V_{0.25})_3Au$	3	$Nb_3Sn_{0.2}Ga_{0.8}$	13.1	$Nb_3Rh_{0.05}Au_{0.95}$	11.0
$(Nb_{0.67}. V_{0.33})_3Au$	2.9	$Nb_3Sn_{0.81}Ge_{0.19}$	14.3	$Nb_3Rh_{0.02}Au_{0.98}$	11.9
$(Nb_{0.57}. V_{0.43})_3Au$	1.6	$Nb_3Sn_{0.73}Ge_{0.27}$	8.2	$Nb_3Rh_{0.98}Co_{0.02}$	2.28
$(Nb_{0.95}. Mo_{0.05})_3Al$	13.4	$Nb_3Sn_{0.56}Ge_{0.44}$	6.8	$Nb_3Rh_{0.95}Co_{0.05}$	1.96

TABLE 24 (Continued)

Composition	T_c, °K	Composition	T_c, °K	Composition	T_c, °K
$(Nb_{0.9}Mo_{0.1})_3Al$	11.4	$Nb_3Sn_{0.5}Ge_{0.5}$	12.6	$Nb_3Rh_{0.9}Co_{0}$,†	1.90
$(Nb_{0.7}Mo_{0.3})_3Al$	8.3	$Nb_3Sn_{0.39}Ge_{0.61}$	7.6	$Nb_3Rh_{0.98}Ru_{0.02}$	2.42
$(Nb_{0.5}Mo_{0.5})_3Al$	4.9	$Nb_3Sn_{0.8}Sb_{0.2}$	18.0	$Nb_3Rh_{0.95}Ru_{0.05}$	2.42
$(Nb_{0.4}Mo_{0.6})_3Al$	4.2	$Nb_3Sn_{0.6}Sb_{0.4}$	15.8	$Nb_3Rh_{0.9}Ru_{0}$,†	2.44
$(Nb_{0.3}Mo_{0.7})_3Al$	2.73	$Nb_3Sn_{0.4}Sb_{0.6}$	12.4	$Nb_3Rh_{0.98}Pd_{0.02}$	2.50
$Nb_{2.75}Ta_{0.25}Sn$	17.8	$Nb_3Sn_{0.35}Sb_{0.65}$	10.5	$Nb_3Rh_{0.95}Pd_{0.05}$	2.49
$Nb_{2.5}Ta_{0.5}Sn$	17.6	$Nb_3Sn_{0.3}Sb_{0.7}$	6.8	$Nb_3Rh_{0.9}Pd_{0}$,†	2.55
Nb_2TaSn	16.4	$Nb_3Sn_{0.25}Sb_{0.75}$	<5	$Nb_3Rh_{0.98}Os_{0.02}$	2.42
$NbTa_2Sn$	10.8	$Nb_3Sn_{0.2}Sb_{0.8}$	<4.2	$Nb_3Rh_{0.95}Os_{0.05}$	2.39
$Nb_3Sn_{0.98}Al_{0.02}$	17.9–17.8*	$Nb_3Sn_{0.1}Sb_{0.9}$	<4.2	$Nb_3Rh_{0.9}Os_{0.1}$	2.30
$Nb_3Sn_{0.96}Al_{0.04}$	18.0			$Nb_3Rh_{0.7}Os_{0.3}$	1.7
$Nb_3Sn_{0.94}Al_{0.06}$	18–17.8*	$Nb_3Al_{0.81}Ge_{0.19}$	17.6	$Nb_3Rh_{0.5}Os_{0.5}$	<1.7
$Nb_3Sn_{0.92}Al_{0.08}$	17.9	$Nb_3Al_{0.62}Ge_{0.38}$	17.2	$Nb_3Rh_{0.3}Os_{0.7}$	<1.7
$Nb_3Sn_{0.9}Al_{0.1}$	18.4–16.2*	$Nb_3Al_{0.55}Ge_{0.45}$	11.7	$Nb_3Rh_{0.1}Os_{0.9}$	<1.7
$Nb_3Sn_{0.88}Al_{0.12}$	17.8–18.4*	$Nb_3Al_{0.5}Ge_{0.5}$	12.6	$Nb_3Rh_{0.98}Pt_{0.02}$	2.52
$Nb_3Sn_{0.81}Al_{0.19}$	14.6	$Nb_3Al_{0.39}Ge_{0.61}$	7.3	$Nb_3Rh_{0.95}Pt_{0.05}$	2.53
$Nb_3Sn_{0.8}Al_{0.2}$	16.9–13.1*	$Nb_3Ge_{0.5}Ga_{0.5}$	7.3	$Nb_3Rh_{0.9}Pt_{0.1}$	2.8
$Nb_3Sn_{0.61}Al_{0.39}$	14.1	$Nb_3In_{0.5}Zr_{0.5}$	6.4	$Nb_3Rh_{0.7}Pt_{0.3}$	5.1
$Nb_3Sn_{0.6}Al_{0.4}$	15.1–15.4*	$Nb_3Si_{1-x}As_x$	17.91–18.0*	$Nb_3Rh_{0.5}Pt_{0.5}$	6.25
$Nb_3Sn_{0.5}Al_{0.5}$	16.3	$Nb_3Si_{0.6}Sn_{0.4}$	6.5	$Nb_3Rh_{0.3}Pt_{0.7}$	7.4
$Nb_3Sn_{0.4}Al_{0.6}$	13.6–15.2*	$Nb_3Si_{0.5}Sn_{0.5}$	8.3	$Nb_3Rh_{0.1}Pt_{0.9}$	7.9
$Nb_3Sn_{0.2}Al_{0.8}$	14.6–15.6*	$Nb_3Rh_{0.98}Au_{0.02}$	2.53	$Nb_3Rh_{0.05}Pt_{0.95}$	8.9
$Nb_3Sn_{0.1}Al_{0.9}$	16.1	$Nb_3Rh_{0.95}Au_{0.05}$	2.52	$Nb_3Rh_{0.02}Pt_{0.98}$	9.6
$Nb_3Sn_{0.5}Zr_{0.5}$	16.5	$Nb_3Rh_{0.9}Au_{0.1}$	2.70		

3. Quaternary solid solutions ‡

Composition	T_c, °K	Composition	T_c, °K	Composition	T_c, °K
$Nb_2Ta_{0.5}V_{0.5}Sn$	12.2	$Nb_{76}Sn_{2.4}Al_{19.2}Ge_{2.4}$	16	$Nb_{79.1}Sn_{9.4}Al_{2.1}Ge_{9.4}$	8.2
$NbTaVSn$	6.2	$Nb_{78.4}Sn_{12.9}Al_{2.2}Ge_{6.5}$	13.3	$Nb_{79.2}Sn_{5.2}Al_{5.2}Ge_{10.4}$	10.9
$Nb_{77.3}Sn_{18.1}Al_{2.3}Ge_{2.3}$	13	$Nb_{77.9}Sn_{11.1}Al_{5.5}Ge_{5.5}$	13	$Nb_{78.6}Sn_{2.2}Al_{9.6}Ge_{9.6}$	12.7
$Nb_{77}Sn_{13.8}Al_{6.9}Ge_{2.3}$	13.3	$Nb_{78.1}Sn_{7.3}Al_{7.3}Ge_{7.3}$	12.5	$Nb_{79.7}Sn_{6.1}Al_{2.0}Ge_{12.2}$	7.15
$Nb_{76.3}Sn_{10.5}Ge_{2.3}Al_{10.9}$	14.4	$Nb_{77.5}Sn_{5.6}Al_{11.3}Ge_{5.6}$	15.2	$Nb_{79.5}Sn_{2.1}Al_{6.1}Ge_{12.3}$	9.4
$Nb_{76.3}Sn_{7.1}Al_{14.3}Ge_{2.3}$	13.9	$Nd_{78.5}Sn_{2.2}Al_{12.8}Ge_{6.5}$	17.0	$Nb_{80.7}Sn_{1.9}Al_{1.9}Ge_{15.5}$	7.8

*Data of various research workers.
† Three-phase alloy.
‡Results of Alekseevskii, Ageev, and Shamrai [176] (except for the first two solid solutions).

TABLE 25. Solubility of Transition Metals in the Compound V_3Si

Dissolving metal	Spatial factor,* %	Difference in electronegativity*	Solubility in V_3Si, at.%		Lattice constant and microhardness of saturated $(V, X)_3 Si$ at 800°C	
			800°C	maximum	a, Å	H_μ, kg/mm²
Ce	35.2	0.64	0.2	†	4.729	1430
La	39.0	0.68	0.1	†	4.727	1500
Ti	8.9	0.23	14	~15	4.762	1480
Zr	19.3	0.37	7.5	†	4.441	1580
Nb	8.2	0.03	18	30 (26 for 1400°C)	4.788	1500
Ta	9.0	0.38	—	5 for 1400°C	—	—
Cr	5.5	0.22	CSSS‡	CSSS for 1300°C	—	—
Mo	3.7	0.20	The same	The same	—	—
Mn	3.07	0.16	<5	49	4.715	1490
Re	2.2	0.475	12	13	4.734	1780
Fe	6.3	0.10	—	23 for 1000°C	—	—
Co	7.2	0.15	—	5	—	—
Ni	8.1	0.05	—	5	—	—
Pd	2.2	0.25	17	†	4.728	1360

*The spatial factor and the difference in electronegativity relative to vanadium are calculated from earlier data [38].
†Solubility almost constant with changing temperature.
‡CSSS = continuous series of solid solutions.

sional and electrochemical factors relative to vanadium (Table 25) leads to the isomorphic replacement of the vanadium atoms in the V_3Si lattice; thus a continuous series of solid solutions is formed in the quasi-binary systems of V_3Si with Mo_3Si and Cr_3Si at all temperatures [174, 180, 181]. The lattice constant of an annealed $(V, Mo)_3Si$ solid solution varies from $a = 4.726$ Å for V_3Si to $a = 4.893$ Å for Mo_3Si [174]. The maximum microhardness (1560 kg/mm²) occurs at 25-35 at.% Mo. On adding chromium the lattice constant of V_3Si diminishes and the microhardness increases ($a = 4.709$ Å, $H_\mu = 1440$ kg/mm² at 4.9 at.% Cr) [174].

Other transition metals not forming compounds of the A_3B type with silicon (Ce, La, Ti, Zr, Nb, Re, Pd [179]) or forming compounds not isostructural with V_3Si (Mn_3Si is a body-centered lattice of the α-iron type [182]) are partially soluble in V_3Si. The limiting solubility of these is directly proportional to the spatial and electrochemical factors (Table 25). The maximum solubility in V_3Si occurs for Mn, Nb, and Ti, the nearest neighbors to vanadium in the periodic table. Among the remaining metals, Pd, Re, and Fe have a considerable solubility in V_3Si, these elements being similar to vanadium in atomic radius. As the difference in atomic radii increases, the solubility of these metals in V_3Si decreases accordingly. The effect of electronegativity is less important in this case.

The solubility of Ce, La, Zr, Pd, Ti, and Re in V_3Si is very low, while that of Nb and Mn varies substantially with changing temperature. Ternary V_3Si-base solid solutions have narrow ranges of homogeneity drawn out along the V_3Si-A_3Si sections. The silicon content varies very little within the range of homogeneity (1-2 at.%) [179, 183].

On dissolving transition metals in V_3Si, in almost all cases there is an appreciable change in the lattice constant of the compound: an increase on introducing Mo, Ti, Zr, Re, Nb, Ce, and La, the atomic radii of which are greater than the atomic radius of vanadium, and a decrease on alloying with Cr and Mn, elements with smaller atomic radii. Only on dissolving palladium, which has an atomic radius very similar to that of vanadium, is there an inappreciable change in the lattice constant of the compound.

Independently of any change in the lattice constant and electron concentration (the latter decreases for Ti, Zr, Ce, and La, but increases for Mn, Re, Pd, Mo, and Cr, remaining unchanged for Nb), the T_c of V_3Si decreases on alloying with transition metals. Certain data relating to the superconducting transition temperature of individual V_3Si-base alloys are presented in Table 24 [57, 58, 174]. The minimum effect occurs in the case of Nb (0.57°K/at.%), i.e., an element of group V (Fig. 57). A sharp decrease in T_c is produced by the rare-earth metals (La by 9, Ce by 6°K/at.%), which are poorly soluble in V_3Si, and also by transition metals of group VII (Mn by 1.8, Re by 2.6°K/at.%). Molybdenum (2.4) and iron (2.3°K/at.%) also sharply reduce the critical temperature of V_3Si, iron having a greater effect in this respect than manganese.

By considering the position of the manganese and rhenium points in Fig. 57 we may well expect that technetium will reduce the critical temperature of V_3Si more than ruthenium, and that among the alloying elements of the second long period the greatest

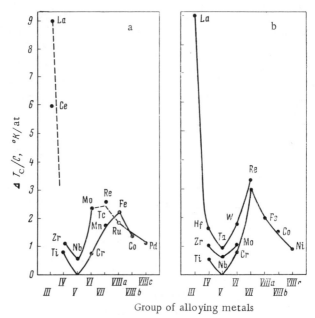

Fig. 57. Reduction in the transition temperature of V_3Si (a) and V_3Ga (b) into the superconducting state in introducing equal atomic proportions of alloying additives within the limits of the corresponding solid solutions.

reduction in critical temperature will probably occur for group VII (or, at any rate, VI-VII).

Data relating to the T_c of V_3Si alloyed with Ru and Pd confirm that in general group VIII transition metals have less effect than those of group VII belonging to analogous periods on the critical temperature of compounds with the Cr_3Si type of structure.

Alloying V_3Ga with transition metals also reduces the value of T_c, in the same way as that of V_3Si (Table 24). The extent of this reduction in T_c differs for equal proportions of different alloying elements. Thus alloying transition metals may be placed in the following order as regards the extent to which they reduce the T_c of V_3Ga (for equal atomic proportions of the alloying additive, within the range of homogeneity): Ti (0.56°K/at.%), Nb (0.7), Cr (0.84), Ta (0.9), Ni (0.94), Mo (1.02), Zr (1.05), Co (1.56), Hf (1.6), W (1.8), Fe (2.0), Mn (3.0), Re (3.3), Ce (~ 8.5), and La (~ 9°K/at.%). The reduction in the T_c of V_3Ga on introducing La or Ce constitutes an approximate estimate. On melting alloys of V_3Ga with lanthanum or cerium no homogeneous alloys resulted. For any specified proportion of the rare-earth additive, two sharply-distinguished layers appeared in the resultant casting, one of these being richer in the alloying additive. The critical superconducting temperature of both layers was much lower than that of the original V_3Ga. By analyzing the layers chemically it was found that the solubility of lanthanum and cerium in V_3Ga was no greater than a few tenths of an atomic percent.

Figure 57b shows the decrease in the critical temperature of V_3Ga for equal atomic proportions of additive in relation to the position of the alloying element in the periodic table. We see that transition metals of the same group in the periodic table reduce the critical temperature of the compound more the greater the mass number of the alloying element. Of all the transition metals of the same period, the smallest reduction in the critical temperature of the compound occurs for elements of group V, i.e., analogs of vanadium. As the difference between the electron structure of the alloying transition metals of groups III-VII and that of the vanadium atom (or its corresponding analogs) increases on passing along the period, so does the decrease in the critical temperature of V_3Ga. Transition elements of group VIII reduce the critical temperature of the compound less than do those of group VII.

The superconducting transition temperature has been studied in more detail in the following systems: V_3Si-Mo_3Si [81], Nb_3Au-V_3Au [184], V_3Sn-Nb_3Sn, V_3Sn-Ta_3Sn, Nb_3Sn-Ta_3Sn [185].

On increasing the amount of molybdenum in a $(V, Mo)_3Si$ solid solution, the T_c of the alloys decreases sharply, particularly for Mo contents of up to 5 at.% (Fig. 58). At 25 at.% Mo the alloys become nonsuperconducting at $T \geq 1.9°K$. Since the T_c of Mo_3Si equals 1.3°K, we see that for compositions with 25-75 at.% Mo there is a gradual decrease in the value of T_c to this temperature.

In the Nb_3Au-V_3Au system a solid solution with a bcc lattice exists at high temperature; after low-temperature annealing these compounds form a continuous series of solid solutions with the Cr_3Si type of structure [184]. The temperature for the formation of a continuous series of solid solutions of the $(V, Nb)_3Au$ type rises from 760° for V_3Au to 1050-1200°C for Nb_3Au. The T_c of

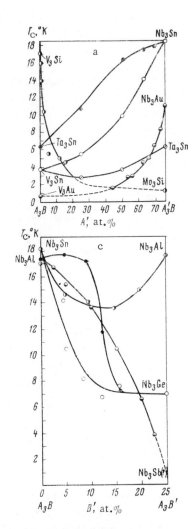
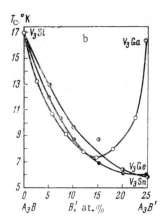

Fig. 58. Variation in the critical superconducting temperature of alloys with the Cr_3Si structure in the following systems: $A_3B-A_3'B$ (a), V_3B-V_3B' (b), and Nb_3B-Nb_3B' (c).

Nb_3Au equals 10.98°K [184] or 11.5°K [186]. An alloy of this composition with a bcc lattice passes into the superconducting state at 1.24°K [184]. The T_c of V_3Au, with a structure of the Cr_3Si type, equals 0.74°K [8]. The alloy of vanadium with 25 at.% Au, having a bcc lattice, hardly passes into the superconducting state at all. The variation in the T_c of solid solutions with a cubic lattice of the Cr_3Si type is illustrated in Fig. 58. Low T_c values occur in the Nb_3Au-V_3Au system even for a composition of $Nb_{1.5}V_{1.5}Au$.

The transformation taking place in this system confirms the effect of crystal structure on superconductivity [20]. For structures of the Cr_3Si type, narrow d bands and a high density of states on the Fermi surface are characteristic features. The density of states on the Fermi surface for Nb_3Au (with a structure of the Cr_3Si type) is twice as great as that of the bcc lattice [20]. This may be concluded from the temperature dependence of the magnetic susceptibility. In annealed samples of Nb_3Au the magnetic susceptibility is 50% greater than in case samples. Taken in conjunction with the BCS theory [130], the foregoing result explains why the T_c values of cast and annealed samples differ by an order of magnitude.

In the pseudobinary systems of V_3Sn with Nb_3Sn and Ta_3Sn, the existence of a continuous series of solid solutions and the fairly high superconducting transition temperatures of all the compounds ensure a smooth change in the critical temperature with changing alloy composition (Fig. 58) [185]. The critical temperature of V_3Sn obtained in one case [185] is much too low, apparently because the alloys were not in a state of equilibrium or were insufficiently pure.

There is also a smooth decrease in T_c in the Nb_3Al-Mo_3Al system (Table 24) [178]. In the V_3Si-Mo_3Si and Nb_3Au-V_3Au systems, however, one of the components has a very low critical temperature. In these systems there is a rather different type of T_c variation: The critical temperature of the compound with the initial high T_c decreases sharply when the atoms of one of its components are replaced by atoms of the third component for small proportions of the additive; subsequently the rate of decrease of T_c becomes less severe.

In the Nb_3Sn-Ta_3Sn system the T_c of the alloys also changes smoothly with changing composition; the critical temperatures of all the alloys are lower than that of Nb_3Sn itself.

Thus in the pseudobinary systems of two compounds with the Cr_3Si type of structure T_c changes smoothly and continuously with composition as one A component is isomorphically replaced by another (the B component remaining the same). In all the systems studied there is a deviation from the additivity law with respect to the composition dependence of T_c. The reasons for the different signs of this deviation remain undiscovered.

Among the A_3B phases with the Cr_3Si type of structure there is a group of compounds in which the B component is a transition metal. The critical temperatures of such compounds of niobium with group VIII transition metals and gold ($Nb_3Rh_{1-x}B_x$, with the Cr_3Si type of structure, in which B = Co, Ru, Pd, Os, Ir, Pt, or Au) were studied in detail [61]. All the alloying additives increased the unit-cell volume of Nb_3Rh. Depending on the composition of the alloys, their critical temperature varied smoothly (with a slight minimum for small proportions of the alloying elements). The T_c of Nb_3Rh decreased for slight additions of alloying elements (x ≤ 0.03). For larger amounts (x > 0.03) of Pt, Au, Pd, and Ru the T_c of the solid solution increased. The lowest critical temperatures (1.8-2.5°K) occurred for an electron concentration of the solid solution equal to 6 electrons/atom, the greatest (11-12°K) for 6.5 electrons/atom. A maximum appears on the composition−T_c curve of annealed alloys of the Nb_3Pt−Nb_3Au system [185a].

2. Effect of B-Subgroup Elements on the Properties of Cr_3Si-Type Compounds

In pseudobinary systems of A_3B compounds with the Cr_3Si type of structure there is an isomorphic replacement of the B components by other B components even in the presence of a favorable spatial factor [174]. The possible formation of solid solutions in pseudobinary alloys with the Cr_3Si structure with the replacement of the B components was studied in particular detail by von Philipsborn [20]. As B components, this author took a variety of elements (transition metals, metals and nonmetals of the main subgroups of the periodic table) and selected from these systems with the same or different valences of the B components (B components from the same or different groups in the periodic table). Among the B components chosen were elements with both large and small atomic radii. The effect of the type of interaction taking place between the B components themselves and between the B and A components on the properties of the alloys based on these compounds was also considered. It was found that continuous series of solid solutions $A_3(B, B')$ were only formed when the B and B' components had physicochemical properties of reasonably similar natures [20, 174, 187].

Philipsborn [20] also studied the possibility of ordering in these systems. The alloys were studied in the cast state and also after annealing at 550-1050°C, it being well known that the ordering temperature of the compounds equalled $0.55-0.63\,T_m$ (T_m = melting point) [188]. No superstructural lines appeared in these systems at any temperature. The absence of superstructure confirms that only weak bond forces act between the B atoms and that chains of A atoms are characteristic features of Cr_3Si-type structures.

In the absence of isomorphic compounds, or if there is a great difference in the atomic radii of the B components in the A_3B-A_3B' systems, limited solid solutions based on the Cr_3Si-type compounds are formed [174].

In accordance with the foregoing principles, in pseudobinary systems of V_3Si with isomorphic compounds such as V_3Ga [47], V_3Ge, or V_3Sn [182] the silicon atoms in the V_3Si lattice are isomorphically replaced by gallium, germanium, or tin over the whole range of concentrations [189-191]. The formation of continuous series of solid solutions has been established in the systems Nb_3Sn-Nb_3Al [176, 177, 192], Nb_3Sn-Nb_3Ge [176, 193], Nb_3Sn-Nb_3Sb [192], Nb_3Al-Nb_3Ge [176], and Nb_3Al-Nb_3Sb [194].

In the V_3Si-V_3Ga system a continuous series of solid solutions $V_3(Si, Ga)$ with the Cr_3Si type of structure (involving the isomorphic replacement of silicon by gallium atoms) is formed at temperatures below 1300°C [19]. The microhardness of the solid solutions reaches its maximum (1680 kg/mm^2) at approximately 7 at.% of Ga, while the lattice constant varies with a slight deviation from the Vegard rule. At high temperatures there is a vanadium-base solid solution in alloys containing more than 15 at.% gallium. The solubility of gallium in V_3Si at high temperatures (above 1300°C) is limited and never exceeds 5 at.%.

The variation in the superconducting critical temperature for annealed alloys of this system is illustrated in Fig. 58 [189-191]. The minimum critical temperature of the solid solution $V_3(Si, Ga)$ with the Cr_3Si structure in this pseudobinary system occurs for approximately equal silicon and gallium contents. The increase in the electron concentration of the alloys on replacing gallium with silicon conversely leads to a decrease in the critical temperature of the corresponding binary compounds with the Cr_3Si type of structure.

Cast alloys of the V_3Si-V_3Ga section containing more than 15 at.% Ga are not superconductors at temperatures above 4.2°K.

The fact that T_c decreases below 4.2°K is due to the formation of a vanadium-base solid solution in this part of the V_3Si-V_3Ga section at high temperatures.

In the V_3Si-V_3Ge and V_3Si-V_3Sn systems the critical superconducting temperature of the alloys decreases, on increasing the proportions of germanium or tin, from the T_c of V_3Si to those of V_3Ge (6.01°K) and V_3Sn (6.00°K) [189, 190]. All the alloys of these three systems exhibit superconductivity, in the same way as the compounds themselves. The electron concentration of the alloys V_3(Si, Ge) and V_3(Si, Sn) remains constant, since the elements in question are being replaced by others belonging to the same subgroup of the periodic table. On replacing silicon by gallium, there is a reduction in the electron concentration of the alloys, and vice versa. We see from Fig. 58 that whether the electron concentration is kept constant or whether it is changed in any way, the critical temperature of the compounds with high T_c values is always reduced in this case.

The dependence of the T_c(°K) of V_3(B, B') solid solutions on the proportion of one of the B components present may be expressed by means of an empirical formula of the following general form:

$$T_c = ae^{-bx} + Ce^{dx},$$

where x is the proportion of the B component in the alloys and a, b, c, d are constants characterizing the particular system under consideration.

For the V_3Si-V_3Ge system in particular

$$T_c = 16.9e^{-0.74x} + 0.37e^{0.088x};$$

for the V_3Si-V_3Sn system

$$T_c = 17.1e^{-0.057x} + 0.069e^{0.13x};$$

for the V_3Si-V_3Ga system

$$T_c = 17.1e^{-0.074x} + 0.059e^{0.216x}.$$

These laws governing the changes in T_c obey the same general empirical formula and are evidently characteristic for all pseudobinary systems of isomorphic compounds of vanadium with elements of the main subgroups of the periodic table (with the

Cr$_3$Si type of structure) forming continuous series of solid solutions.

On isomorphically replacing the vanadium in compounds of the Cr$_3$Si type by other transition metals, in the case of certain systems there is an analogous law governing the changes in the T$_c$ of the alloys (see Fig. 58). In certain pseudobinary systems formed by niobium compounds with a structure of the Cr$_3$Si type there is a rather different composition dependence of the critical temperature when a continuous series of solid solutions is formed [191, 191a].

Thus the Nb$_3$Sn–Nb$_3$Al system contains two compounds with high critical temperatures; when the proportion of the B component is varied, T$_c$ changes smoothly, with a minimum at a ratio of Sn:Al = 1:1 (Fig. 58) [176, 177]. In one case [192], however, it was suggested that the solubility of the components in this system was only limited, the Nb$_3$Sn-base solid solution extending only up to the composition Nb$_3$Sn$_{0.4}$Al$_{0.6}$. For larger aluminum contents a second phase appears. The critical temperature of metalloceramic Nb$_3$Sn$_{1-x}$Al$_x$ alloys prepared at 1200°C first rises slightly with increasing x, then passes through a minimum, and finally approaches the value characteristic of Nb$_3$Al (Fig. 59). However, in the case of alloys prepared at 1000°C, no initial rise in the transition temperature occurs. These results [192] may clearly be attributed to the nonequilibrium nature of the alloys prepared by the metalloceramic technique. In the Nb$_3$Sn–Nb$_3$Ge system, as in the case of

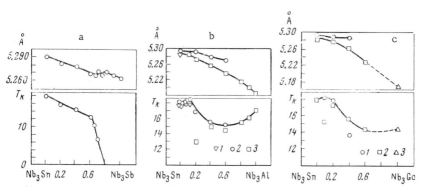

Fig. 59. Composition dependence of the lattice constant and critical superconducting temperature of niobium alloys: a) alloys Nb$_3$Sb$_x$Sn$_{1-x}$; b) alloys Nb$_3$Al$_x$Sn$_{1-x}$ obtained at 950 (1), 1200 (2), and 1500°C (3); c) alloys Nb$_3$Ga$_x$Sn$_{1-x}$ obtained at 1200 (1) and 1500°C (2), and from the results of other authors (3).

V_3Si-V_3Ge and V_3Si-V_3Sn, (these being systems comprising compounds with high and low T_c values respectively), the critical temperature of the alloys decreases along a smooth curve without any very obvious minimum (Fig. 58b) [176, 193]. It is true that in the system formed by the niobium compounds there is a sharp fall in T_c after adding even the slightest trace of germanium.

In the Nb_3Al-Nb_3Ge system the critical temperature of the alloys generally decreases on increasing the germanium content. However, in contrast to the systems of vanadium compounds, for slight additions of germanium there is an initial small rise in T_c, and a maximum appears on the curve for an Al:Ge ratio of 4:1 (Fig. 58) [176]. The subsequent sharp decrease in T_c coincides with the region in which the degree of ordering of the system diminishes rapidly. The presence of the maximum T_c is evidently associated with different degrees of ordering in the crystal structure of the solid solutions.

Recently the creation of a material with a T_c of 20.98°K on the basis of this system was described [2b, 195]. This material constitutes a combination of the compounds Nb_3Al and Nb_3Ge with a complex tubular structure in the longitudinal direction. The existence of such a directional structure in this material underlines the necessity of having a specific ordering of the crystal lattice of the alloy in order to produce high superconducting characteristics.

The addition of traces (2-3 at.%) of Ga, In, Tl, Pb, and Bi, according to Hagner [177], slightly raises the T_c of the compound Nb_3Sn (Fig. 60). Alloys of this containing up to 7.5 at.% of Ga, In, Tl, Ge, Pb, As, Sb, and Bi were also prepared by the metalloceramic method (1200°C, 6 h). Hagner's values of the critical temperature of the alloyed compound Nb_3Sn (18-18.3°K) lie within the wider limits of the fluctuations in the critical temperature of the pure compound arising from fluctuations in its chemical composition and conditions of production (17.5-18.5°K) [1]. Hence the slight rise in the critical temperature of the compound Nb_3Sn may be associated with local deviations from stoichiometric composition in the metalloceramic samples studied, with the diffusion method of producing the sample, or with the purity of the original compound Nb_3Sn [287], all the more so in view of the fact that in another investigation [196] no rise in the transition temperature

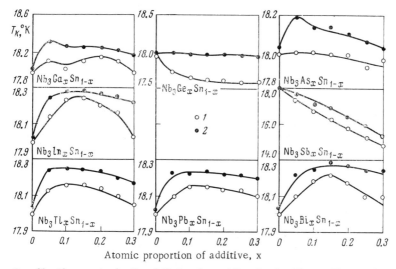

Fig. 60. Changes in the T_c of Nb_3Sn after adding Ga, In, Tl, Ge, Pb, As, Sb, or Bi: 1) T_c determined by measuring the electrical resistance of the alloys; 2) determined by the magnetic method.

of Nb_3Sn was observed on alloying this compound with exactly the same elements.

An initial slight rise in the T_c of Nb_3Sn was also observed in the Nb_3Sn-Nb_3Ga system [192]. According to these results, the Nb_3Sn-base solid solution extended up to a composition of $Nb_3Sn_{0.6}Ga_{0.4}$. For larger proportions of gallium a second phase appeared in the alloys. The critical temperature of metalloceramic $Nb_3Sn_{1-x}Ga_x$ alloys prepared at 1200°C first increases slightly with increasing x, then decreases to a minimum, and finally tends toward the values characteristic of Nb_3Ga (Fig. 59). In the case of the alloys prepared at 1000°C no initial rise in the transition temperature was observed.

According to Reed et al. [192], the initial rise in T_c is similar to the slight increase in the transition temperature observed for an excess of niobium in the compound Nb_3Sn. Since there is a relatively slight reduction in the lattice constant of these alloys for x < 0.2, it is quite possible that the aluminum does not at this stage enter into the crystal lattice of the compound with the Cr_3Si-type structure, so that there is an excess of niobium in this compound in the alloys in question. On reaching the solubility limit of niobium in Nb_3Sn, the aluminum atoms begin to occupy the positions of the B components in the lattice of the compound [192]. The niobium chains therefore become curved, and T_c accordingly decreases until the properties of the compound Nb_3Al begin to dominate the situation. The low values of T_c in certain alloys are evidently associated with the severe disordering of the crystal structure of the compound. In our own opinion, both the experimental results quoted [192] and the underlying mechanism just discussed suggest that the alloys under consideration were not in an equilibrium state.

The Nb_3Sn-Nb_3Sb system forms a continuous series of solid solutions [192]. The transition temperature of the $Nb_3(Sn, Sb)$ solid solution first decreases slowly with increasing antimony content (to $Nb_3Sn_{0.4}Sb_{0.6}$) and then more rapidly (Fig. 59). Alloys of composition $Nb_3Sn_{0-0.25}Sb_{1-0.75}$ are not superconducting at 4.2°K. A similar effect is found [197] in $Nb_3Al_{1-x}Sb_x$ alloys, in which T_c decreases sharply at x = 0.5. According to other data [194], however, the composition dependence of T_c in the Nb_3Al-Nb_3Sb system has quite a different character. The critical temperature of the ternary solid solution with the Cr_3Si structure decreases smoothly from 17.17°K for Nb_3Al to 3.92°K for $Nb_3Al_{0.1}Sb_{0.9}$ (Fig. 58) [194]. The value of T_c obeys a parabolic relationship throughout the concentration range studied:

$$T_c(x) = [T_c(Nb_3Al) - 0.55](1 - x^2) + 0.55.$$

In this formula the value of 0.55°K corresponds to the extrapolated value of T_c for Nb_3Sb. However, the question as to whether Nb_3Sb is a superconductor remains open. It was in fact noted [194] that in structure, lattice contant, and T_c the alloy $Nb_3Al_{0.5}Sb_{0.5}$ differed sharply from the compound Nb_3Sb.

Otto [197a] found a T_c maximum in the Nb_3Al-Nb_3Ga, Nb_3Al-Nb_3Ge, Nb_3Sn-Nb_3Ga, Nb_3Sn-Nb_3Ge, Nb_3Sn-Nb_3In systems. The density of the valence electrons of these alloys, except for those of the Nb_3Sn-Nb_3In system, was 0.257-0.262 electrons/$Å^3$. This value practically coincides with the electron density favoring high T_c values for niobium compounds with the Cr_3Si type of structure. A minimum T_c was observed in the systems Nb_3Al-Nb_3Sn, Nb_3Al-Nb_3In, Nb_3Ga-Nb_3Ge; in these systems the optimum electron density was never attained for any compositions.

Alloys of composition $Nb_3Ge_{0.5}B_{0.5}$, in which B = Sn, Al, or Ga were also studied in detail [192] (see Table 24).

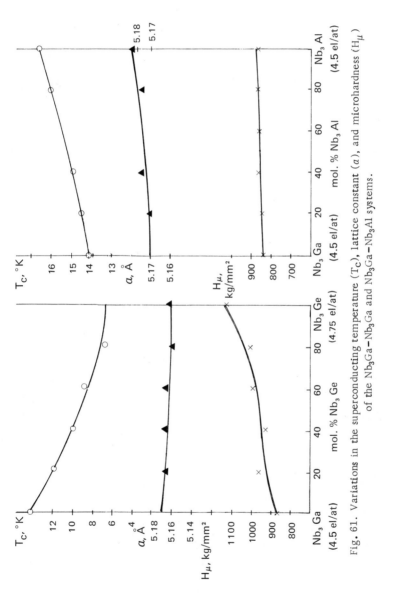

Fig. 61. Variations in the superconducting temperature (T_C), lattice constant (a), and microhardness (H_μ) of the Nb_3Ga–Nb_3Ga and Nb_3Ga–Nb_3Al systems.

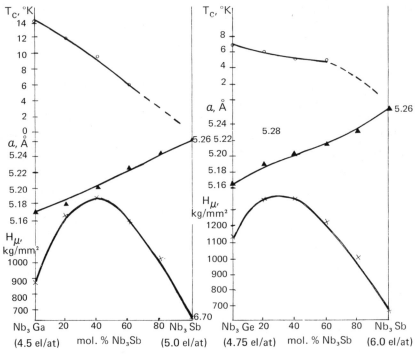

Fig. 62. Variations in the superconducting temperature (T_c), lattice constant (a), and microhardness (H_μ) of the Nb_3Ga-Nb_3Sb and Nb_3Ge-Nb_3Sb systems.

We studied alloys of a further series of pseudobinary systems formed by various compounds of niobium and vanadium with a structure of the Cr_3Si type [197b]. All the alloys of the systems Nb_3Ga-Nb_3Ge, Nb_3Ga-Nb_3Al (Fig. 61), Nb_3Ga-Nb_3Sb, Nb_3Ge-Nb_3Sb (Fig. 62) constitute continuous series of solid solutions. We see from Figs. 61 and 62 that after homogenization at 1400°C for 13 h these pseudobinary systems reveal no alloys with a transition temperature exceeding the T_c of the original compounds. The lattice constant and T_c vary almost additively with the composition of the alloys. It is interesting to note that, in the systems Nb_3Ga-Nb_3Ge and Nb_3Ga-Nb_3Al (Fig. 61), in which all the alloys exhibit superconductivity, the microhardness also varies additively with composition, while in the systems Nb_3Ge-Nb_3Sb and Nb_3Ga-Nb_3Sb (Fig. 62) the microhardness-composition curve passes through a maximum. The same systems are distinguished by the absence of superconductivity in some Nb_3Sb-base alloys. Prolonged annealing at lower temperatures (800°C, 170 h) raises the T_c of Nb_3Al and Nb_3Ga by 0.5°K, and in the alloy of composition $Nb_3Al_{0.8}Ga_{0.2}$ raises T_c above that of pure Nb_3Al (Fig. 63). Analogous results were obtained for the Nb_3Al-Nb_3Ga system by Blaugher et al. [192a], who also observed a rise in the T_c of Nb_3Sn-Nb_3Al alloys after annealing. The rise in T_c was attributed to ordering.

Limited solid solutions based on V_3Si are formed on alloying this compound with aluminum and beryllium [174]. In contrast to germanium, tin, and gallium, aluminum and beryllium have a less favorable difference in atomic radii and electronegativity with respect to silicon ($\Delta r = 0.113$ and $\Delta E_n = 0.34$ for aluminum, $\Delta r = 0.191$ and $\Delta E_n = 0.4$ for beryllium). Furthermore no A_3B compound is formed in the V-Be system.

The solubility of aluminum in V_3Si at 800°C is about 8 at.% [174]. At saturation the lattice constant of the $V_3(Si, Al)$ solid solution is $a = 4.745$ Å and the microhardness $H_\mu = 1500$ kg/mm². The phase $V_3(Si, Al)$ with a cubic structure of the Cr_3Si type was found in alloys containing up to 20 at.% Al. For 20 at.% Al, in addition to the $V_3(Si, Al)$ and a vanadium-base solid solution, the alloy also contained a third phase X, which also existed in small

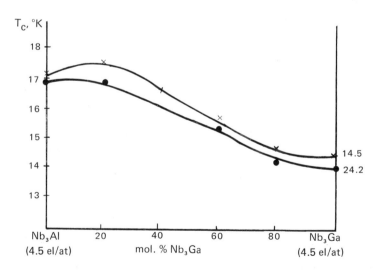

Fig. 63. Variation in the superconducting temperature of Nb_3Al-Nb_3Ga alloys (● cast, × annealed 800°C, 170 h).

quantities in the binary alloy of vanadium with 25 at.% Al. The phase X remained unidentified.

The lattice constant and microhardness of V_3Si in alloys with beryllium practically coincide with the values characterizing the pure compound [174]. All the alloys with beryllium were heterogeneous. This indicates that beryllium is either insoluble in V_3Si or has a very low solubility.

The introduction of aluminum and beryllium into the alloys also reduces the T_c of V_3Si (Table 24) [50e, 174]. For the limited solubility the decrease in T_c is particularly sharp within the range of homogeneity of the solid solution of $V_3(Si, Al)$, over which there is a sharp rise in the lattice constant of the compound. In the two-phase regions of the alloys of V_3Si with aluminum and beryllium, in which the lattice constant of the solid solution $V_3(Al, element)$ remains unaltered, the change in the critical temperature is far less marked. The decrease in the critical temperature of V_3Si for equal atomic proportions of Sb, Be, Ge, Al, Ga, and Sn respectively equals 19; 15; 1.02; 0.95; 0.9, and 0.82°K/at.% [174, 189-191].

On replacing the gallium atoms in the crystal lattice of V_3Ga by elements from the main subgroups of the periodic table there is also a decrease in the critical superconducting temperature (Table 24) [175]. The sharpest decrease in the T_c of V_3Ga is produced by elements of group II. The introduction of 0.027 at.% of Zn reduces the T_c by 6.6°K (244°K/at.%) and that of 0.016 at.% of Cd by 5.8°K (300°K/at.%). The remaining alloying elements from the main subgroups of the periodic table may be placed in the following order as regards the effect of equal atomic proportions (small total concentrations) of these on the decrease in critical temperature: tin (1.8°K/at.%), antimony (1.9), silicon (2.4), bismuth (3.3), germanium (3.4), indium (3.4), and lead (5.6°K/at.%). Aluminum also raises the T_c of V_3Ga [50e].

A comparison of the changes taking place in the critical temperatures of the compounds V_3Ga and V_3Si on alloying these with elements belonging to the main subgroups of the periodic table shows that the same alloying elements have different effects on different bases. This is evidently due to the great difference in the electron structure of the alloying elements belonging to the B subgroups and the elements replacing them (gallium and silicon) as compared with the difference in the electron structures of the transition metals. Alloying the compounds V_3Ge, V_3Sn, and V_3Sb with aluminum leads to a rise in their T_c [50d, 50e].

According to Roberts [62] the optimum conditions for high critical temperatures are realized in binary compounds with 4.7 electrons/atom and a mean atomic volume of over 21 Å^3, in which the mean electron density is over 0.25 electron/Å^3. Furthermore, all the sites for component A in the lattice of the binary compound should be completely occupied by the atoms of this component.

The extensive experimental material discussed in this chapter shows that, strictly speaking, the main causes of the high superconducting properties of compounds with the Cr_3Si type of structure are still not completely understood, and that in particular we cannot yet specify the optimum relationships between the physicochemical properties of the components required in order to raise the superconducting characteristics of such compounds. There is still a wide scope for experimental work and creative discoveries.

3. Influence of the Simultaneous Replacement of the A and B Components on the Properties of Compounds of the Cr_3Si Type

In the V_3Ge-Nb_3Al and V_3Ge-Nb_3Sn systems [197b] we have the simultaneous replacement of the vanadium by niobium and the germanium by aluminum or tin atoms. In both systems continuous series of solid solutions are formed. The T_c of alloys of the V_3Ge-Nb_3Al system decreases smoothly from Nb_3Al to V_3Ge (Fig. 64). The T_c curve of the V_3Ge-Nb_3Sn system (Fig. 65) passes through a minimum. The T_c of Nb_3Sn decreases very sharply even for the slightest traces of vanadium and germanium.

Low-temperature annealing of the alloys of these systems does not lead to any rise in their T_c.

4. Effect of Interstitial Impurities on the Properties of Cr_3Si-Type Compounds

We have shown earlier that even slight traces of various metals and nonmetals exert an extremely strong influence on the superconducting properties of compounds. Compounds with the

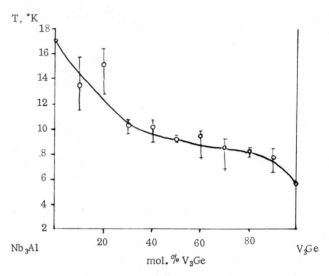

Fig. 64. Variation in the superconducting temperature of Nb_3Al-V_3Ge alloys.

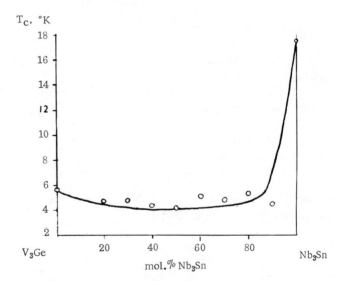

Fig. 65. Variation in the superconducting temperature of V_3Ge-Nb_3Sn alloys.

Cr_3Si-type structure often incorporate chemically-active, rare, and refractory metals. The most harmful impurities for these elements are interstitial impurities (hydrogen, nitrogen, oxygen, and carbon) [198]. Even minute traces of these (hundredth and thousandth parts of a percent) radically alter the properties of the rare metals and their alloys. This makes the problem of interactions taking place between the compounds under consideration and atmospheric gases and carbon one of the utmost importance.

Data relating to the phase diagrams formed by these elements with niobium are presented in Chap. IV. The interaction of these elements with other metals of group VA such as vanadium are approximately the same [198a]. As yet no data have been published relating to the solubility of interstitial impurities in many superconducting compounds.

The substantial effect of interstitial impurities on superconducting properties explains, for example, the fact that different opinions exist as to the T_c of V_3Ga; some consider this to be 16.5°K [57, 77] and others between 14.4 and 14.45°K [199, 200]. This difference is evidently associated with the purity of the original materials and the methods employed for producing the alloys. According to our own data, increasing the amount of interstitial impurity from 0.1 to 0.45 wt.% (total) reduces the T_c of V_3Ga from 15.8 to 14.2°K [54].

The crystal structure of the Cr_3Si type is characterized by the close packing of atoms with different sizes. The small size of the vacancies in this packing (r = 0.33 Å) limits the formation of interstitial solid solutions.

The binary single-phase alloys V_3Si obtained by some authors [54, 174] contained 0.2 wt.% C and 0.1% O in solid solution. Increasing the carbon content to 0.4 wt.% (1.4 at.%) or oxygen to 0.56 wt.% (1.6 at.%) led to the formation of a second phase. The lattice constant of V_3Si changes very little on dissolving carbon, oxygen, and boron (Table 26). However, the sharp increase in the microhardness of V_3Si in alloys containing carbon and oxygen indicates the formation of interstitial solid solutions. The limiting solubility of carbon in V_3Si is about 0.3 wt.% (< 1 at.%) and oxygen 0.5 wt.% (< 1.5 at.%); the solubility of boron is under 0.07 wt.% (< 0.3 at.%).

TABLE 26. Alloys of V_3Si with Carbon, Oxygen, and Boron [54, 174]

Composition of the alloys, at.%			Ratio V : Si	Properties of the phase of the Cr_3Si structural type		
V	Si	third component		lattice constant a, Å	microhardness H_μ, kg/mm^2	T_c, °K
75.57	23.04	1.39 C	3.29	4.725	1780	—
74.65	23.80	1.55 C	3.14	4.724	1780	16.5
75.00	22.70	2.30 C	3.30	4.725	1780	—
75.00	22.50	2.25 C	3.34	—	—	16.40
75.15	23.27	1.58 O	3.22	4.726	1690	15.94
75.96	23.75	0.29 B	3.20	4.725	1500	—
69.43	21.75	8.82 B	3.19	4.726	1500	16.20
75.00	22.50	2.25 B	3.34	—	—	15.80

The introduction of carbon, oxygen, boron, and nitrogen atoms into the crystal lattice of the compound leads to a decrease in its superconducting transition temperature (Table 26). A decrease in the value of T_c was also noted on alloying V_3Si with boron [58], the latter being introduced into the alloy in place of silicon.

We noticed a similar effect on alloying V_3Ga with carbon or boron. Over the range of the solid solution of carbon or boron in V_3Ga there was a sharp decrease in the critical superconducting temperature of the compound. In two-phase alloys the T_c of the compound fell much less.

The dissolution of interstitial impurities thus leads to a decrease in the T_c of a compound with the Cr_3Si type of structure. The critical temperature decreases particularly sharply over the range of homogeneity of the interstitial solid solution. In the two-phase regions the T_c of the compounds also diminishes, but to a far lesser extent.

The results obtained in the foregoing experiments agree with theoretical conclusions relating to dilute superconducting solid solutions containing nonmagnetic impurities [201]. According to the Anderson model of contaminated superconductors, in the presence of impurities the anisotropy of the energy gap becomes blurred. This should primarily lead to a reduction in T_c. We have seen that this does in fact also happen in the case of compounds with the Cr_3Si structure.

Traces of oxygen, nitrogen, and carbon may completely change the crystal structure of compounds of the Cr_3Si type. On adding 15-20 at.% (3-4 wt.%) of oxygen, the compounds Ti_3Au, V_3Au, and V_3Pt assume an ordered structure of the Cu_3Au type [20]. For smaller traces of interstitial impurities a second phase with the Cu_3Au type of lattice appears in addition to the Cr_3Si lattice [202]. It is interesting that the atoms of these impurities are very strongly connected to the atoms of the Cu_3Au-type lattice; electron-beam remelting of the Ti_3AuO phase hardly removes any of its oxygen.

Other compounds of the Cr_3Si type (for example, V_3Rh, V_3Ir, and others) do not form any phases of the Cu_3Au type on alloying with oxygen [20]. On introducing oxygen or nitrogen into the Cr_3Si-type structure of Ti_3Sb the lattice is converted into body-centered tetragonal form [203].

All these structures are interstitial phases. The metal atoms form an ordered structure of the Cu_3Au type, the oxygen, nitrogen, or carbon atoms lie in the octahedral vacancies in the position $(\frac{1}{2}, \frac{1}{2}, \frac{1}{2})$. The structure with completely-occupied vacancies is of the perovskite type (Fig. 45) [204]. A complete conversion of a Cr_3Si-type phase into one of the Cu_3Au type occurs, for example, at the composition $V_3AuO_{0.7}$, where 70% of the vacancies are occupied by oxygen atoms.

The introduction of oxygen impurities leads to the vanishing of superconductivity. The phases Ti_3AuO, $V_3AuO_{0.7}$, and $V_3PtO_{0.7}$ do not pass into the superconducting state at temperatures above 1.2°K [20].

The influence of hydrogen on the structure and superconducting properties of compounds of the Cr_3Si type has hardly been considered. It is known that the compound Nb_3Ge containing 22.1 at.% Ge and 16 at.% H has no superconductivity above 4.2°K [192]. After removing the hydrogen (by vacuum-annealing at 1200°C) the T_c of this compound equals 5.3°K. On introducing hydrogen the T_c of Nb_3Sn after first increasing by ~ 0.2°K decreases sharply [204a]. The hydrides Ti_3AuH and Ti_3AuH_2 (Cr_3Si type) show no superconductivity right down to 1.6°K [204b].

An analogous sharp decrease in the T_c of compounds of the Cr_3Si type occurs on introducing nitrogen into the crystal lattice. According to our own data, for 0.154 wt.% N (practically the solubility limit) the T_c of V_3Ga equals 15.1°K. The simultaneous introduction of several elements leads to a sharper decrease in the T_c of this kind of compound than the introduction of one element at a time. According to our results, the T_c of V_3Si falls to 14.9°K for 0.49 wt.% of oxygen and 0.009 wt.% of hydrogen.

Nothing has been said as to the effect of interstitial impurities on the critical magnetic field and critical current.

5. Effect of Alloying on the Properties of Compounds with Other Types of Crystal Structure

Alloys of carbides and nitrides with the NaCl-type structure form an interesting group of superconductors with high critical parameters. These alloys with a common crystal structure enable us to follow the changes taking place in the superconducting characteristics with changing composition and physical properties. In certain carbide and carbonitride systems with continuous series of solid solutions, the T_c of the alloys is higher than that of the original binary carbides and nitrides (Table 27). The maximum on the T_c curve of the NbC–NbN, HfC–MoC, and TaC–MoC systems occurs at an electron concentration equal to 4.85 electrons/atom (Fig. 66a,b) [102]. A particularly sharp rise in T_c appears on forming niobium carbonitride (17.9°K) [102, 113, 205, 207]. The

TABLE 27. Critical Temperature Solid Solutions with the NaCl Structure [1, 102, 105, 113, 205, 206]

Material	T_c, °K	Material	T_c, °K
$NbC_{0.28}N_{0.78}$	17.9	$V_xMo_yC_{0.50}$	9.3
$NbC_{0.3}N_{0.7}$	17.8—17.38	$Ta_{0.95-0.99}Co_{0.05-0.01}C$	8.5
$NbC_{0.9}N_{0.1}$	16.97	(Hf, Ta) C	7.9—7.3
$Mo_{0.95}Hf_{0.05}C_{0.75}$	14.2	(Hf, Nb) C	7.5—6.0
$Nb_{0.8}W_{0.2}C$	12.7	$Ti_{0.5}W_{0.5}C$	6.7
$Nb_xMo_yC_{0.50}$	12.5 (max)	(Hf, Zr) C	1.68—1.38
Hf_xZr_yN	10.7—6.2	(Hf, Ti) C	1.38
$Ta_xW_yC_{0.50}$	10.5 (max)	$Nb_{0.99}Cr_{0.01}N$	<1.2
$Ti_xMo_yC_{0.50}$	10.2	$Nb_{0.97}Cr_{0.03}N$	<1.2
$Zr_xMo_yC_{0.50}$	9.5		

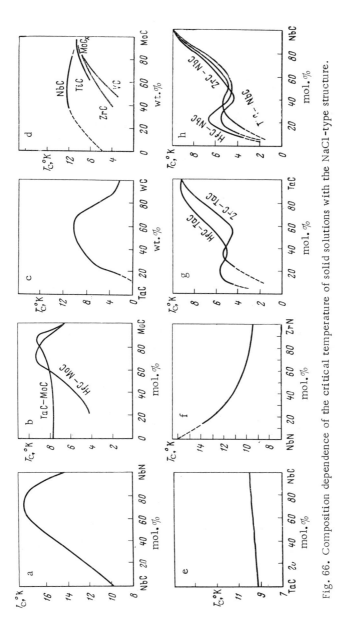

Fig. 66. Composition dependence of the critical temperature of solid solutions with the NaCl-type structure.

maximum critical field H_c of these alloys at 4.2°K equals 110 kOe [208]. Darnell et al. [208] noted that the critical current of NbC, NbN, and Nb(C, N) whiskers, being superconductors of the second kind, was proportional to $d^{0.6}$, where d is the diameter of the whiskers (between 10 and 100 μ). The magnetic field created by this current on the surface of the whiskers was 10-100 times smaller than H_{c1}.

A rise in the critical temperature of NbC and TiC occurs on alloying these with tungsten (the T_c of $Ti_{0.5}W_{0.5}$ equals 6.7°K, that of $Nb_{0.8}W_{0.2}C$ — 12.7°K [102]), although tungsten monocarbide and titanium monocarbide are not superconducting above 1.28°K. Chromium nitride, which is not a superconductor, sharply reduces the T_c of niobium nitride. The alloys $Nb_{0.99}Cr_{0.01}N$ and $Nb_{0.97}Cr_{0.03}N$ exhibit no superconductivity at temperatures above 1.2°K [113].

In approximately the same range of electron concentrations, at 50-70% WC, a T_c maximum appears in the TaC–WC system (Fig. 66c) [209]. It would appear that the smooth change with composition in the critical temperature in a system formed by components of different crystal structure (TaC is a cubic crystal of the NaCl type, while WC is hexagonal, similar to MoC) is associated with the suppression of the hexagonal form of WC in these alloys. The smooth change in the systems of MoC with NbC, TiC, ZrC, and VC (Fig. 66d) [209], which is a characteristic of continuous series of solid solutions, was also explained as being due to the suppression of the stable γ form (hexagonal structure) and to the proposed existence of a hypothetical α form of MoC (cubic lattice) in the alloys. In the MoC–NbC system there is a flat maximum (12.5°K) at about 50% NbC. The isomorphic replacement of molybdenum with titanium, zirconium, and vanadium leads to a smooth fall in the critical superconducting temperature of the alloys. The replacement of molybdenum by hafnium, however, with a deficiency of carbon in the compound, leads to a substantial increase in the critical temperature of MoC (T_c of the alloy $Mo_{0.95}Hf_{0.05}C_{0.75}$ = 14.2°K).

A smooth, almost additive variation in T_c was observed in the TaC–NbC (Fig. 66e) and HfN–ZrN systems [102, 104]. In these systems all the alloys have a constant electron concentration. A monotonic decrease in T_c from 15.7 to 10°K is also observed on isomorphic substitution for niobium in the NbN–ZrN system (Fig.

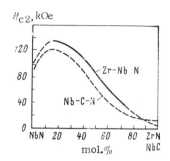

Fig. 67. Changes in the H_c of (Nb, Zr)N and Nb(C, N) solid solutions with the NaCl type of structure in relation to the composition of the alloys at 4.2°K.

66f) [209]. The critical magnetic field of this ternary solid solution has a maximum (135 kOe) in the region of 20 mol.% ZrN (Fig. 67). These alloys possess high critical current densities (up to $6 \cdot 10^3$ A/cm² in a field of 120 kOe for the composition $Nb_{0.85}Zr_{0.15}N$, with a 3-7% deficiency of nitrogen in the alloys). The variation in the critical current of several alloys at 4.2°K is illustrated in Fig. 68.

In solid solutions of the TiC–NbC, ZrC–NbC, HfC–NbC, ZrC–TaC, and HfC–TaC systems (Fig. 66g,h) [102, 210] there is a minimum of T_c in the region of equimolar compositions (~ 4.25 electrons/atom). However, the value of this minimum in each system is greater than the corresponding T_c of one of the components. For lower electron concentrations (~ 4.1 electrons/atom), slight maxima appear in these systems (the maxima lie below the T_c of one of the components of the system). An increase in T_c also occurs on isomorphically substituting titanium or zirconium for hafnium in the monocarbide (Table 27) [102]. This is particularly interesting in view of the fact that hafnium carbide is not a superconductor at temperatures above 1.23°K (Table 13). The presence

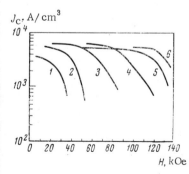

Fig. 68. Critical current density of several alloys of the NbN–ZrN system at 4.2°K in relation to the the magnetic field. 1) $Nb_{0.25}Zr_{0.75}N$; 2) $Nb_{0.35}Zr_{0.65}N$; 3) $Nb_{0.5}Zr_{0.5}N$; 4) $Nb_{0.67}Zr_{0.33}N$; 5) $Nb_{0.75}Zr_{0.25}N$; 6) $Nb_{0.85}Zr_{0.15}N$.

of several extrema in the latter systems (Fig. 66) still lacks an explanation. These results are not in accordance with the general laws governing changes in the physical properties of systems with continuous series of solid solutions.

Bilz [211] calculated the change in the density of states for carbides and nitrides with the NaCl type of structure. The value of T_c was related to the density of states on the Fermi surface in the following manner:

$$T_c = c \exp(-1/N(0)v),$$

where the coefficient c is proportional to the Debye temperature, while the exponent contains the product of the density of states on the Fermi surface N(0) times the interaction constant v [130]. Comparison of these results [211] with the composition dependence of T_c, or more precisely the dependence of T_c on the mean electron concentration of the alloys, demonstrates the similar character of the T_c curves in systems with continuous series of solid solutions (the position of the Fermi level on the calculated curves [211] is slightly displaced in the high-energy direction for compounds with valence electrons and in the low-energy direction for those with ten electrons). There is no doubt that the change in the critical temperature of molecular solid solutions based on compounds with the cubic NaCl structure is mainly a consequence of the change in the density of states following a change in composition.

In systems with limited solid solutions, NbC–WC and TiC–WC, the T_c of the solid solution with the NaCl structure increases on increasing the proportion of WC, reaching 13 and 7°K, respectively, for the saturated solutions. The carbides WC and TiC themselves are superconductors with critical temperatures below 1.20°K. It has been shown [212] that the addition of 1% Co raises the T_c of TaC from 2.4-3.6 to 4.2-5.3°K.

It is interesting to note that, in contrast to compounds with the Cr_3Si structure, the alloying of carbides and nitrides often leads to an increase in their superconducting characteristics. Even replacing the nitrogen atoms in the crystal lattice of niobium nitride by oxygen atoms raises the critical superconducting temperature (up to 16.97°K for 5 at.% of oxygen) [206].

A considerable amount of information has been published regarding the effect of alloying on the critical temperature of the silicides and germanides of transition metals containing more than 25 at.% of the nonmetal. The microalloying of A_5B_3 silicides (Mn_5Si_3 structure) with carbon fails to produce superconductivity. Like the corresponding binary compounds, the alloys $V_5Si_3C_{0.05}$, $Nb_5Si_3C_{0.05}$, and $V_5Ga_3C_{0.05}$ are not superconducting at temperatures down to 0.3-1.02°K [1]. The alloying of the superconducting cobalt disilicide (1.22°K), however, leads to a rise in its critical temperature.

The critical temperature of solid solutions with the $MgCu_2$ type of structure was studied by Bozorth et al. [213]. The $GdRu_2$–$ThRu_2$ system forms a continuous series of solid solutions with the $MgCu_2$ cubic type of structure. The compound $ThRu_2$ is a superconductor with T_c = 3.56°K [214]. The compound $GdRu_2$ shows no superconductivity, at any rate above 1.2°K [213]. On isomorphically replacing thorium by gadolinium the T_c of the alloys first decreases smoothly (Fig. 69); at 9-10 mol.% of $GdRu_2$ there is a sharp decrease in T_c to temperatures below 1.2°K; at the same concentrations the Curie point of $GdRu_2$ falls to zero. Analogous results were obtained for the $CeRu_2$–$GdRu_2$ and $CeRu_2$–$PrRu_2$ systems, which also form continuous series of solid solutions with the $MgCu_2$ structure [62]. The replacement of vanadium by niobium in the compound ZrV_2 ($MgCu_2$ type) leads to a decrease in T_c [214a]. The superconductivity of the phase $LaAl_2$ (the same type) vanishes on adding 0.59 at.% Gd [214b].

Zhuravlev et al. [215] studied the formation of solid solutions in systems of compounds of the nickel–arsenide type (PtBi–

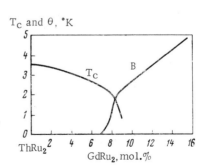

Fig. 69. Composition dependence of the critical superconducting temperature and the Curie point in the $ThRu_2$–$GdRu_2$ system.

TABLE 28. Mutual Solubility of Superconducting
Compounds of the Nickel Arsenide Type [215]

System	Solubility, mol.%		T_c, °K
	in the first compound	in the second compound	
PtBi — PtSb	CSSS[1*]		1.21—2.1 [2*]
PtBi — PtPb	CSSS		1.29—? [2*]
NiBi — NiSb	CSSS		4.26—1.0 [2*]
NiBi — RhBi	CSSS		4.2—2.06 [2*]
PtBi — PdBi	55	<10	1.21—3.7 [3*]
PdSb — PdBi	45	~8	1.5—3.7 [3*]
NiBi — PtBi	~5 [4*]	~5 [4*]	4.26—3.7 [3*]
NiBi — MnBi	~5 [4*]	~5 [4*]	<4.26 [3*]
PtBi — MnBi	~5 [4*]	~5 [4*]	<3.7 [3*]
PtBi — PtSn	<10	~20	<3.7 [3*]
PtSb — PdSb	0	0	2.1—1.5 [3*]
PtSb — CoSb	0	<20	<2.1 [3*]

[1*] CSSS = continuous series of solid solutions.
[2*] Limits of variation of T_c on increasing the content of the second component.
[3*] Proposed limits of T_c for alloys of the eutectic systems in question.
[4*] After homogenization at a temperature close to that of the formation of NiBi and MnBi.

PtSb, NiBi–RhBi, PdSb–PdBi, PtBi–PdBi, NiBi–NiSb, and PtBi–PtPb). A number of such systems form continuous series of solid solutions based on superconducting compounds (Table 27). In the formation of the continuous series of solid solutions no linear change in critical temperature appears. On varying the composition the T_c of the alloys never exceeds the T_c of the superconducting components of these systems.

EFFECT OF HEAT TREATMENT AND OTHER FACTORS ON THE SUPERCONDUCTING CHARACTERISTICS OF COMPOUNDS

Thermal effects influence the superconducting properties of compounds in so far as they modify the composition of these (and in changing the phase composition of an alloy the crystal structure of the compounds as well). The critical temperature of compounds with the Cr_3Si type of structure is directly related to the degree of ordering in these [62, 192, 216, 217]. A phase in the superconducting state, having a higher entropy than the phase in the normal state for the same temperature, is more ordered [218]. Experimental data relating to the composition dependence of T_c show that the majority of superconducting compounds belong to the daltonide class, and a single maximum appears on the T_c-composition curves (see Chap. IV). A high degree of order ensures the wholeness of the chains of A atoms and high critical temperatures [192]. In the course of disordering, the chains of A atoms are disrupted (B atoms lie in the A positions) and the critical temperature of the compounds diminishes. Thus the lower T_c values of Nb-Sn alloys (Cr_3Si structure) obtained by diffusion from the gas phase [129] as compared with the T_c of sintered samples [220] are associated with the high degree of disorder due to the method of producing the first samples [62]. Figure 70 [62] shows the degree of occupation of the niobium atom positions in the crystal lattice of Nb_3Sn in relation to niobium content (from 75 to 100 at.%). In the case of an

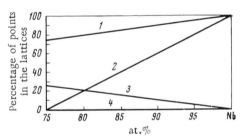

Fig. 70. Diagram to illustrate the degree of occupation of the crystal-lattice points in relation to the chemical composition of the ordered and disordered Nb_3Sn phase: 1) percentage of tin points in the disordered lattice of the compound occupied by niobium; 2) percentage of tin points in the ordered lattice of the compound occupied by niobium; 3) percentage of niobium points in the disordered lattice of the compound occupied by tin; 4) percentage of niobium points in the ordered lattice of the compound occupied by tin.

ordered lattice [220] in which some of the tin atom positions (from 0 to 100%) are occupied by niobium atoms, the stoichiometric composition of Nb$_3$Sn is disrupted, but the positions of the niobium atoms are not occupied by tin. The second case assumes the occupation of all points in the lattice by niobium or tin atoms; the percentage of the tin positions occupied by niobium atoms varies from 75 to 100%, while the percentage of the niobium positions occupied by tin atoms varies from 25 to 0% on approaching pure niobium. The latter scheme assumes that a structure of the Cr$_3$Si type exists continuously even for a niobium content of over 75%. However, it is found experimentally that alloys with A concentrations of over 90% do not have a structure of the Cr$_3$Si type. Figure 71 [62] shows the dependence of the critical temperature of the percentage of tin positions occupied by niobium. This percentage was determined from the known composition on the assumption that over 80% of the occupied positions in the ordered lattice could be described in accordance with Fig. 70, and that the material was single-phased. In this case we find an almost linear dependence of T_c on the percentage of tin positions occupied by niobium atoms in sintered or diffusion samples of the compound. The increase in the T_c of diffusion samples on subsequent annealing agrees with the relationship so established.

On increasing the degree of ordering in the crystal lattice of V$_3$Au from 0.92 to 0.99 the T_c increases from 0.86 to 1.78°K [64b].

The transition temperature of arc-melted Nb$_3$Sn samples varies with annealing temperature, from 18.1°K for annealing at 900°C to 15.6°K for annealing at 1500°C (Fig. 72) [221]. According to subsequent data [222], the T_c of cast niobium alloys containing 23-40 at.% Sn equals 17.5-17.7°K ($\Delta T = 0.1$-0.7°K); after heat treat-

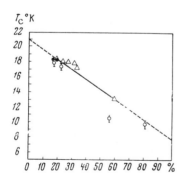

Fig. 71. Dependence of the transition temperature of Nb$_3$Sn on the number of tin positions occupied by the excess niobium atoms.

Fig. 72. Superconducting transition temperature of Nb_3Sn as a function of annealing temperature.

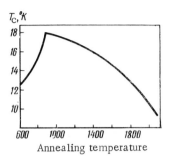

ment at 900-1000°C the critical temperatures of the alloys increase (to 17.9-18.0°K for an alloy containing 20 at.% Sn). Reed and colleagues [192, 223] showed that heat treatment at temperatures above optimum disrupted the ordered arrangement of the atoms in the crystal structure of Cr_3Si-type compounds and reduced their superconducting transition temperature. Thus the T_c of an alloy of niobium with 20 at.% Sn is 5.6°K after annealing at 1800°C for 3 h, and 18°K after annealing at 1200°C for the same period; holding for 16 h at 1200°C raises T_c to 18.5°K. Disorder may be introduced into the lattice structure by sintering above the optimum temperature, by preparing the samples in such a way as to produce a nonequilibrium state (e.g., diffusion coatings), or by having an excess of atoms of one particular type over and above that required for the stoichiometric composition of the sample [62]. It is true that a slight excess of niobium at the expense of tin in the compound Nb_3Sn sometimes produces a slight increase in the critical temperature (with a subsequent reduction on increasing the excess of niobium) [62]. The reasons for the increase in T_c in the presence of an excess of niobium are none too clear. It is possible that the excess niobium atoms, occupying the A positions, lead to the development of tin vacancies. At the same time it is quite possible that they will occupy the B positions; this will give rise to a network of niobium chains which may very well be more favorable toward superconductivity than the normal arrangement of the chains in a compound of stoichiometric composition [223].

Analogous results were obtained elsewhere [63] when studying the effect of sintering time and temperature on T_c, H_c, and the degree of long-range order of the compound Ta_3Sn. The maximum degree of order in Ta_3Sn equals 0.68; this gives $T_c = 8.35°K$. As a result of prolonged vacuum treatment at 1425-1600°C the tin is partly removed from the Ta_3Sn. The positions of the tin atoms at first

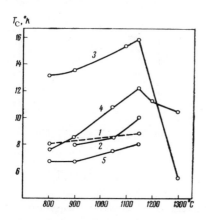

Fig. 73. Critical temperature of vanadium–gallium alloys containing the β phase in relation to their homogenization temperature. Gallium content, at. %: 1) 17.5; 2) 21.3; 3) 25.9; 4) 31.8; 5) 33.9.

remain vacant. On reducing the amount of tin by 15% (with respect to the stoichiometric composition) the tantalum atoms occupy the tin positions. The degree of order of the crystal structure falls sharply, and the superconducting properties thereupon become worse.

The effect of heat treatment on the T_c of V_3Ga was studied in another investigation [54, 168]. In the cast state the compound V_3Ga (Cr_3Si structure) has a fairly low transition temperature (13.1°K). On increasing the homogenization temperature T_c increases (Fig. 73). The superconducting transition of the alloys also becomes sharper (Fig. 74). The increase in the T_c of the alloys containing the compound V_3Ga and homogenized at the optimum temperature is clearly associated with the closer approximation of the equilibrium state and the occurrence of ordering processes [54, 200]. On increasing the homogenization temperature, the mobility of the atoms increases, diffusion processes are accelerated, and so is the rearrangement of the crystal lattice.

The substantial influence of annealing on the width of the superconducting transition of Nb_3Sn diffusion layers (Fig. 75) was mentioned elsewhere [224].

The reasons for the instability of the superconducting properties of Nb_3Sn have been repeatedly discussed [225-229], various hypotheses having been set forward.

In one case [229] an attempt was made at explaining the decrease in the critical current density of Nb_3Sn on increasing the

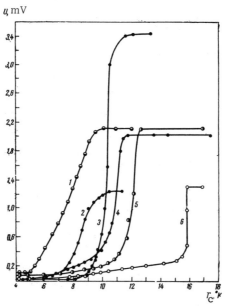

Fig. 74. Transition curves of vanadium–gallium alloys containing the β phase of the compound V_3Ga. Gallium content, at.%: 1) 31.8 (cast); 2) 21.5 (ditto); 3) 26.3 (homogenized at 800°); 4) 31.8 (at 1000°); 5) ditto (at 1150°); 6) 25.9 (at 1150°).

Fig. 75. Transition curves (in terms of resistance) for the superconducting state of diffusion layers of Nb_3Sn produced on a niobium surface at 700 (1), 800 (2), 830 (3), 850 (4), 990 (5), and 1250°C (6).

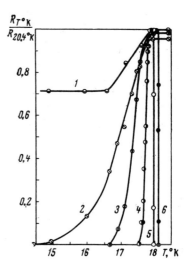

temperature of heat treatment by way of a dislocation mechanism; the critical current density depends on the dislocation density, and on raising the annealing temperature above 900-1000°C the dislocation density in the Nb_3Sn grains decreases sharply. It was found elsewhere [227] that Nb_3Sn might occur in a disordered state when a certain number of the niobium atoms lay in the "tin" points and tin atoms in some of the "niobium" points. An attempt was made at associating the transition into a disordered state with the instability of the superconducting properties; however, it was shown in [225] that no order–disorder transformation occurred at temperatures up to 1400°C.

Enstrom and others [226, 228] came to the conclusion that the high superconducting parameters of Nb_3Sn (and probably other compounds with the Cr_3Si structure) were associated with the presence of close-packed rows (chains) of niobium atoms. The transition temperature and the values of the other superconducting parameters decrease when these chains are broken. It was found that the evaporation of tin during high-temperature vacuum annealing and also a deviation from stoichiometric composition in the niobium-rich direction led to a reduction in lattice constant and a corresponding reduction in T_c and H_c. This effect was explained as being due to the creation of vacancies in the tin positions. After reaching a certain critical concentration the vacancies pass over into the niobium positions and break the chains of niobium atoms. The disordering of the crystal lattice of A_3B compounds with the Cr_3Si structure not only leads to a severe decrease in T_c but also reduces the critical current of the samples [192].

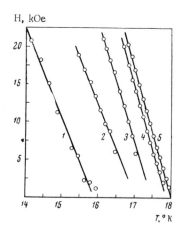

Fig. 76. Temperature dependence of the critical magnetic field of diffusion layers of Nb_3Sn produced on a niobium surface at 800 (1), 830 (2), 850 (3), 990 (4), and 1250°C (5).

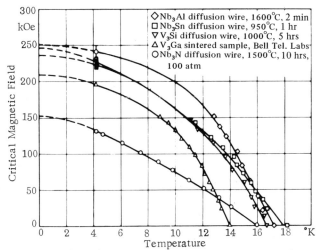

Fig. 77. Effect of temperature on the critical magnetic fields of the compounds Nb_3Al, Nb_3Sn, V_3Si, V_3Ga, and Nb_3N.

The effect of annealing temperature on the T_c of the Laves phase ZrV_2 was studied in another paper [230]. The critical temperature of cast samples varies from 7.35 to 7.05°K over the narrow range of homogeneity of the compound. Annealing at high temperatures with subsequent quenching substantially raises the superconducting transition temperature of the Laves phase. After annealing at 1200°C, the T_c of ZrV_2 attains a value of 9.64°K.

Heat treatment has a considerable influence on the critical magnetic field of the compounds as well. The effect of annealing temperature on the temperature dependence of the critical magnetic field of diffusion-type Nb_3Sn samples close to the T_c point is illustrated in Fig. 76 [224].

The temperature dependence of the critical magnetic field of Nb_3Sn, V_3Ga, V_3Si, and NbN in transverse magnetic fields (up to 240 kOe) as determined by Saur and Wizgall [89] is shown in Fig. 77. The critical field decreases particularly sharply close to the temperature corresponding to the transition into the normal state (Fig. 78) [231]. The value of $dH_c/dT = -44$ kOe/deg agrees closely with existing data [88, 93]. At lower temperatures (11-14.5°K) $dH_c/dT = -17$-26 kOe/deg [200].

The critical current density of bulk samples of the compounds in question may reach 10^5 A/cm^2 [232]. Lower values of critical

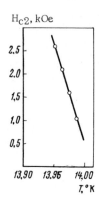

Fig. 78. Temperature dependence of the critical magnetic field of V_3Ga near the T_C point.

current density (Fig. 79) were obtained for case samples of Nb_3Sn, V_3Ga, V_3Si, and Nb_3Ga in one case [92]. However, this may have been due to the nonequilibrium state of the alloys and also to the effects of impurities, since the samples were prepared in aluminum oxide or zirconium oxide crucibles. It is of interest that a "peak effect" appeared in V_3Ga and NbN subjected to strong magnetic fields (Fig. 80) [89], the nature of this as yet being uncertain. For NbN samples held for a long time at high temperatures under high nitrogen pressures the density of the critical current in fields of 110 kOe is usually of the order of 10^2 A/cm², while at 110-120 kOe a sharp peak occurs: $5 \cdot 10^3$ A/cm² [89a]. In samples containing zirconium the "peak effect" is less marked [89a].

Data are now available in relation to the critical current density of the "record" superconductor $Nb_3(Al, Ge)$. One paper [2b] suggests that the critical current is very low, so that the practical

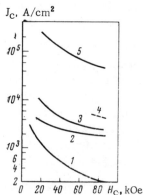

Fig. 79. Critical current density of cast samples of Nb_3Ga (1), Nb_3Sn (2), V_3Si (3), V_3Ga (4), and wires with a Nb_3Sn core (5) at 4.2°K in relation to the magnetic-field strength.

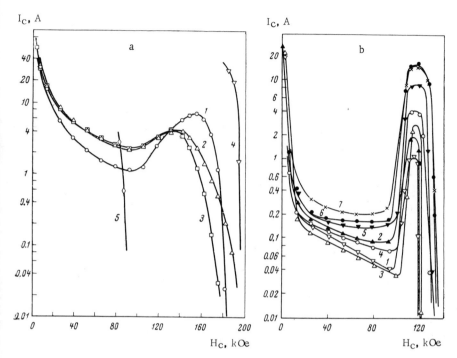

Fig. 80. Critical current of V_3Ga (a) and NbN (b) samples at 4.2°K in relation to the magnetic field. a: 1) Diffusion sample (1200°C, 10 h, d = 0.5 mm); 2) wire with a superconducting core (1485°C, 2 h, d = 1 mm); 3) ditto (1440°C, 2 h, d = 1 mm); 4) sintered sample (cross-sectional area 22 mm^2); 5) ditto (area 0.6 m^2); b: diffusion samples produced at 1500°C (22 h) and the following pressures (in atm): 1) 10; 2) 15; 3) 20; 4) 30; 5) 65; 6) 100; 7) 200.

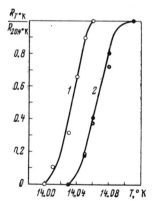

Fig. 81. Superconducting transition curves of V_3Ga under pressure (1) and without additional pressure (2).

use of this material is in doubt. Some Japanese research workers however [232a] have obtained a current density of $I_c = 10^3$ A/cm^2 in this superconductor under pulse conditions at 4.2°K in a field of 350 kOe and 10^2 A/cm^2 at H_{c2}. It would appear that the latter is associated with the "peak effect" in the alloy Nb$_3$(Al, Ge).

External pressure has a certain effect on the superconducting characteristics of the compounds. Figure 81 shows the superconducting transition curves of V$_3$Ga plotted in the absence of pressure (T_c = 14.06°K) and again at a pressure of 1730 kg/cm^2 [231]. The transition temperature decreases under pressure ($dT_c/dP = -1.7$-$3.4 \cdot 10^{-5}$ deg/atm). The effect of pressure on V$_3$Ga is the same in sign and order of magnitude as on the majority of other superconductors [233]. The transition temperature of Nb$_3$Sn also decreases on increasing the pressure ($dT_c/dP = -2.5 \cdot 10^{-5}$ deg/atm) [234, 235, 235a, 235b].

A negative derivative of the critical temperature with respect to pressure also occurs for Nb$_3$Al [235b] and V$_3$Si [81]. On applying a stress to a V$_3$Si sample in the [100] crystallographic direction the critical superconducting temperature decreases at a rate of about $5 \cdot 10^{-4}$ deg/atm [236]; a stress along the [111] axis does not lead to any appreciable change in T_c. The severe anisotropy of T_c is apparently associated with the anisotropy of the elastic properties of the crystal lattice in Cr$_3$Si-type compounds [237-239]. Under hydrostatic compression the change in the T_c of V$_3$Si equals $-2 \cdot 10^{-5}$ deg/atm [240]. Weger et al. [236] considered that the coefficient was even a little smaller than this. The fact that the change in T_c on applying a pressure along the (100) axis was more than an order greater than that obtained under hydrostatic compression indicates that the critical temperature is affected not only by changes in volume but also by shear strains in the lattice [71].

The uniaxial compression of a V$_3$Si single crystal under a stress of 4 kg/mm^2 completely suppresses the transformation of a Cr$_3$Si-type structure into the tetragonal form [239, 241]. Under a stress of 2 kg/mm^2 the temperature of the transformation in question remains unaltered, but the degree of the transformation diminishes substantially.

The fall in the T_c of the foregoing superconducting compounds is, according to Smith [241a], associated with the inhomogeneity of the applied pressure. According to these data, the T_c of V$_3$Si, V$_3$Ga, and V_3Ge rises under a uniform hydrostatic pressure ($dT_c/dp = 3.7$, 1.0, and $8.1 \cdot 10^{-5}$ °K/bar, respectively). A rise in T_c was also observed elsewhere [235b] for V_3Si and niobium nitride under pressure.

Thus in the case of the compound $CaSi_2$, which remains in the normal state down to temperatures of 0.32°K under ordinary pressures, high pressures induce a structural transformation (the rhombohedral structure being transformed into a body-centered tetragonal structure of the $ThSi_2$ type with $a = 2.2832$, $c = 13.542$ Å), and the compound becomes a superconductor with $T_c = 1.58$°K (in the absence of a field) [242]. On applying a magnetic field the critical temperature decreases to 1°K at 300 Oe and 0.35°K at 1000 Oe.

High pressure has a similar effect on InSb [243, 244]. On slowly raising the pressure (2 kbar/day) between 20 and 30 kbar a phase with an orthorhombic structure ($a = 2.928$, $b = 5.623$, $c = 3.07$ Å) is formed [243]. A rapid pressure rise to over 30 kbar (in approximately 30 sec) leads to the simultaneous formation of an orthorhombic phase with the structure of β-tin [243]. The proportion of the latter rises from 20% at 35 kbar to 70% at 150 kbar. Both phases are preserved in a metastable state at temperatures below 77°K on removing the pressure. The superconducting transition temperature of these phases depends on the pressure at which they were formed. The T_c of the first orthorhombic phase formed at pressures of 30, 35, 70, and 150 kbar, respectively, equals 3.3-3.5; 3.6-4.7; 4.0-5.0, and 3.5-5.0°K; for the phase with the β tin type of structure the T_c values at the corresponding pressures equal 2.4-2.9; 2.6-3.3; 3.1-3.8, and 2.6-3.1°K [243]. The critical temperature of the forms of InSb developing at high pressures also depend on the temperatures at which these pressures were imposed [244]; on reducing the temperature of pressure application the value of T_c increases. It was considered that one of the chief reasons for the changes in the T_c of this compound lay in a change in the density of states of the electrons on the Fermi surface [244].

Data relating to the effect of pressure on the characteristics of superconducting elements were also presented in Chap. II.

The critical temperature of compounds of the Cr_3Si type falls slightly on irradiating the samples with neutrons, while the critical current rises sharply [245, 246]. After irradiation with 15-MeV deuterons at 30°K at an intensity of 10^{17} deuterons/cm^2 the critical temperature of Nb$_3$Sn decreases by tenths of a degree Kelvin [247]. After annealing the irradiated sample at 300°K the T_c value is only partially restored. The critical current increases or decreases according to its initial value. After irradiation the critical magnetic field (H_c) decreases in proportion to the change in T_c.

We see from the experimental data of this section that by varying thermodynamic factors (concentration, temperature, and pressure) or imposing magnetic, radiation, or other fields the structure and properties of superconducting compounds may to a certain extent be controlled.

In conclusion, it should be mentioned that only compounds with a high density of states on the Fermi surface possess high values of T_c. The band structure of superconductors remained uncertain for a very long time. Recently [249, 249a] the structure of the energy spectrum was calculated theoretically for several compounds of the V$_3$X type (X = Ga, Si, Ge, As) and the density-of-states curve was plotted for V$_3$Ga. Nemnonov and Kurmaev [250, 250a] used an x-ray spectrographic method to confirm experimentally the distribution of the density of states with respect to energies for the compounds V$_3$Si and Cr$_3$Si. A characteristic of superconducting compounds with high T_c values is the positioning of the Fermi boundary within the limits of the highest maximum on the N(E) curve relating the density of states to the energy [250, 251]. The absence of superconductivity in the compound Cr$_3$Si itself is a consequence of the low density of states at the Fermi level [250].

Savitskii and colleagues [252] carried out some work on predicting the properties of metallic (including superconducting) compounds from data relating to the electron structure of the component atoms, using an electronic computer. This work showed that at least 3000 compounds with the stoichiometric formula A$_3$B might be expected to exist, some of these being superconductors.

The results obtained with the aid of an electronic computer deepen our knowledge of the nature of matter and demand the refinement, or in some cases a complete review, of a large number of concepts and viewpoints in physical chemistry, metal physics, and metallography. The authors' own method [253-255] of using computers in this field enables data relating to the electron structure of the atoms, together with a detailed knowledge of specific test cases, to be converted into predictions regarding the properties of new compounds and the conditions of their formation (type of reaction, range of homogeneity, phase diagrams describing their formation, etc.), and provides useful estimates of their physical (including superconducting) properties. Computer mathematics offers wide possibilities for the multilateral pursuit of superconducting compounds.

LITERATURE CITED

1. B. W. Roberts, in: New Materials and Methods of Studying Metals and Alloys [Russian translation], Izd. Metallurgiya, Moscow (1966), p. 9.
2. R. A. Hein, Phys. Letts., 23:435 (1966).
2a. B. T. Matthias, T. H. Geballe, L. D. Longinott, E. Corenzwit, G. W. Hull, R. H. Willens, and J. P. Maita, Science 156(3775):645 (1967).
2b. S. Foner, E. J. McNiff, B. T. Matthias, and E. Corenzwit, J. Appl. Phys., 40(2): 2010 (1969).
2c. B. T. Matthias, Science, 168(3927):103 (1970).
3. E. M. Savitskii and V. V. Baron, Izv. Akad. Nauk SSSR, Metallurgiya i Gornoe Delo, No. 5, p. 3 (1963).
4. M. Laue, Ann. Phys., 6:40 (1948).
5. J. Daunt, Progr. Low-Temperature Phys., 2:194 (1957).
6. I. B. Borovskii, Collected Works of the Institute of Metallurgy, No. 15, Izd. AN SSSR, Moscow (1963), p. 79.
7. I. B. Borovskii, Collected Works of the Institute of Metallurgy, No. 6, Izd. AN SSSR, Moscow (1960), p. 41.
8. B. T. Matthias, T. H. Geballe, and V. B. Compton, Rev. Mod. Phys., 35(1):1 (1963).
9. B. T. Matthias, Progr. Low-Temperature Phys., 2:138 (1957).
10. B. Boren, Arkiv Kemi, Mineral., Geol., No. 10, 11 (1933).
11. G. B. Bokii, Introduction to Crystal Chemistry, Izd. MGU (1954).
12. H. Hartmann, F. Ebert, and O. Bretschneider, Z. Anorg. und Allgemein. Chem., 198:116 (1931).
13. A. J. Hegedus, T. Millner, J. Neugebauer, and K. Sasvari, Z. Anorg. und Allgemein. Chem., 281:64 (1955).
14. T. Millner, A. J. Hegedus, K. Sasvari, and J. Neugebauer, Z. Anorg. und Allgemein. Chem., 289:288 (1957).
15. J. Neugebauer, A. J. Hegedus, and T. Millner, Z. Anorg. und Allgemein. Chem., 293:241 (1958).
16. G. Hagg and N. Schonberg, Acta Crystallogr., 7:351 (1954).
17. F. Laves, Theory of Phases in Alloys [Russian translation], Metallurgizdat, Moscow (1961), p. 111.
18. L. Kihlborg, Acta Chem. Scand., 16:2458 (1962).
19. N. Schonberg, Acta Chem. Scand., 8:221 (1954).

20. H. von Philipsborn, Mischsysteme von Verbindungen des Cr_3Si Typs under deren Polimorphia-Erscheinungen, Juris Verlag, Zurich (1964).
21. K. Schubert, Kristallstrukturen zweikomponentiger Phasen, Springer-Verlag, Berlin (1964).
22. P. Greenfield and P. A. Beck, Trans. AIME, 206:265 (1956).
23. T. Millner, Z. Anorg. und Allgemein. Chem., 292:25 (1957).
24. S. Geller, Acta Crystallogr., 9:885 (1956).
25. S. Geller, Acta Crystallogr., 10:380 (1957).
26. S. Geller, Acta Crystallogr., 10:678 (1957).
27. L. Pauling, Acta Crystallogr., 10:374 (1957).
28. L. Pauling, Acta Crystallogr., 10:685 (1957).
29. M. V. Nevitt, Trans. AIME, 212:350 (1958).
30. H. J. Wallbaum, Z. Metallkunde, 31:362 (1939).
31. H. J. Wallbaum, Naturwissenschaften, 32:76 (1944).
32. D. H. Templeton and C. H. Dauber, Acta Crystallogr., 3:261 (1950).
33. E. Raub and P. Walter, Heraeus Festschrift, p. 124 (1951).
34. P. Duwez, Trans. AIME, 191:564 (1951).
35. P. Duwez and C. B. Jordan, Acta Crystallogr., 5:213 (1952).
36. W. Rostoker and A. Yamamoto, Trans. ASM, 46:1136 (1954).
37. E. M. Savitskii, V. V. Baron, and Yu. V. Efimov, Dokl. Akad. Nauk SSSR, Tekh. Fiz., 171(2):84 (1967).
38. E. Teatum, K. Gschneider, and J. Waber, Los Alamos Scient. Lab. Rept., No. 23, p. 45 (1960).
39. J. B. Darby, D. J. Lam, L. J. Norton, and J. W. Downey, J. Less-Common Metals, 7(4):558 (1962).
40. J. B. Darby and S. T. Zegler, J. Phys. Chem. Solids, 23:1825 (1962).
41. K. Schubert, T. R. Anantharaman, H. O. K. Ata, H. C. Meissner, M. Potzschke, W. Rossteutscher, and E. Stolz, Naturwissenschaften, 47:512 (1960).
42. R. M. Watersfrat and E. C. van Reuth, Trans. Metallurg. Soc. AIME, 236(8):1232 (1966).
43. D. H. Killpatrick, J. Phys. Chem. Solids, 25(1):1213 (1964).
44. New Scientist, 31:512, 556 (1966).
45. S. Rosen, J. Goebel, and J. A. Mullins, J. Less-Common Metals, 12(6):510 (1967).
46. H. Holleck, H. Nowotny, and F. Benesovsky, Monatsh. Chem., 94(2):473 (1963).
47. M. V. Nevitt, Electronic Structure and Alloy Chemistry of the Transition Elements, New York (1963), p. 101.
48. T. Millner, Zh. Neorg. Khim., 3(4):946 (1958).
49. D. H. Killpatrick, J. Metals, 16:98 (1964).
50. M. D. Banus, T. B. Reed, and H. C. Gatos, J. Phys. Chem. Solids, 23:971 (1962).
50a. H. L. Luo, E. Vielhaber, and E. Corenzwit, Z. Phys., 230(5):443 (1970).
50b. Yu. V. Efimov, in: Phase Diagrams of Metallic Systems, Izd. Nauka, Moscow (1968), p. 12.
50c. H. Holleck, H. Nowotny, and F. Benesovsky, Mh. Chem., 94(2):473 (1963).
50d. V. I. Surikov, M. K. Borzhitskaya, A. K. Shtol'ts, V. L. Zagryazhskii, and P. V. Gel'd, Fiz. Met. Metallov, 30(6):1167 (1970).
50e. T. Asada, T. Horiuchi, and M. Uchida, Japan J. Appl. Phys., 8(7):958 (1969).

50f. A. Müller, Z. Naturforsch., 24a(7):1134 (1969).
51. Y. L. Yao, Trans. AIME 224:1146 (1962).
52. Yu. V. Efimov, Izv. Akad. Nauk SSSR, Neorgan. Mat., 2(4):598 (1966).
53. I. I. Hauser and H. C. Theurerer, Phys. Rev., 129(1):103 (1963).
54. E. M. Savitskii, P. I. Kripyakevich, V. V. Baron, Yu. V. Efimov, Izv. Akad. Nauk SSSR, Neorgan. Mat., 3(1):45 (1967).
55. E. Raub and E. Roschel, Z. Metallkunde, 57(6):470 (1966).
56. V. B. Compton, Phys. Rev., 123:1567 (1961).
57. G. F. Hardy and J. K. Hulm, Phys. Rev., 89:884 (1953).
58. G. F. Hardy and J. K. Hulm, Phys. Rev., 93:1004 (1954).
59. B. T. Matthias, Phys. Rev., 97:74 (1955).
60. B. T. Matthias, E. A. Wood, E. Corenzwit, and V. B. Bala, J. Phys. Chem. Solids, 17(1):188 (1956).
61. S. T. Zegler, Phys. Rev., 137(5A):1437 (1965).
62. B. W. Roberts, Intermetallic Compounds (J. H. Westbrook, Ed.), John Wiley and Sons, New York–London–Sydney (1966), p. 581.
63. T. H. Courtney, G. W. Pearsall, and J. Wulff, J. Appl. Phys., 36(10):3256 (1965).
64. V. Sadagopan, H. C. Gatos, and B. C. Giessen, J. Phys. Chem. Solids. 26(11): 1687 (1965).
64a. G. Meyer, Naturwissenschaften, 54(18):489 (1967).
64b. E. C. Reuth, R. M. Waterstrat, R. D. Blaugher, R. A. Hein, and J. E. Cox, Transactions of the Tenth International Conference on Low-Temperature Physics, Izd. VINITI, Moscow (1967), Vol. 2B, p. 137.
64c. B. W. Roberts, Superconductive Materials and Some of Their Properties, Note 408 (1967).
64d. H. L. Luo, E. Vielhaber, and E. Corenzwit, Z. Phys., 230:443 (1970).
64e. E. M. Savitskii, V. V. Baron, and Yu. V. Efimov, Fiz. Met. Metallov., 25(6):1126 (1968).
65. B. W. Batterman and C. S. Barrett, Phys. Rev. Lett., 13:390 (1964).
65a. S. A. Medvedov, K. V. Kiseleva, and V. V. Mikhailov, Fiz. Tverd. Tela, 10(3):746 (1968).
65b. H. W. King, F. H. Cocks, and J. T. A. Pollock, Phys. Lett., A26(2):77 (1967).
65c. L. J. Vieland, R. W. Cohen, and W. Rehwald, Phys. Rev. Lett., 26(7):373 (1971).
65d. E. Nembach, K. Tachikawa, and S. Takano, Phil. Mag., 21(172):869 (1970).
66. M. J. Goringe and U. Valdre, Phys. Rev. Lett., 14:823 (1965).
67. J. E. Kunzler, J. P. Maita, E. J. Ryder, and H. J. Levinstein, Phys. Rev., 143:390 (1966).
68. J. J. Hauser, Phys. Rev. Lett., 13:470 (1964).
69. J. J. Hauser, Phys. Rev. Lett., 14:422 (1965).
69a. J. C. F. Brook, Solid-State Communs., 7(24):1789 (1969).
70. P. W. Anderson and E. I. Blount, Phys. Rev. Lett., 14:217 (1965).
71. A. P. Levanyuk and R. A. Suris, Uspekhi Fiz. Nauk, 91(1):113 (1967).
72. G. Otto and E. Saur, Z. Naturforsch., 20a(7):975 (1965).
72a. L. J. Vieland, J. Phys. Chem. Solids, 31:1449 (1970).
73. M. Weger, Rev. Mod. Phys., 36:175 (1964).
74. A. M. Clogston, A. C. Gassard, V. Jaccarino, and Y. Yafet, Rev. Mod. Phys., 36:170 (1964).

75. H. J. Williams and R. C. Sherwood, Bull. Amer. Phys. Soc., 5:430 (1960).
76. J. S. Shier and R. D. Taylor, Solid-State Communs., 5(2):147 (1967).
76a. H. Krebs, Z. Naturforsch., 23a(2):332 (1968).
77. E. A. Wood, V. B. Compton, B. T. Matthias, and E. Corenzwit, Acta Crystallogr., 11:604 (1958).
78. E. Corenzwit, J. Phys. Chem. Solids, 20(9):98 (1959).
79. K. Raetz and E. Z. Saur, Physik, 169:315 (1962).
80. P. S. Swartz, Phys. Rev. Lett., 9:448 (1962).
81. N. E. Alekseevskii, E. M. Savitskii, V. V. Baron, and Yu. V. Efimov, Dokl. Akad. Nauk SSSR, 145(1):82 (1962).
82. J. K. Hulm and R. D. Blaugher, Phys. Rev., 123(5):1569 (1961).
83. B. T. Matthias, J. Phys. Chem. Solids, 20(10):342 (1959).
84. R. M. Bozorth, A. J. Williams, and D. D. Davis, Phys. Rev. Lett., 5:148 (1960).
85. J. E. Kunzler, E. Buchler, F. S. L. Hsu, and J. H. Wernick, Phys. Rev. Lett., 6:89 (1961).
86. V. D. Arp, R. H. Kropschot, J. H. Wilson, W. F. Love, and R. Phelan, Phys. Rev. Lett., 6:452 (1961).
87. J. O. Betterton, R. W. Boom, G. D. Kneip, and R. E. Worsham, Phys. Rev. Lett., 6:532 (1961).
88. F. J. Morin, J. P. Maita, H. J. Williams, R. C. Sherwood, J. H. Wernick, and J. E. Kunzler, Phys. Rev. Lett., 8:275 (1962).
88a. Yasukochi Ko, Akihama Ryozo, and Usui Nobumitsu, Japan J. Appl. Fiz., 9(7):845 (1970).
89. E. Saur and H. Wizgall, Les Champs Magnetiques Intenses, Colloque Internat., Grenoble (1966), p. 223.
89a. K. Hechler, E. Saur, and H. Wizgall, Z. Phys., 205(4):400 (1967).
90. D. B. Montgomery, Bull. Amer. Phys. Soc., 10(3):359 (1965).
91. H. R. Hart, J. Jacobs, C. L. Kolbe, and P. E. Lawrence, High Magnetic Fields, New York (1962), p. 584.
92. J. Wernick, in: Superconducting Materials, Izd. Mir, Moscow (1965), p. 64.
93. J. H. Wernick, F. J. Morin, F. S. L. Hsu, D. Dorsi, J. R. Maita, and J. E. Kunzler, High Magnetic Fields, Technol. Press, New York–London–Cambridge (Mass.) (1962), p. 609.
94. Chemical Week, 89(20):175 (1963).
95. C. B. Chandraseker, Appl. Phys. Lett., 1:7 (1962).
96. D. B. Montgomery and H. Wizgall, Phys. Letts., 22(1):48 (1966).
97. J. J. Hauser, D. D. Bacon, and W. H. Haemmerle, Phys. Rev., 151(1):296 (1966).
98. J. J. Hauser, Transactions of the Tenth International Conference on Low-Temperature Physics, Izv. VINITI, Moscow (1967), p. 111.
99. H. J. Levinstein and J. E. Kunzler, Phys. Letts., 20(6):581 (1966).
100. Asayama Kunisuke and Yamagata Hidekim, J. Phys. Soc. Japan, 22(1):347 (1967).
100a. A. G. Shepelev, Uspekhi Fiz. Nauk, 96(2):217 (1968).
101. G. V. Samsonov, in: Metallography and Metallurgy of Superconductors, Izd. Nauka, Moscow (1965), p. 65.
102. O. I. Shulishova, Superconductivity of Carbides and Nitrides of Transition Metals and Their Solid Solutions with the NaCl Structure, Author's abstract of Dissertation, Inst. Metallofiz. Akad. Nauk Ukr. SSR, Kiev (1966).

103. O. I. Shulishova and I. A. Shcherbak, Izv. Akad. Nauk SSSR, Neorgan. Mat., 3(8):1495 (1967).
104. A. L. Giorgi and E. G. Szklarz, J. Less-Common Metals, 11(6):455 (1966).
104a. A; L. Giorgi, E. G. Szklartz, M. C. Krupka, T. C. Wallace, and N. H. Krikorian, J. Less-Common Metals, 14(2):247 (1968).
105. V. Sadagapan and H. C. Gatos, J. Phys. Chem. Solids, 27(2):235 (1966).
105a. A. L. Giorgi, E. G. Szklarz, M. C. Krupka, T. C. Wallace, and N. H. Krikorian, J. Less-Common Metals, 14(2):247 (1968).
106. G. Dorfman and I. K. Kikoin, Physics of Metals, GTTI (1934), p. 405.
107. H. J. Fink, A. C. Therson, E. Parker, V. F. Zackay, and L. Toth, Phys. Rev., 138(4A):1170 (1955).
108. N. Pessall, C. K. Jones, H. A. Johansen, and J. K. Hulm, Appl. Phys. Lett., 7:38 (1965).
109. G. V. Samsonov and Ya. S. Umanskii, Hard Compounds of Refractory Metals, Metallurgizdat, Moscow (1957).
110. A. Giorgi, E. Szklarz, E. Storms, A. Bowman, and B. Matthias, Phys. Rev., 125:837 (1962).
111. N. E. Alekseevskii, G. V. Samsonov, and O. I. Shulishova, Zh. Éksp. Teor. Fiz., 44:1413 (1963).
112. J. S. Rajput and A. K. Gupta, J. Phys. Soc. Japan, 21(10):2075 (1966).
113. T. H. Geballe, B. T. Matthias, J. P. Remeika, A. M. Clogston, V. B. Compton, J. P. Maita, and H. J. Williams, Physics, 2(6):293 (1966).
114. T. Z. Jurrianse, Z. Kristallogr., 90:322 (1935).
115. T. H. Geballe, Transactions of the Tenth International Conference on Low-Temperature Physics, Izd. VINITI, Moscow (1967), p. 196.
116. L. Gold, Phys. Stat. Solidi, 4:261 (1964).
117. H. Nowotny, Freiberger Forschung., 123:7 (1967).
117a. A. L. Giorgi et al., J. Less-Common Metals, 17(1):121 (1969).
117b. H. C. Krupka et al., J. Less-Common Metals, 17(1):91 (1969).
117c. M. C. Krupka and M. G. Bowman, Colloq. Int. CNRS, 188:409 (1970).
117d. R. E. Intra et al., J. Less-Common Metals, 22:149 (1970).
117e. M. C. Krupka et al., J. Less-Common Metals, 19(2):113 (1969).
117f. A. L. Giorgi et al., J. Less-Common Metals, 22(1):131 (1970).
118. G. Hagg, Z. Phys. Chem., 12:33 (1931).
119. T. V. Massal'skii, in: Theory of Phases in Alloys, Metallurgizdat, Moscow (1961), p. 49.
120. H. Nowotny, W. Jeitschko, and F. Benesovsky, Planseeber. Pulvermetallurgie, 12:31 (1964).
121. L. E. Toth, W. Jeitschko, and C. M. Yen, J. Less-Common Metals, 10(1):29 (1966).
121a. N. B. Hannag, T. H. Geballe, B. T. Matthias, K. Andres, P. Schmidt, and D. MacNair, Phys. Rev. Lett., 14:225 (1965).
121b. B. T. Matthias, Trans. of the Tenth Internat. Conf. on Low-Temperature Physics [Russian translation], Izd. VINITI, Moscow (1967), Vol. 2B, p. 77.
122. Ch. J. Raub, V. B. Compton, T. H. Geballe, B. T. Matthias, J. P. Maita, and G. W. Hull, J. Phys. Chem. Solids, 26(12):2051 (1965).
123. R. D. Blaugher, J. K. Hulm, and P. N. Yocom, J. Phys. Chem. Solids, 26(12):2037 (1965).

124. G. L. Guthrie and R. L. Palmer, Phys. Rev., 141(1):346 (1966).
125. M. H. Maaren and G. M. Schaeffer, Phys. Letts., 20(2):131 (1966).
126. V. P. Zhuze, S. S. Shalyt, V. A. Noskin, and V. M. Sergeeva, ZhETF, Pis. Red., 3(5):217 (1966).
127. R. M. Bozorth, F. Holtzberg, and S. Methfessel, Phys. Rev. Lett., 14:952 (1965).
128. T. A. Bither, C. T. Prewitt, J. L. Gillson, P. E. Bierstedt, R. B. Flippen, and H. S. Young, Solid State Communs., 4(10):533 (1966).
129. P. E. Zeiden and F. Gol'tsberg, Transactions of the Tenth International Conference on Low-Temperature Physics [Russian translation], Izd. VINITI, Moscow (1967).
130. J. Bardeen, L. N. Cooper, and J. R. Schrieffer, Phys. Rev., 108:1175 (1957).
131. B. Reiss and H. Wagini, Z. Naturforsch., 21a(11):2008 (1966).
132. B. T. Matthias and J. K. Hulm, Phys. Rev., 87:799 (1952).
133. M. D. Banus, L. B. Farrell, and A. J. Stauss, Solid State Res. Lincoln Lab. MIT, 27(4):33 (1964/1965).
134. B. B. Goodman and S. G. Marcucci, Suomalais. Tiedeakat. Tiomituks, Sar. AVI, 210:86 (1966).
135. L. Finegold, Phys. Rev. Lett., 13:233 (1964).
136. S. Geller, A. Jagaraman, and G. W. Hull, Appl. Phys. Lett., 4:35 (1964).
137. S. Geller and G. W. Hull, Phys. Rev. Lett., 13:127 (1964).
138. J. K. Hulm, C. K. Jones, R. Mazelsky, R. C. Miller, R. A. Hein, and J. W. Gibson, Low-Temperature Physics, LT-9, Part A, Plenum Press, New York (1965), p. 600.
139. K. Andres, N. A. Kuebler, and M. B. Robin, J. Phys. Chem. Solids, 27(11-12):1747 (1966).
140. H. P. R. Frederikes, J. E. Schooley, W. R. Thurber, E. Pfeiffer, and W. R. Hosler, Phys. Rev. Lett., 16(13):579 (1966).
141. E. Ambler, J. H. Colwell, W. R. Hosler, and J. E. Schooley, Phys. Rev., 148(1):280 (1966).
142. P. E. Bierstedt, T. A. Bither, and F. J. Darnell, Solid State Communs., 4(1):25 (1966).
143. A. R. Sweedler, J. K. Hulm, B. T. Matthias, and T. H. Geballe, Phys. Letts., 19(2):82 (1965).
143a. M. Robin, K. Andres, T. H. Geballe, N. A. Kuebler, and D. B. McWhan, Phys. Rev. Lett., 17:917 (1966).
143b. N. M. Builova and V. B. Sandomirskii, Uspekhi Fiz. Nauk, 97(1):119 (1969).
144. C. W. Tucker, Science, 112:448 (1950).
145. G. J. Dickins, A. M. Douglas, and W. H. Taylor, J. Iron Steel Inst., 167:27 (1951).
146. J. Thewliss, Acta Crystallogr., No. 7, p. 323 (1954).
147. P. Duwez, in: Theory of Phases in Alloys [Russian translation], Metallurgizdat, Moscow (1961), p. 225.
148. D. S. Bloom and N. J. Grant, Trans. AIME, 197:88 (1953).
149. E. Bucher, F. Heiniger, and J. Muller, Phys. Kondensierten Materie, 2(3):210 (1964).
150. P. Greenfield and P. A. Beck, Trans. AIME, 200:253 (1954).
151. A. J. Lena, Metal Progr., 66:122 (1954).
152. R. D. Blaugher and J. K. Hulm, J. Phys. Chem. Solids, 19(1-2):134 (1961).

153. S. H. Autler, J. K. Hulm, and R. S. Kemper, Phys. Rev., 140A:1117 (1965).
154. V. Sadagopan and H. C. Gatos, Phys. Stat. Solidi, 13(2):423 (1966).
154a. E. M. Savitskii and O. Kh. Khamidov, Izv. Akad. Nauk SSSR, Metally, 5:51 (1969).
155. F. Laves and H. Witte, Metallwirtschaft, 14:645 (1935).
156. R. L. Berry and G. V. Raynor, Acta Crystallogr., 6:178 (1953).
157. H. Witte, Zur Struktur und Materia des Festkorper, Springer Verlag, Berlin (1952).
158. H. Klee and H. Witte, Z. Phys. Chem., 202:352 (1954).
159. H. Wuhe, Z. Phys., 197(3):276 (1966).
160. R. L. Falge and R. A. Hein, Phys. Rev., 148(2):940 (1966).
161. D. C. Hamilton, Ch. J. Raub, and B. T. Matthias, J. Phys. Chem. Solids, 26(3):665 (1965).
161a. Rapp Östen, J. Less-Common Metals, 21(1):27 (1970).
161b. B. Hillenbrand and M. Wilhelm, Phys. Lett., A33(2):61 (1970).
162. A. Scharri, Phys. Letts., 20(6):619 (1966).
163. J. S. Kasper, Acta Metallurgica, 2:456 (1954).
164. D. K. Das, S. P. Rideout, and P. A. Beck, Trans. AIME, 194:1071 (1952).
165. N. E. Alekseevskii and N. N. Mikhailov, Zh. Éksp. Teor. Fiz., 43(6):2110 (1962).
166. E. M. Savitskii, M. A. Tylkina, and K. B. Povarova, Rhenium Alloys, Izd. Nauka, Moscow (1965), pp. 158, 245.
167. O. Kh. Khamidov, Physicochemical Interaction of Rhenium with Transition Metals of Groups III-VI in the Periodic System, Author's abstract of Dissertation, IMET, Moscow (1967).
167a. T. Claeson, Phys. Stat. Solidi, 25(2):K95 (1968).
168. E. M. Savitskii, V. V. Baron, and Yu. V. Efimov, Izv. Akad. Nauk SSSR, Neorgan. Mat., 12:2170 (1967).
169. J. Muller, E. Bucher, R. Burton, F. Heiniger, and G. Zambelli, Transactions of the Tenth International Conference on Low-Temperature Physics [Russian translation], VINITI, Moscow (1967), p. 211.
170. E. Bucher, F. Heiniger, J. Muller, and P. Spitzli, Phys. Letts., 19(4):263 (1965).
171. V. M. Pan, V. V. Pet'kov, and O. G. Kulik, in: Metallography, Physical Chemistry, and Metal Physics of Superconductors, Izd. Nauka, Moscow (1967), p. 161.
172. E. M. Savitskii, in: Metallography and Metal Physics of Superconductors, Izd. Nauka, Moscow (1965), p. 3.
173. E. M. Savitskii, Vestnik Akad. Nauk SSSR, No. 3 (1966).
174. Yu. V. Efimov, V. V. Baron, E. M. Savitskii, and E. I. Gladyshevskii, in: Metallography and Metal Physics of Superconductors, Izd. Nauka, Moscow (1965), p. 91.
175. Yu. V. Efimov, V. V. Baron, and E. M. Savitskii, in: Physical Chemistry, Metallography, and Metal Physics of Superconducting Materials, Izd. Nauka, Moscow (1968), p. 108.
176. N. E. Alekseevskii, N. V. Ageev, and V. F. Shamrai, Izv. Akad. Nauk SSSR, Neorgan. Mat., 2(12):2150 (1966).
177. R. Hagner, Z. Phys., 177(1):10 (1964).
177a. E. Ronald, J. Appl. Phys., 37(13):4880 (1966).
178. N. R. Alekseevskii, I. I. Kornilov, N. M. Matveeva, and Yu. A. Maksimov, Dokl. Akad. Nauk SSSR, 173(3):553 (1967).

179. M. Hansen and K. Anderko, Constitution of Binary Alloys [Russian translation], Metallurgizdat, Moscow (1962).
180. H. Nowotny, R. Maschenschalk, H. Kieffer, and F. Benesovsky, Monatsch. Chem., 85:241 (1954).
181. E. M. Savitskii, V. V. Baron, Yu. V. Efimov, and E. I. Gladyshevskii, Zh. Neorgan. Khim., 9(7):1655 (1964).
182. W. B. Pearson, Handbook of Lattice Spacings and Structures of Metals and Alloys, Pergamon Press, London (1958), p. 1.
183. E. M. Savitskii, V. V. Baron, Yu. V. Efimov, and E. I. Gladyshevskii, Trudy Inst. Met., 12, 166 (1962).
184. E. Bucher, F. Laves, J. Muller, and H. von Philipsborn, Phys. Letts., 8:27 (1964).
185. C. D. Cody, J. J. Hanak, G. T. McConville, and F. D. Rosi, Rev. RCA, 25(3):338 (1964).
185a. R. Flükiger, Phys. Lett., A29(7):407 (1969).
186. E. A. Wood and B. T. Matthias, Acta Crystallogr., 9:534 (1956).
187. S. T. Zegler and J. W. Downey, Trans. AIME 227:1407 (1963).
188. R. A. Oriani, Acta Metallurgica, 2:343 (1954).
189. E. M. Savitskii, V. V. Baron, Yu. V. Efimov, V. R. Karasik, T. V. Vylegzhanina, and E. I. Gladyshevskii, Zh. Neorgan. Khim., 9(8):2045 (1964).
190. E. M. Savitskii, V. V. Baron, Yu. V. Efimov, and E. I. Gladyshevskii, Izv. Akad. Nauk SSSR, Neorgan. Mat., 1(2):208 (1965).
191. E. M. Savitskii, V. V. Baron, and Yu. V. Efimov, Izv. Akad. Nauk SSSR, Neorgan. Mat., 2(8):1444 (1966).
191a. C. Susz, R. Flukiger, and J. Muller, Helv. Phys. Acta, 43(5):476 (1970).
192. T. B. Reed, H. C. Gatos, W.-J. La Fleur, and G. T. Goddy, Metallurgy of Advanced Electronic Materials, 19:71 (1963).
192a. R. Blaugher, N. Ressall, and A. Patterson, J. Appl. Phys., 40(5):2000 (1969).
193. H. Holleck, F. Benesovsky, and H. Nowotny, Monatsh. Chem., 93:996 (1962).
194. F. Rothwarf, J. A. Schwitz, C. C. Dickson, R. C. Thiel, H. Boller, and E. Parth, Phys. Rev., 152(1):341 (1966).
195. Sci. News, 91(20):475 (1967).
196. R. Hagner and E. Saur, Naturwissenschaften, 49:444 (1962).
197. F. Rothwarf, C. C. Dickson, E. Parthe, and H. Boller, Bull. Amer. Phys. Soc., 7:322 (1962).
197a. H. Otto, Z. Phys., 215(4):323 (1968).
197b. E. M. Savitskii, V. V. Baron, M. I. Bychkova, S. D. Gindina, Yu. V. Efimov, N. D. Kozlova, L. F. Martynova, B. P. Mikhailov, L. F. Myzenkova, and V. A. Frolov, Summaries of Contributions to the Sixteenth All-Union Conference on Low-Temperature Physics, Leningrad (1970), p. 198.
198. E. M. Savitskii, New Metallic Alloys, Izd. Znanie, Moscow (1967).
198a. E. M. Savitskii and G. S. Burkhanov, Metallography of Refractory Metals and Alloys, Izd. Nauka, Moscow (1967).
199. J. H. N. Van Vucht, H. A. C. M. Brunning, H. C. Donkersloot, and A. H. Gomes de Mesquito, Philips Res. Rpts., 19(5):407 (1964).
200. H. J. Levinstein, J. H. Wernick, and C. D. Capio, J. Phys. Chem. Solids, 26(7):1111 (1965).

201. E. A. Linton, Superconductivity [Russian translation], Izd. Mir, Moscow (1964).
202. H. von Philipsborn and F. Laves, Acta Crystallogr., 17:213 (1964).
203. A. Kjekshus, F. Gronvold, and J. Thorbjornsen, Acta Chem. Scand., 16:1493 (1962).
204. H. von Philipsborn, Tafeln zum Bestimmen der Minerale nach äusseren Kennzeichen, Stuttgart (1953).
204a. P. R. Sahm, Phys. Lett., A26(10):459 (1968).
204b. J. B. Vetrano, G. L. Guthrie, and H. E. Kissinger, Phys. Lett., A26(1):45 (1967).
205. M. W. Williams, K. M. Ralls, and M. R. Pickus, J. Phys. Chem. Solids, 28(2):333 (1967).
206. G. K. Gaulé, J. T. Breslin, R. L. Ross, J. R. Pastore, and J. R. Shappirio, Low Temperature Physics, LT-9, Part A, Plenum Press, New York (1965), p. 612. Press, New York (1965), p. 612.
207. B. T. Matthias and J. K. Hulm, Phys. Rev., 92:874 (1953).
208. F. J. Darnell, P. E. Bierstedt, W. O. Forshey, and R. K. Waring, Phys. Rev., 140(5A):1581 (1965).
209. L. E. Toth, C. M. Yen, L. G. Rosner, and D. E. Anderson, J. Phys. Chem. Solids, 27(11-12):1815 (1966).
210. N. E. Alekseevskii, G. V. Samsonov, and O. I. Shulishova, Izv. Akad. Nauk SSSR, Neorgan. Mat., 3(1):61 (1967).
211. H. Bilz, Z. Phys., 153:338 (1958).
212. G. Lautz and D. Schneider, Z. Naturforsch., 172A: 54 (1962).
213. R. M. Bozorth, B. T. Matthias, and D. D. Davis, Transactions of the Seventh International Conference on Low-Temperature Physics (Canada 1960), Univ. Toronto Press (1961), p. 385.
214. B. T. Matthias, V. B. Compton, and E. Corenzwit, J. Phys. Chem. Solids, 19:130 (1961).
214a. N. E. Alekseevskii, L. N. Guseva, and N. M. Matveeva, Dokl. Akad. Nauk SSSR, 178(5):1047 (1968).
214b. M. B. Maple, Phys. Lett., A26(10):513 (1968).
215. N. N. Zhuravlev, G. S. Zhdanov, and E. M. Smirnova, Fiz. Met. Metallov., 13(1):62 (1962).
216. G. G. Hanak and G. R. Gody, High Magnetic Fields, New York (1961), p. 592.
217. R. Eustrom, F. Courtney, and G. Pearsall, Metallurgy of Advanced Electronic Materials, Vol. 19, New York (1963), p. 60.
218. W. A. Rachinger, J. Austral. Inst. Metals, 1(4):231 (1966).
219. H. G. Jansen and E. J. Saur, Transactions of the Seventh International Conference on Low-Temperature Physics (Canada 1960), Univ. Toronto Press (1961), p. 185.
220. J. J. Hanak, G. D. Cody, J. L. Cooper, and M. Rayl, Transactions of the Eighth International Conference on Low-Temperature Physics (London, 1962), Washington (1963), p. 353.
221. V. N. Svechnikov, V. M. Pan, and Yu. I. Beletskii, in: Metallography, Physical Chemistry, and Metal Physics of Superconductors, Izd. Nauka, Moscow (1967), p. 100.
222. W. Kunz and E. Saur, Z. Phys., 189(4):401 (1966).
223. M. Tanenbaum and W. V. Wright (eds.), Superconductors, New York (1962), p. 143.

224. W. S. Kogan, A. I. Krivko, B. G. Lazarev, L. S. Lazareva, A. A. Matsakova, and O. N. Ovcharenko, in: Metallography and Metal Physics of Superconductors, Izd. Nauka, Moscow (1965), p. 76.
225. R. Enstrom, T. Courtney, G. Pearsall, and J. Wulff, Metallurgy of Advanced Electronic Materials, Vol. 19, Interscience, New York (1962), p. 121.
226. R. E. Enstrom, N. Y. Pearsall, G. W. Pearsall, and J. Wulff, J. Metals, 16:97 (1964).
227. J. J. Hanak, G. D. Cody, P. R. Aron, and H. C. Hitchcock, High Magnetic Fields, Technol. Press, New York–London–Cambridge (Mass.) (1962), p. 592.
228. T. H. Courtney, G. W. Pearsall, and J. Wulff, Trans. Metallurg. Soc. of AIME, 233(1):212 (1965).
229. E. Buchler, J. H. Wernick, K. M. Olsen, F. S. L. Hsu, and J. E. Kunzler, Metallurgy of Advanced Electronic Materials, Vol. 19, Interscience, New York (1963), p. 105.
230. N. M. Matveeva and T. O. Malakhova, in Metallography, Physical Chemistry, and Metal Physics of Superconductors, Izd. Nauka, Moscow (1968), p. 141.
231. B. G. Lazarev, L. S. Lazareva, A. A. Matsakova, and O. N. Ovcharenko, in Metallography and Metal Physics of Superconductors, Izd. Nauka, Moscow (1965), p. 89.
232. V. R. Karasik, Physics and Techniques of Strong Magnetic Fields, Izd. Nauka, Moscow (1964).
232a. Ko Yasukochi, Ryozo Akihama, and Nobumitsu Usui, Japan J. Appl. Phys. 9(7):845 (1970).
233. L. S. Kan, B. G. Lazarev, and V. I. Makarov, Éksp. Teor. Fiz., 40:457 (1961).
234. B. G. Lazarev, L. S. Lazareva, O. N. Ovcharenko, and A. A. Matsakova, Zh. Éksp. Teor. Fiz., 43:2309 (1962).
235. E. S. Itskevich, M. A. Il'ina, and V. A. Sukhoparov, Zh. Éksp. Teor. Fiz., 45:1378 (1963).
235a. J. P. McEvey, Colloq. Int. CNRS, No. 188, p. 101 (1970).
235b. H. Neubauer, Z. Phys., 226:211 (1969).
236. M. Weger, B. G. Silbernagel, and E. S. Greiner, Phys. Rev. Lett., 13:521 (1964).
237. L. R. Testardi, T. B. Bateman, W. A. Reed, and V. G. Chirba, Phys. Rev. Lett., 15:250 (1965).
238. K. R. Kieler and D. D. Hanak, Transactions of the Tenth International Conference on Low-Temperature Physics [Russian translation], Izd. VINITI, Moscow (1967), p. 391.
239. J. R. Patel and B. W. Batterman, Phys. Rev., 148(2):662 (1966).
240. C. B. Muller and E. L. Saur, Rev. Mod. Phys., 36:103 (1964).
241. J. R. Patel and B. W. Batterman, J. Appl. Phys., 37(9):3347 (1966).
241a. T. F. Smith, Phys. Rev. Lett., 25(21):1483 (1971).
242. D. B. McWhan, V. B. Compton, M. S. Silverman, and J. R. Soulen, J. Less-Common Metals, 12(1):75 (1967).
243. D. B. McWhan and M. Marezio, J. Chem. Phys., 45(7):2508 (1966).
244. S. Minomura, B. Okai, Y. Onoda, and S. Tanuma, Phys. Letts., 23(11):641 (1966).
245. J. P. McEvoy, R. F. Decell, and R. L. Novak, Appl. Phys. Lett., 4:43 (1964).
246. P. S. Swartz, H. R. Hart, and R. R. Fleischer, Appl. Phys. Letts., 4:71 (1964).

247. E. L. Keller, H. T. Coffey, A. Patterson, and S. H. Autler, Trans. Amer. Nucl. Soc., 9(1):55 (1966).
248. V. F. Shamrai, Superconductivity of the Beta Phase in the Nb−Sn−Al−Ge System, Author's abstract of Dissertation, A. A. Baikov Inst. Metallurgy (1966).
249. L. F. Mattheis, Phys. Rev., 138:A112 (1965).
249a. S. A. Nemnonov, É. Z. Kurmaev, and V. P. Belash, Phys. Stat. Sol., 39:39 (1970).
250. S. A. Nemnonov and E. Z. Kurmaev, Phys. Stat. Solidi, 24:K43 (1967).
250a. S. A. Nemnonov, É. Z. Kurmaev, V. I. Minin, V. G. Zyryanov, and I. A. Brytov, Fiz. Met. Metallov., 30(3):659 (1970).
251. C. B. Muller and E. J. Saur, Rev. Mod. Phys., 36:103 (1964).
252. E. M. Savitskii, Yu. V. Devingtal', and V. B. Gribulya, Dokl. Akad. Nauk SSSR, 183(5):11 (1968).
253. E. M. Savitskii and V. B. Gribulya, Izv. Akad. Nauk SSSR, Neorgan. Mat., 7(7):1097 (1971).
254. E. M. Savitskii, Yu. V. Devingtal', and V. B. Gribulya, Dokl. Akad. Nauk SSSR, 185:561 (1969).
255. E. M. Savitskii, Yu. V. Devingtal', and V. B. Gribulya, in: Problems of Superconducting Materials, Izd. Nauka, Moscow (1970), p. 39.

Chapter IV

Physicochemical Analysis of Superconducting Systems

The study of binary, ternary, and more complex systems of chemically-active refractory metals, including "hard" superconductors" only became possible in the second half of the present century after pure, rare, and refractory metals had been prepared and the apparatus and techniques developed for their vacuum melting, "neutral" melting, and heat treatment [1].

After this, investigations into the phase diagrams of refractory metals developed with increasing rapidity, chiefly in connection with the search for new superconducting and heat-resistant materials. At the present time almost all possible binary systems, several dozen ternary systems, and individual quaternary systems have been investigated. The study of composition–properties diagrams has lagged seriously. There have been approaches to the solution of metal-science problems with the aid of electronic computers [1a]. Data relating to the superconducting properties of alloys are presented in this chapter.

The characteristics of superconducting alloys (T_c and H_c) are mainly determined by the properties of the original components and the nature of their interactions; they depend on the chemical composition and structure of these materials.

The BCS theory and the empirical Matthias rule [2] signifying the relation between the development of superconductivity and the number of valence electrons indicate that the electron structure has a major influence on the very existence of superconductivity as well as on the level of the superconducting properties of metals and alloys. The general structure of the phase diagrams of alloys and the formation of various intermediate phases and solid solutions in these are determined by the interatomic interaction, which depends on the position of the atoms in the periodic table, on their electron structure, and on the tendency of the component atoms to give or receive valence electrons [3]. It is for this reason that the Mendeleev Table constitutes the basis for the science of metallic alloys. However, the properties of individual atoms cannot explain the whole array of properties of the vast numbers of atoms constituting a metal crystal. Crystalline material may incorporate all kinds of chemical bonds due to the interaction of the outer valence electron.

A metal differs sharply from nonmetallic substances in the nature of the forces involved in the interatomic bond in the liquid and solid states. A typical metal constitutes an assembly of ions immersed in a "cloud" of electrons; the electrons are in perpetual motion, not being linked to any particular ion, but appertaining to the whole crystal. This form of bond determines the luster, the high ductility, and the electrical conductivity of metals [1].

In transition metals and metallic compounds other forms of bond (covalent, ionic) also occur. The strength of the interatomic bonds in metals and alloys is at the moment approximated from the heat of sublimation, the melting point, the recrystallization temperature, the heat content, the activation energies of diffusion and self-diffusion, the elastic modulus, the thermal-expansion coefficient, and so on.

The development of methods for directly measuring the interatomic bond forces in metal crystals is one of the chief problems in the physical chemistry of metals [1, 1b].

Alloys may be: a) solid solutions with a disordered distribution of the component atoms (mainly with a metallic interatomic bond); b) chemical compounds between metals (metallic compounds) — with various laws governing the disposition of the component atoms (crystal structure) — having a mixed (metallic, covalent, and ionic) bond between the atoms; c) a mixture of phases.

Metallic solid solutions are characterized by high ductility and mechanical strength; they constitute the basis for the majority of technical construction alloys. The creation of metallic solid solutions is one of the chief ways of increasing the hardness and strength of metals (for example, the strength of niobium may be increased by a factor of five in this way).

The mutual solubility of elements in one another is determined by three factors: firstly, the structural factor, i.e., the similarity or difference between the crystal lattices of the solvent and dissolved metal; secondly, the dimensional factor (the difference between the atomic radii of the components); thirdly, the chemical factor (the electronegativity, the position of the elements in the periodic table, the bond energy between the components).

Metals dissolve easily in one another when they are close to one another in the Mendeleev Table (i.e., when they have a similar structure of the outer electron shells of the atoms), when they are isomorphic as regards crystal structure, when they differ by no more than 8-12% in atomic diameter, and when the difference in electronegativity is 0.2 or less. If the metals differ substantially in any of these respects, their mutual solubility will be poor and they will tend to form chemical compounds [1].

Graphs exist expressing the dependence of the maximum solid-state solubility of various alloying elements on the atomic diameters of the solvent and dissolved additive, and also on the electronegativity of these elements. The atomic radii of the elements are plotted along the horizontal axes of these graphs and the electronegativity along the vertical axes. In order to determine the solubility limits, auxiliary ellipses are plotted: an inner one (with a major axis representing ±0.2 units of electronegativity and a minor axis representing ± 8% difference in atomic radii), and an outer one (±0.4 units of electronegativity, ± 15% difference in atomic radii). The graphs are plotted for the solvent metals and may be given the name of solubility graphs. Within the smaller ellipse we most frequently encounter metals forming unlimited solid solutions with the base metal. Between the smaller and larger ellipses we find metals with a limited solubility in the base. Outside the larger ellipse the spatial and valence factors are unfavorable toward the formation of solid solutions. The niobium solubility graph is shown in Fig. 82; others have been plotted for other metals.

It should be noted that in the interaction of various elements there are a number of exceptions to the foregoing semiempirical solubility rules.

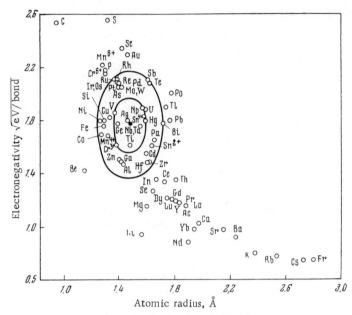

Fig. 82. Solubility graph for niobium.

The phase diagrams (diagrams of state) describe the structure of alloys; however, it is also important to know how the structure is related to the corresponding chemical and physical properties. N. S. Kurnakov discovered the relations governing the changes taking place in many physical properties of binary equilibrium systems (Kurnakov laws). The formation of metallic solid solutions is accompanied by an increase in hardness, mechanical strength, and electrical resistance relative to the values of these parameters in the original components. The formation of a metallic compound is always accompanied by an increase in hardness and electrical resistance. In alloys constituting mixtures of phases the properties vary additively.

The use of physicochemical analysis when studying superconducting alloys led to the establishment of some important laws governing changes in the properties of superconducting systems. It was found, for example, that all alloys of the metallic systems formed by superconductors were themselves superconductors, independently of the phase composition and crystal structure of the alloys. In binary systems involving the formation of a continuous series of solid solutions, the transition temperature varies continuously and often additively. In the presence of polymorphism in one or both of the superconducting components, a maximum appears on the T_c-composition curve. Superconducting compounds which may be formed by both superconducting and nonsuperconducting metals in a number of systems are distinguished by singular points. In alloys of mixtures formed by two superconductors, superconductivity is also preserved over the whole concentration range, while T_c varies linearly with composition [1b, 4, 5]. If one of the components is not superconducting, then in such mixtures superconductivity is preserved until the proportion of the superconductor becomes immeasurably small. In all cases the T_c of alloys and compounds is greater than the T_c of at least one of the components.

A systematic study of the composition–property diagrams of superconducting systems will undoubtedly enable us to predict the superconducting properties of alloys yet unstudied and to correct existing results, discrepancies in which may be partly attributed to variations in the nature of the samples and the experimental conditions, differences in the purity of the material, the degree of equilibrium of the alloys, and other metallurgical factors. An example of the substantial and sometimes decisive influence of the metallurgy of superconducting metals and alloys on their characteristic properties is the production of a superconducting wire of high parameters suitable for making superconducting magnets and solenoids, which only became possible after the development of methods for melting and thermomechanically processing a number of rare and refractory metals and alloys in vacuum and in protective media [5].

In view of their special physical properties, superconductors have a number of characteristic features (high structural sensitivity of the critical current, serious effect of chemical and phase inhomogeneity on T_c and H_c, the influence of the "proximity effect" on T_c, and so on) which make it vital to study the superconducting and other physical properties at helium temperatures. In other words, we really need a special branch of metal science dealing solely with the properties of superconductors.

The study of the superconducting properties of alloys and the determination of their parameters will in turn promote inves-

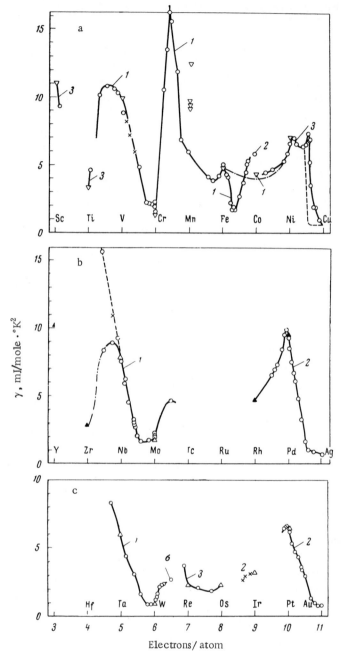

Fig. 83. Dependence of the electron specific heat on the number of valence electrons per atom in alloys of transition metals from the fourth (a), fifth (b), and sixth (c) periods. Crystal lattice: 1) bcc; 2) fcc; 3) hcp.

tigations into the electron structure of metallic materials, since these very parameters (the existence of superconductivity, the temperature of the superconducting transition) are explicitly related to the electron structure, the density of states near the Fermi surface (proportional to the electron specific heat), the width of the energy gap, and the Debye temperature.

Figure 83, based on existing data [306], illustrates the specific heat γ in relation to the number of valence electrons per atom in alloys of transition metals belonging to the fourth, fifth, and sixth periods; the same figure presents information relating to the crystal structure of the alloys. On comparing existing data relating to the temperature of the superconducting transition we notice that alloys with the highest T_c values also have the highest specific heats.

Many investigations have taken place in this field over the last ten years, revealing a large number of alloys and compounds with superconducting properties [4-6], and this number is still growing. In this chapter we shall systematize existing data relating to binary, ternary, and more complex superconducting systems.

BINARY SUPERCONDUCTING SYSTEMS

1. Systems with Unlimited Solubility in the Liquid and Solid States

In systems with unlimited solubility in the liquid and solid states at all concentrations and temperatures, superconductivity is only found when transition metals of the VA and VIA groups are alloyed with each other. Superconductivity occurs over the whole range of concentrations in systems of these elements in the case of Nb–V, Nb–Ta, and Nb–Mo alloys.

Figure 84 presents the phase diagram [7] and the T_c–composition relationship [8, 9] of Nb–V alloys. The liquidus and solidus curves in this system have a minimum in the range 70-80 at.% V, which suggests the possibility of ordering in the solid state after appropriate heat treatment. However, this has never yet been realized. All the alloys have the bcc lattice of the solid solution in the solid state. The critical temperature decreases from 9.2 to 5.4°K on adding vanadium to niobium, with a slight minimum at

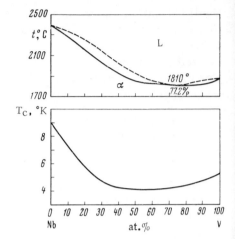

Fig. 84. Phase diagram and superconducting transition temperature of Nb−V alloys.

50-70 at.% V. Information as to the critical magnetic field of Nb−V alloys is given in [9b].

The phase diagram of Nb−Ta alloys [10] is presented together with their transition temperature and upper critical magnetic field [8, 9, 11, 12] in Fig. 85. Niobium−tantalum alloys have a narrow range of crystallization and an almost linear composition dependence of the solidus temperature. It is also interesting to

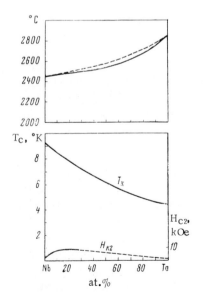

Fig. 85. Phase diagram, superconducting transition temperature, and critical magnetic field of Nb−Ta alloys.

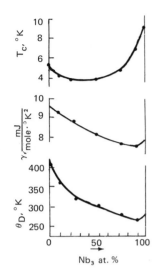

Fig. 86. Transition temperature (T_c), electron specific heat (γ) and Debye temperature (θ) as functions of composition in the V–Nb system.

note the almost linear dependence of T_c on the changing composition of the alloys. In [12a] the dependence of T_c on the proportion of the components in a sample of variable composition was studied; the values of T_c coincide with those shown in Fig. 85. The critical magnetic field (H_{c2}) increases slightly on adding 20 at.% Ta, then falls again; however, even at 20 at.% Ta it fails to rise above 7-8 kOe. According to [13] H_{c2} reaches 5 kOe in an alloy with 45 at.% Ta. The temperature dependence of the H_{c2} of Nb–Ta alloys was studied elsewhere [14] and the value of H_{c2} extrapolated to 0°K was determined. The composition–$H_{c2(0)}$ has a maximum (9.5 kOe) at 40% Ta. The changes in the electron specific heat and the Debye temperature are presented in Figs. 86 and 87 [9b].

Figure 88 shows the phase diagram of Nb–Mo alloys obtained for samples prepared by sintering [15] and casting [16]; the same figure shows the T_c–composition diagram of these alloys [17]. On adding molybdenum to niobium, the T_c of the alloys decreases linearly to 45 at.% Mo; alloys containing up to 80% Mo have extremely low transition temperatures (around 0.03°K), and then T_c rises to 0.9°K for molybdenum itself. The lowest T_c was found [17] in an alloy containing 70 at.% Mo (0.016°K). On measuring the Hall constants of this system, the alloy containing 70 at.% Mo [18] yielded a maximum value; this suggested the possible formation of an intermediate phase or the development of ordering. On comparing the T_c data with the Matthias rule [17], it was found

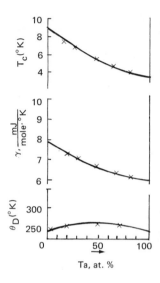

Fig. 87. Transition temperature (T_C), electron specific heat (γ) and Debye temperature (θ) as functions of composition in the Nb–Ta system.

that the T_c minimum (which occurred at 5.6-5.7 electrons/atom) coincided with the electron concentration corresponding to the absence of conductivity predicted by Matthias; however, the alloy did remain superconducting even in this case. The same authors attempted to compare the density of states at the Fermi surface, $N(0)$, and the electron–electron interaction coefficient (V) with the measured values of T_c; subject to certain simplifying assumptions in the calculations, they deduced that the product $N(0) \cdot V$ passed through a minimum at 70% Mo while the parameter V rose monotonically on passing from vanadium to molybdenum (Fig. 89).

The values of γ and θ_D were also determined for the Nb–Mo system in [9b]. In contrast to the Nb–Ta system, in this case the

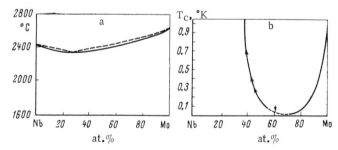

Fig. 88. Phase diagram (a) and superconducting transition temperature (b) of Nb–Mo alloys.

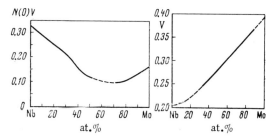

Fig. 89. Dependence of the product $N(0) \cdot V$ and the electron-electron interaction parameter V on the composition of alloys in the Nb-Mo system.

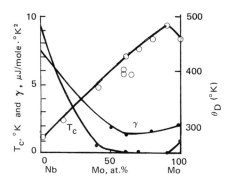

Fig. 90. Transition temperature (T_c), electron specific heat (γ), and Debye temperature (θ) as functions of composition in the Nb-Mo system.

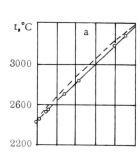
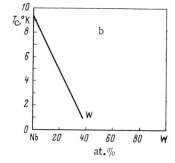

Fig. 91. Phase diagram (a) and superconducting transition temperature (b) of Nb-W alloys.

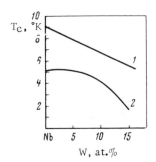

Fig. 92. Superconducting transition temperature of Nb—W alloys in relation to impurity content: 1) total interstitial impurities amounting to hundredths of a percent; 2) tenths of a percent.

γ-composition curves showed a flat maximum and the θ_D-composition curve a maximum at ~ 10 at.% Nb. These are shown in Fig. 90.

The Nb—W system also exhibits continuous solubility [19, 19a]; however, superconductivity only appears in alloys containing up to 37-38 at.% W (Fig. 91). On adding tungsten to niobium T_c decreases linearly and at 37-38 at.% W is no greater than 1°K [8, 9]. Data relating to the effect of the purity of the original metal on the critical temperature of Nb—W alloys [8] were presented by Hulm. On studying niobium containing 0.01% oxygen, 0.018% nitrogen, and less than 0.02% carbon (1 in Fig. 92) the T_c of the alloys was found to be considerably higher than on using metalloceramic niobium containing impurities of the order of tenths of a percent (2 in Fig. 92). At the same time metallic impurities, even ferromagnetics, had little effect on T_c if their concentration were no greater than a few tenths of a percent.

Chromium and molybdenum have a still worse effect on the superconducting transition temperature of vanadium than that of tungsten (Fig. 93) [8, 9, 20-23]; equally serious is the effect of tungsten on tantalum (Fig. 94)[8, 9, 24, 25]. In both cases T_c decreases sharply. In the V—Mo system the fall in T_c with increasing Mo corresponds to a fall in γ (Fig. 93), while θ_D increases from V to Mo [9b]. In the Ta—W system the alloys become non-superconducting above 1°K for a content of over 20% W. The sharpest fall in γ occurs in the Ta—W system in W-base alloys (Fig. 95) [9b]. In contrast to the Nb—Mo system, alloys of the Nb—W and V—W systems have never been studied at temperatures below 1°K. By invoking the foregoing laws based on the physicochemical analysis of superconducting systems, we may neverthe-

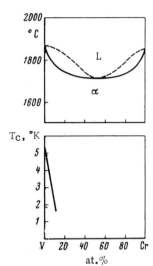

Fig. 93a. Phase diagram and superconducting transition temperature of V−Cr alloys.

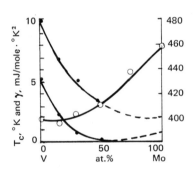

Fig. 93b. Transition temperature (T_c), electron specific heat (γ), and Debye temperature (θ) as functions of composition in the V−Mo system.

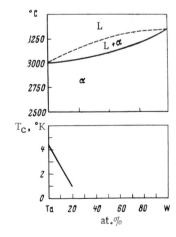

Fig. 94. Phase diagram and T_C of Ta−W alloys.

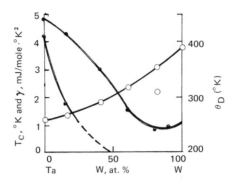

Fig. 95. Transition temperature (T_c), electron specific heat (γ), and Debye temperature (θ) as functions of composition in the Ta-W system.

less assert with complete confidence that superconductivity will exist in these systems at all concentrations, although only at extremely low temperatures.

In all the cases which we have considered in systems with continuous solubility in the liquid and solid states, T_c decreases on adding a second component with a lower transition temperature, this effect being the greater, the lower the transition temperature in question; this indicates that a fundamental law governing the changes in T_c with composition in systems of the type under consideration is that of additivity. However, the quantitative values of T_c for these systems require further refinement over the whole concentration range.

2. Systems with Unlimited Solubility and a Polymorphic Transformation of the Components

A considerable number of superconducting systems with continuous solubility in the liquid and solid states are known in which the solid solution decomposes at lower temperatures as a result of the polymorphism of one or both of the components. Such systems are formed by the alloying of group VA and VIA elements with polymorphic metals of group IVA, or by the alloying of group IVA elements with each other. Alloys of niobium with titanium and zirconium have had their superconducting properties studied most fully.

Many alloys of niobium with zirconium and titanium possess advanced superconducting characteristics. It should of course be remembered that, owing to the formation of a number of metastable phases during the decomposition of the solid solution and difficulties arising in connection with establishing a state of equilibrium, the diagrams representing these characteristics cannot yet be regarded as final. The great affinity of these refractory transition metals and their alloys toward interstitial impurities (the concentrations of which fundamentally influence solid-state diffusion processes and the boundaries of the phase regions) also makes it more difficult to secure comparative data relating to the structure and properties of these alloys, including the superconducting versions.

Figure 96 shows the phase diagram of the Nb-Ti system of alloys as well as their transition temperature and critical magnetic field. The Nb-Ti phase diagram has been studied by a number of authors [26-31]. According to the diagram illustrated in Fig. 96a, this system is characterized by complete solubility in the solid and

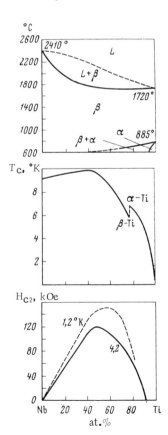

Fig. 96. Phase diagram, superconducting transition temperature, and critical magnetic field of Nb-Ti alloys.

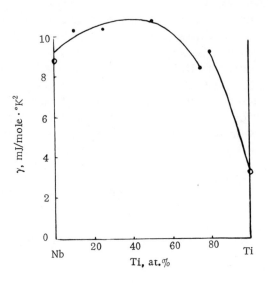
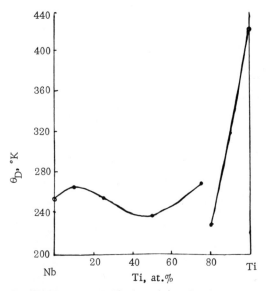

Fig. 97. Electron specific heat (γ) and Debye temperature (θ) as functions of composition in the Nb-Ti system.

liquid states; on reducing the temperature of the alloys below the polymorphic transformation point of titanium, however, with a niobium content of about 36 at.%, decomposition of the solid solution sets in. The boundary of the two-phase $\alpha + \beta$ region thus formed has not been established precisely below 40 at.%. According to certain data [28], a compound of the NbTi type is formed in this system, while the boundary of the two-phase $\alpha + \beta$ region extends only as far as 12 at.% Nb. The first version of the Nb−Ti phase diagram is supported by results derived from materials of much greater purity [30], however, and this accordingly appears more likely to be correct. Other work carried out with titanium-rich alloys [32] indicated that the decomposition of the β solid solution resulted in the formation of several metastable low-temperature phases, α', α'', ω.

Figure 96b shows the results obtained by measuring the critical temperatures of alloys belonging to this system after homogenization and quenching from 1000°C [8, 33]. X-ray analysis showed that all the alloys containing up to 80 at.% Ti had the structure of the β solid solution with a bcc lattice after this form of heat treatment [32]. We ourselves noted that in cast alloys of the same compositions melted in an electric-arc furnace with a water-cooled copper sole quenching of the β solid solution took place. The T_c-composition curve of these alloys is continuous up to 80 at.% Ti and has a flat maximum (about 10.2°K) for a content of 40-50 at.% Ti. Figure 97 represents certain data [32a] relating to changes in γ and θ_D for the Nb−Ti system; the γ-composition relationship is similar in form to the T_c-composition curve. For over 80% Ti the second phase (α) separates out, this being accompanied by a change in the behavior of the T_c-, γ-, and θ_D-composition curves. The effect of different phase compositions of the titanium-rich alloys on their transition temperature was studied by Hulm et al. [8]. In the case of the quenching of the alloys, after the bend on the T_c-composition curve (at 80% Ti), the transition temperature decreases sharply on further reducing the niobium content, and for zero niobium extrapolates to values agreeing with the T_c of titanium (0.49°K). After slow cooling from 1000°C the transition temperature of titanium-rich alloys decreases much less with increasing titanium content, and the alloy containing 99% Ti has a T_c of about 6°K instead of the 1°K for the same alloy in the quenched state. This effect may be attributed to the decomposition of the β solid solution as a result of slow cooling, the precipitation of the titanium-rich α phase (with $T_c < 1°K$), and the enrichment of the β solid solution with niobium. In alloys richer with respect to niobium the difference between the T_c values of the alloys subjected to the two kinds of heat treatment is much less, and indeed is nonexistent in the alloy containing 50% Nb. Hence no decomposition of the β solid solution occurs in the Nb-rich alloys.

The behavior of a Ti–25%-Nb alloy after quenching and aging at various temperatures was studied in more detail [34]. The solid-solution decomposition processes leading to the change in the phase composition of this alloy also had a major effect on its T_c value. After homogenization, quenching from 700°C, and cold working (in order to accelerate the decomposition processes in subsequent annealing) the transition temperature equalled 7°K. In this case the alloy had the structure of the β solid solution with a bcc lattice. Annealing at 500, 450, and 500°C led to the decomposition of the β solid solution (in accordance with the phase diagram) into two phases: $β_1$ (with a higher niobium content than the original β phase) and α (with an hcp lattice). This process was accompanied by an increase in the T_c of the alloy, which was determined by the chemical composition of the β phase (the α phase as indicated earlier had a transition temperature of under 1°K). Knowing the dependence of T_c on the composition of Nb–Ti alloys, we may to a certain accuracy estimate the chemical composition of the β phase. After annealing for 1 h at 400-500°C the amount of niobium in the β solid solution was about 30%, as compared with 25 at.% (39 wt.%) before annealing (Figs. 96-99).

The degree of equilibrium of the alloys also affected the value of T_c. On isothermally annealing the same alloys (1-100 h)

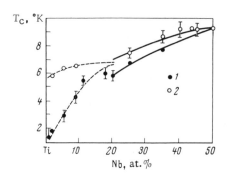

Fig. 98. Superconducting transition temperature as a function of the composition of Ti–Nb alloys: 1) after quenching from 1000°C; 2) after annealing.

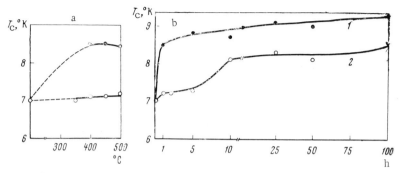

Fig. 99. Dependence of the superconducting transition temperature of Ti−25% Nb on the annealing temperature (a) and the period of annealing at 450°C (b); 1) worked samples; 2) recrystallized samples.

an increase occurred in T_c. After holding for 1 h, T_c reached 8.5°K, then it increased more slowly, and after reaching 100 h and thus achieving a higher degree of equilibrium the transition temperature was 9.4°K. Figure 99 shows the effect of annealing temperature and holding time on the T_c of worked and recrystallized samples of Ti−25% Nb. It is interesting to note that the tensile strength and specific electrical resistance of the same samples experienced their greatest change in roughly the same range of annealing temperatures (Fig. 100). Owing to the greater diffusion velocity and the faster decomposition of the β solid solution, these properties changed more rapidly at lower temperatures and for shorter annealing periods in the worked material.

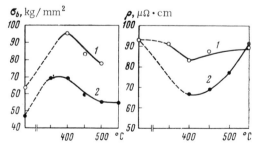

Fig. 100. Dependence of the tensile strength and electrical resistance of a Ti−25%-Nb alloy on the annealing temperature. 1) Worked samples; 2) recrystallized samples.

The critical magnetic field of the alloys of this system was measured at 1.2 and 4.2°K in a pulsed magnetic field by observing the restoration of the electrical resistance [35-37]. The measurements were carried out on cold-worked samples. The maximum critical magnetic field of Nb-Ti alloys at 1.2°K reached 150 kOe (60 at.% Ti), while averaged over alloys containing 30-70% Ti it equalled approximately 120 kOe; at 4.2°K the maximum critical field of about 120 kOe corresponded to an alloy containing 45-47 at.% Ti.

In the Nb-Zr system the superconducting transition temperature, the critical magnetic field, and the electron specific heat vary along curves containing a maximum. The dependence of the Debye temperature on the composition is characterized by the presence of a minimum in the middle of the diagram (Fig. 101) [9b]. In this case also the transition temperature depends on both the chemical and the phase composition of the alloys. According to a number of investigations [38-44] the phase diagram of the Nb-Zr system (Fig. 102) is characterized by a continuous series of solid solutions. On reducing the temperature the high-temperature β solid solution decomposes by a monotectoid reaction over a wide range of concentrations. Data relating to the decomposition temperature of the β solid solution, the temperature of the monotectoid transformation, and the extent of the two-phase regions differ for different authors. Thus the temperature of the monotectoid transforma-

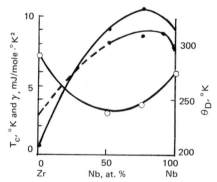

Fig. 101. Transition temperature (T_c), electron specific heat (γ), and Debye temperature (θ) as functions of composition in the Zr-Nb system.

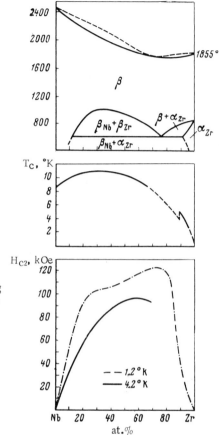

Fig. 102. Phase diagram, superconducting transition temperature, and critical magnetic field of Nb–Zr alloys.

tion varies between 800 and 560°C and the onset of the decomposition of the β solid solution between 1000 and 1180°C; there is also disagreement as to the monotectoid point and the solubility of the components in each other. In addition to the foregoing type of Nb–Zr phase diagram, the use of very pure materials leads to a phase diagram without any $\beta \rightarrow \beta_1 + \beta_2$ decomposition (Fig. 103) [45]; on the other hand, on using metals containing appreciable amounts of oxygen, the system develops a three-phase region typical of a three-component diagram, and the temperature of the monotectoid point is displaced [46].

In view of the serious effect of the impurity content of the original materials and the alloys prepared from these on the ulti-

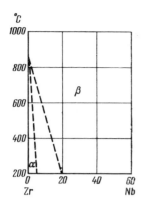

Fig. 103. Phase diagram of Nb-Zr alloys melted from pure original materials.

mate structure, a recent investigation into the kinetics of the decomposition of the β solid solution in this system for various oxygen contents and various grain sizes is of particular interest [47]. Special attention was paid in this investigation to the homogeneity of the samples, both as regards composition and as regards grain size, factors which had a considerable effect on the rate of decomposition of alloys of the same basic composition, as special tests showed. By applying a newly-developed electrolytic-etching technique to the alloys in question, changes in structure during the decomposition of the β solid solution were readily observed in the metallographic microscope. The microstructure of the alloys was studied after heat treatment (at 550-950°C) using both cold-worked (90% deformation) and previously-recrystallized samples; after recrystallization samples with extremely different grain sizes (from 4-8 to 80-180 μ and even 0.5-1 mm) were obtained.

After studying the microstructure and carrying out x-ray and electron-microscope analyses, Love et al. [47] concluded that since decomposition of the β solid solution occurred in every case, even when using extremely pure materials, the generally-accepted structure of the Nb–Zr phase diagram (Fig. 102) was correct; they concluded that the homogeneity and grain size of the annealed alloy had the greatest effect on the decomposition kinetics for the same thermal conditions. For a grain size of 5 μ in an alloy containing 33% Zr, decomposition started after holding for about 1.5 h, while for a grain size of 800 μ 100 h were required in order to initiate decomposition. The temperature required to initiate the reaction, however, changed little, remaining at about 720-750°C. Figure 104 illustrates these results; the curves on the graph re-

Fig. 104. Onset of the decomposition of the solid solution in a Nb-33%-Zr alloy with various grain sizes in relation to the annealing time and temperature.

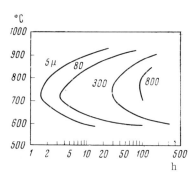

lating to different grain sizes indicate the holding time and temperature required to start the decomposition process in each alloy (to produce 5% decomposition). The kinetics of the $\beta \rightarrow \beta_1 + \beta_2$ decomposition in a fine-grained alloy (grain size 3 μ after recrystallization) with 33% Zr and a low oxygen content (under 0.003 wt.%) are illustrated in Fig. 105. The curves plotted for samples in which the transformation in the solid state had proceeded to the extent of 5, 50, and 95%, respectively, exhibit "projections" at 700°C. This means that the most favorable annealing temperature in the present case is 700°C. The graph shows the times required to initiate and complete the decomposition of the β solid solution at various temperatures (this time is shortest for a temperature of 700°C: from 5 to approximately 150 h). In alloys containing 25% Nb the results of similar measurements differ very little, although the process takes place rather more rapidly (which we had hardly expected).

The annealing of a cold-worked alloy containing 33% Zr of the same purity but without preliminary recrystallization accelerates the decomposition of the solid solution, which begins after

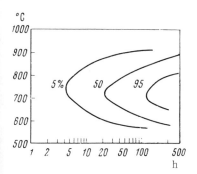

Fig. 105. Degree of decomposition of the β solid solution in a recrystallized fine-grained Nb-33%-Zr alloy in relation to the annealing time and temperature.

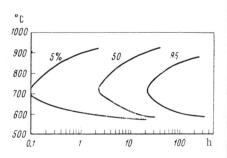

Fig. 106. Effect of annealing time and temperature on the degree of decomposition of the β solid solution in a cold-worked (90% deformation) Nb−33%-Zr alloy.

holding for 10 min, i.e., some 30 times more rapidly; however, 700°C is still the best temperature (Fig. 106).

The effect of various oxygen contents on the time–temperature-transformation curves was also studied; it was found that an oxygen content of 0.02 wt.% (as compared with an oxygen content of 0.003% or under in the "pure" alloys) had no marked effect on the $\beta \rightarrow \beta_1 + \beta_2$ decomposition processes.

A microstructural analysis of the alloys after annealing showed that in the alloys cold-worked and recrystallized to produce a fine grain a pearlitic type of decomposition occurred, this appearing in particular along the grain boundaries.

An earlier review [48] presents data illustrating the substantial effects of interstitial impurities in Nb–Zr alloys on the T_c, H_c, and I_c values. For high oxygen contents in a Nb–25-at.%-Zr alloy, in which the limiting oxygen solubility is under 0.01 wt.% between 1000 and 1200°C, a second phase, identified as α zirconium saturated with oxygen, appears.

The effect of phase composition on the T_c of Nb–Zr alloys (25 and 50% Zr) was also studied elsewhere [49]. The alloys were melted in a vacuum arc furnace using electron-beam-melted niobium and zirconium refined by the iodide process, and contained some 0.02% of oxygen, 0.001% of hydrogen, and 0.02% of carbon. The T_c–composition curve has a maximum at 25-27% Zr (Fig. 102). If the composition of the original alloy with the structure of the β solid solution is similar to the compositions of those alloys of this system having the highest transition temperature (25-27% Zr), then as a rule the decomposition processes tend to reduce T_c because the composition of the matrix deviates from the optimum. In certain cases, after annealing quenched samples of such alloys at temperatures close to the onset of the monotectoid decomposition (900-

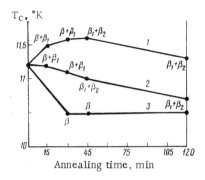

Fig. 107. Superconducting transition temperature and phase composition of a quenched Nb−25%-Zr alloy as a function of annealing time at 900°C (1), 800°C (2), and 600°C (3).

950°C) there may at first be a slight increase in T_c (as compared with its original value), and then, as the alloy develops an equilibrium state, T_c decreases again slightly. The rise in the critical temperature in this case may be attributed to the development of internal stresses as phases differing in chemical composition from the original alloy precipitate from the solid solution. An increase in the annealing time leads to a reduction in these stresses, and T_c resumes a value close to the original. Annealing at lower temperatures reduces T_c, since the composition of the precipitating phases differs more from the original, the more so the nearer the sample is to a state of equilibrium at the temperature in question. A similar effect occurs (to a greater degree) at still lower annealing temperatures (600°C), although in this case the precipitation of the second phase cannot be detected by the x-ray method, as the sample has not yet reached equilibrium. Figure 107 shows the effect of annealing time on the T_c of a Nb−25%-Zr alloy and the phase composition of the corresponding sample. A longer holding period at 700°C (up to 100 h) still further reduces the transition temperature (Fig. 108) of an Nb−25%-Zr alloy, which falls to 10.2 as compared

Fig. 108. T_c and phase composition of a quenched Nb−25%-Zr alloy as a function of annealing time at 700°C.

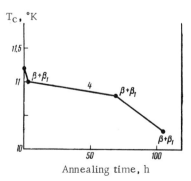

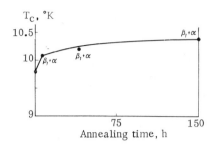

Fig. 109. T_c and phase composition of a quenched Nb–50%-Zr alloy as a function of annealing time at 500°C.

with 11.2°K before annealing. In this case a state of equilibrium was never reached, judging from the phase composition of the sample. Annealing an Nb–50%-Zr alloy, which after quenching had the structure of the β solid solution and a superconducting transition temperature of 10.2°K, led to a rise in T_c: After holding for 100 h at 500°C, T_c rose to about 10.7°K (Fig. 109).

The effect of heat treatment on the transition temperature of Nb–Zr alloys with a high Zr content was studied in another investigation [50]. After annealing a quenched and cold-worked alloy containing 20% Nb at 450°C, the transition temperature rose from 8.2 to 8.7°K. The structure of the alloy also changed in this process; the Zr-rich β solid solution decomposed into β + ω (the transition temperature of the ω phase was about 0.6°K, it was formed in the initial stage of decomposition). The alloy containing 25% Nb had the highest transition temperature (10.5°K) after annealing at 800°C; the structure comprised a mixture of α and β solid solutions. In both cases the increase in the transition temperature of the heat-treated alloys was also associated with the fact that the phases precipitating during the decomposition of the solid solution became rich in niobium, their composition moving in the direction of the optimum niobium content.

In another paper concerned with an alloy containing a still greater proportion of zirconium (4% Nb) [51] it was found that only after annealing a quenched sample of this composition and thus forming a β phase by virtue of a decomposition process expressing itself in the form $\alpha' \to \alpha + \alpha' \to \alpha + \beta$ did any superconductivity appear at temperatures above 4.2°K.

The foregoing information relating to the effect of the chemical and phase compositions on the transition temperature demonstrates the necessity of studying the structure and properties of

Nb–Zr alloys in much greater detail, particularly over the concentration range corresponding to the most promising compositions (20-80% Zr), using materials of semiconductor purity (millionth parts of impurities). A further study of the effects of impurities should enable us to establish the phase structure of this system more precisely as well as to determine the effect of impurity content on the structure and superconducting properties of the alloys.

The critical magnetic field of Nb–Zr alloys at 1.2°K, according to [35], reaches its maximum at 20-30 at.% Nb, while for the remaining alloys in the middle of the system H_{c2} is about 100 kOe.

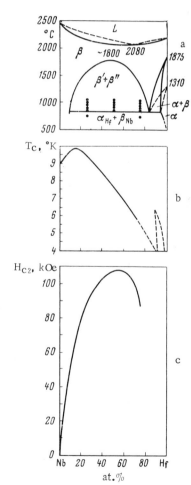

Fig. 110. Phase diagram, superconducting transition temperature, and critical magnetic field of Nb–Hf alloys.

At 4.2°K for alloys of this system $H_{c2} \leq 100$ kOe, and the maximum on the curve moves in the direction of higher niobium contents (45-55% Nb). On passing a measuring current of 10 A/cm^2 density through the samples, in the case of the alloys with the highest transition temperature (25-27 at.% Zr) the H_{c2} at 1.2°K was about 90 kOe, while at 4.2°K it decreased to 60-70 kOe.

The superconducting properties of the Nb–Hf system have been studied much less than those of the Nb–Zr or Nb–Ti systems. The structure of the Nb–Hf diagram has only recently been studied [52, 53], and the results of the two papers in question disagree. Figure 110a shows the phase diagram indicated in the first paper. The structure is similar to that of the Nb–Zr diagram. Here also we have a decomposition of the β solid solution with the bcc lattice, based on a monotectoid reaction over a wide temperature range. Owing to the fact that the polymorphic transformation of hafnium takes place at a higher temperature than that of zirconium, the temperature at the onset of the monotectoid reaction is much higher in the present case. In the second paper [53] it was suggested that the decomposition of the β solid solution in the system took place in accordance with a reaction of the $\beta \rightarrow \beta_2 + \alpha$ type, and the form of the phase diagram was similar to that of the Nb–Ti system. However, later investigations [54] confirmed the existence of the monotectoid reaction over a wide range of concentrations, and the phase diagram of Fig. 110a therefore appeared more probable. A study of a number of alloys belonging to this system showed that on alloying niobium with hafnium the critical temperature rose to 10°K for a hafnium content of 20%, while for 75% Hf it was 6°K [53, 54]. The critical magnetic field in alloys of this system is 100 kOe at 1.2°K [55] (there are no data relating to H_{c2} at 4.2°K). We may therefore conclude that both the T_c and the H_{c2} of alloys belonging to this system are lower than those of Nb–Zr alloys; they are also inferior to alloys of the Nb–Ti system as regards the critical magnetic field. However, we must still remember that the properties of these alloys have not yet been studied adequately.

Figures 110b and c give the values of T_c and H_{c2} of the alloys derived from two of the foregoing papers [54, 55]. A study of the effect of changes taking place in the phase composition after heat treatment on the T_c of niobium alloys containing 25, 50, and 75% Hf showed that this characteristic of the alloys was sensitive to the presence of different phases. We see from the results sum-

TABLE 29. Critical Temperature of Niobium Alloys with 25, 50, and 75% Hafnium as a Function of Annealing Temperature

Composition, Hf, at.%	Treatment	Transition temperature, °K
25	Cold work (99.8%)	9.6
50	Ditto	7.2
75	Ditto	6.1
25	Cold work (99.8%) and annealing at 750°C	9.7
25	Ditto, at 850°C	9.2
25	Ditto, at 900°C	9.1-7.5
25	Ditto, at 950°C	9.3
25	Ditto, at 1000°C	9.3-8.2
50	Cold work (99.8%) and annealing at 750°C	7.4
50	Ditto, at 850°C	7.5
50	Ditto, at 900°C	7.5
50	Ditto, at 1000°C	7.5
50	Ditto, at 1050°C	8.6-7.5
75	Cold work (99.8%) and annealing at 750°C	7.4
75	Ditto, at 850°C	7.5
75	Ditto, at 900°C	7.1-6.3

marized in Table 29 that there is a considerable change in T_c after annealing alloys containing 50 and 75% Hf, quenched to the β phase. On annealing between 750 and 1050°C, T_c increases in every case. The greatest increment occurs after annealing the alloy containing 50% Hf at 1060°C (from 7.2 to 8.6°K) and that containing 75% Hf at 450° (from 6.1 to 7.5°K). X-ray analysis of the alloys showed that the structure consisted of two phases. In the presence of a fair amount of the second phase, its presence is indicated by an additional step on the T_c curves (second values of T_c given in Table 29).

In the Nb–U system there is a polymorphic transformation of the uranium at about 770°C and the solid solution accordingly also decomposes by a monotectoid reaction. The structure of this phase diagram differs from those of the analogous systems (Nb–Zr, Nb–Hf) with respect to the wide range of crystallization of the alloys and the considerable solubility of uranium in niobium, which is also retained on further cooling the alloys (Fig. 111) [56]. There is little information regarding the superconducting properties of

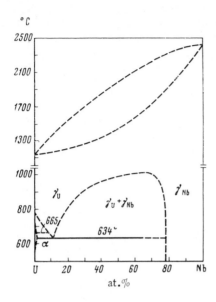

Fig. 111. Phase diagram of Nb–U alloys.

these alloys. It is only known that alloys with the structure of the high-temperature U-base solid solution (74-77 at.% U) have a transition temperature of 1.9-1.8°K and a critical field of 25 kOe [57].

The phase diagram of the V–Ti system is analogous to the Nb–Ti diagram. The system also contains continuous series of liquid and solid solutions (the latter between vanadium and β-Ti). Below the temperature of the polymorphic transformation of titanium (885°C) the β solid solution with a high proportion of titanium decomposes in accordance with a reaction of the $\beta \rightarrow \beta_1 + \alpha$ type similar to the reaction in the Nb–Ti system. The results of different authors relating to the boundaries of the ($\alpha + \beta$) region and the solubility of vanadium in α-Ti fail to agree. The boundaries of these regions have not been established precisely below 600°C.
The temperature and concentration ranges of the liquidus and solidus minima forming in this system also differ slightly. Figure 112 shows the structure of this system according to the latest determinations, due allowance being made for earlier work [58]. The solid solutions of vanadium with titanium (bcc lattice) have a fairly high superconducting transition temperature. The critical temperature of the transition for vanadium is 5.3°K and for β-Ti 4.6°K; the critical temperature of α-Ti is much lower: 0.49°K. When titanium dissolves in vanadium there is an increase in the critical tempera-

ture in the middle of the system; according to [59] the change in the T_c of vanadium with changing concentration (dT_c/dC) equals ±0.8. The T_c-composition curve has a maximum at 35-45 at.% Ti [8, 59-61]. The critical magnetic field also passes through a maximum and reaches 110 kOe for alloys containing 50% Ti at 1.2°K [11, 35, 62]; at 1.2°K the alloy containing 18% Ti has an H_{c2} of 60 kOe, and at 4.2°K 50 kOe. The effect of phase composition on the T_c of V-Ti alloys has hardly been studied at all; it would appear likely, however, that it will be analogous to that found in the Nb-Ti system.

There is little information as to the superconductivity of the Fe-V system, which also forms continuous series of solid solutions in the liquid and solid (above 1234°C) states as a result of the alloying of vanadium with this polymorphic ferromagnetic nonsuperconductor (iron) [58]. At 50% Fe and 1234°C an intermediate phase

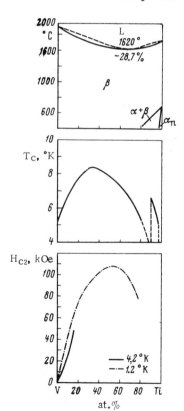

Fig. 112. Phase diagram, superconducting transition temperature, and critical magnetic field of V-Ti alloys.

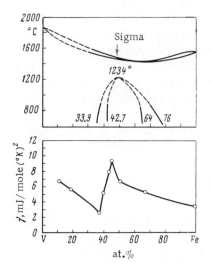

Fig. 113. Phase diagram of the V–Fe alloys, together with their electron specific heat.

isomorphic with the σ-phase (of the FeCr type) is formed. The boundaries of this phase have not been established below 700°C (Fig. 113). In two papers [63, 64] it was indicated that the T_c of alloys containing 1 and 15% of Fe respectively equalled 2.8 and 2.3°K. In the solid solution with the composition of the metallic compound FeV no superconductivity was observed at 0.3°K [64].

In the Ta–Ti system a continuous series of solid and liquid solutions is also formed between tantalum and β-Ti, while below the temperature of the polymorphic transformation of titanium the β solid solution decomposes in alloys containing about 50 at.% Ti [65, 66]. The transition temperature of these alloys was studied by several authors [8, 37, 67]. The T_c of alloys containing 45-50 at.% Ti is considerably higher than that of the original components; the maximum value (in the 50 at.% alloy) reaches 8.5-8.8°K. The transition temperature of these alloys and its sensitivity to working and annealing were studied in greatest detail by Blaugher et al. [68]. Figure 114 shows the phase diagrams of the alloys and their transition temperature for up to 70 at.% Ti. Before measuring the T_c value, the alloys (prepared from 99.96% Ta and iodide-type titanium) were cold-worked after a homogenizing anneal at 1000°C in a vacuum of 10^{-5} mm Hg. The alloys containing 31.5-53.8 at.% Ti were subjected to a subsequent anneal at 1200°C, as the worked samples exhibited a wide temperate range of the superconducting transformation (from 0.2 to 1.2°K). After annealing the transition became sharper in every case, while at the same time there was a slight decrease in T_c (by 0.25-0.75°K). The effect of annealing temperature on T_c was studied in [69] in a 50% Ti alloy containing a rather greater proportion of oxygen (0.25 wt.%). For annealing temperatures of 500-600°C there was a slight decrease in T_c, due to the precipitation of the α-phase from the β-solid solution (detected by x-ray analysis). A further reduction in T_c after heat treatment at temperatures above the boundary of the $\beta \to \beta_1 + \alpha$

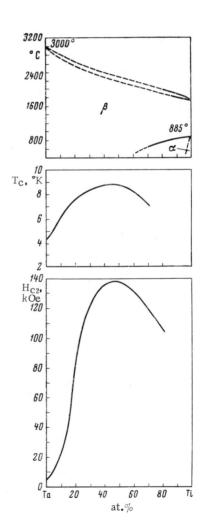

Fig. 114. Phase diagram, superconducting transition temperature, and critical magnetic field of the Ta–Ti system.

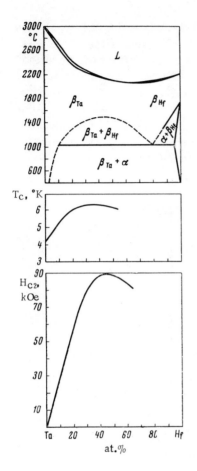

Fig. 115. Phase diagram, superconducting transition temperature, and critical magnetic field of the Ta-Hf system.

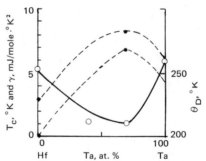

Fig. 116. Transition temperature (T_C), electron specific heat (γ), and Debye temperature (Θ) as functions of composition in the Hf-Ta system.

decomposition was also attributed to the decomposition of the β solid solution as a result of the stabilization of the hexagonal α phase in the presence of oxygen. The transition temperature of the cold-worked 50 at.% Ti alloy heat-treated at 350-1200°C was higher than that of an alloy of similar composition containing less oxygen (9.6 instead of 8.8°K); the width of the transition was nevertheless much greater in the former case (1.6°K). The critical magnetic field was measured for a larger number of alloys than T_c (see Fig. 114). The H_{c2} of the 50%-Ti alloys at 1.2°K was about 140 kOe [8, 70]; at 4.2°K it was much lower (85 kOe for an alloy containing 30% Ti as against 100 kOe at 1.2°K); in both cases however the critical magnetic field of the alloys was quite high.

In the Ta–Hf system continuous series of liquid and solid solutions (between tantalum and β-Hf) are also formed. Below the

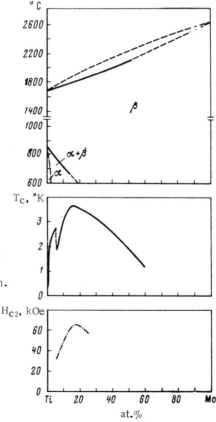

Fig. 117. Phase diagram, superconducting transition temperature, and critical magnetic field (at 1.2°K) of the Mo–Ti system.

polymorphic transformation of hafnium the β solid solution with the bcc lattice decomposes [71, 72]. The critical temperature and the critical magnetic field of Ta–Hf alloys are lower than those of Ta–Ti alloys. The T_c value reaches a maximum (about 6.3°K) for alloys with 40-50% Hf. The critical magnetic field is also lower, and at 1.2°K it is no greater than 90 kOe (40-50% Hf). The phase diagram and all available T_c and H_{c2} data for this system appear in Fig. 115 [8, 9]. Figure 116 represents data [9b] relating to the Debye temperature and the electron specific heat; the maximum T_c in this system corresponds to a maximum γ and a minimum θ_D.

In the Ti–Mo system the β solid solution decomposes for alloys containing up to 20 at.% Mo (Fig. 117) [73], and the transition temperature is much lower, although it rises on adding molybdenum to titanium. The T_c of the Mo-rich alloys has never been measured but would appear to be low. Thus even in an alloy containing 60% Mo the T_c is no greater than 1°K [74, 75]. The critical magnetic field H_{c2} at 1.2°K has been measured for alloys containing 10-25% Mo; it reaches 65 kOe for the 20 at.% Mo alloy [11, 37, 55].

So far we have been considering the superconducting properties of binary alloys involving one polymorphic metal, forming continuous solid solutions of the high-temperature form of this component. Superconductivity also appears in alloys of systems formed by two polymorphic metals in which continuous series of solid solutions are also formed. These systems include alloys of polymorphic metals situated in the same subgroup of the periodic table (IVA), and in all these cases continuous mutual solubility exists between the isomorphic low- and high-temperatures forms of the metals in question.

The phase diagram of the Ti–Zr system is shown in Fig. 118 [76]. The transition temperature of the alloys, the Debye temperature (θ_D), and the electron specific heat (γ) were taken from other papers [9b, 77]. A detailed study of the properties of these alloys revealed possible effects of changing composition on their characteristics, even without taking account of the corresponding changes in the number of valence electrons per atom.

The critical temperature of these alloys passes through a maximum in the same way as that of alloys containing only one polymorphic component. The form of the T_c and γ curves was exactly the same, the latter being proportional to the density of

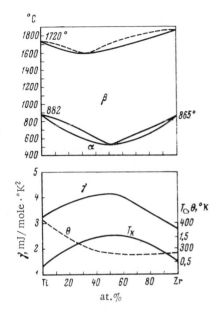

Fig. 118. Phase diagram, superconducting transition temperature, Debye temperature, and electron specific heat of the Ti-Zr system.

states at the Fermi surface, $N(0)$. The values of γ and T_c were used to calculate the electron–electron interaction parameter V; it was found that in this case V increased linearly from titanium to zirconium. An interesting point here was that heat treatment affected T_c but not γ.

The Ti-Hf system also forms two series of continuous solid solutions of the substitutional type between the α and β forms, separated by a $\alpha + \beta$ two-phase region [78]. There is little information as to the occurrence of superconductivity in this system. In one of the alloys of the Zr-Hf system, which has an analogous structure, superconductivity occurs at a very low temperature (0.37°K in the alloy containing 99% Hf) [79]. We may expect superconductivity to occur in both the latter two systems at lower temperatures if materials of high purity are employed. The T_c of zirconium containing 5 at.% of scandium equals ~ 0.06°K [79a].

It should be noted that the effect of different phase compositions, the purity of the materials involved, and the methods of preparation on the superconducting characteristics was hardly studied at all in many of the investigations here cited; more careful research into alloys of the foregoing systems may therefore result in data appreciably differing from that presented.

3. Systems of the Eutectic, Peritectic, and Monotectic Types

Simple eutectic superconducting systems are chiefly formed by nontransition metals. Only two eutectic systems formed by refractory superconducting metals of the transition type are known: Nb–Th and V–Th. The superconductivity of alloys belonging to the latter system has never been studied.

Despite the fact that superconductivity was observed in a large number of alloys of nontransition metals (Pb, Zn, Sn, Cd, Bi, Tl) considerably earlier than in alloys of transition metals, there is only a small number of such systems in which the superconducting characteristics have been studied over a wide concentration range. This is clearly due to the low level of the superconducting properties (T_c and H_c) of these alloys, so that they are less likely to find a practical use than alloys of the transition metals. Published data are sparse and require verification.

It is known from existing data relating to the superconducting properties of alloys containing elements of group IA forming eutectic systems (Cu–Pb and Ag–Pb) that on adding copper or silver to lead the transition temperature of the latter either remains quite constant up to the eutectic composition (Pb +4.7 at.% Ag) or even rises slightly (by 0.5°K) (Pb + 0.18 at.% Cu) [80, 81]. The addition of up to 10% of tin to lead reduces the T_c value to about 3.7°K; the critical magnetic field thereupon rises (from 600 Oe for lead to 1000 Oe for 14 at.% Sn) [82]. Data relating to the magnetic properties of a Pb–7-wt.%-Sn alloys were presented later [83]. The critical magnetic field of this alloy after quenching from the region of the solid solution, and also after annealing, equalled 900 Oe. The different heat treatment only affected the shape of the magnetization curve and the hysteresis. At the same time, the addition of about 10% indium to a Pb–Sn alloy of a similar composition to that just mentioned produced a substantial rise in the critical magnetic field; in such an alloy H_{c2} = 2600 Oe.

In the Pb–As system the transition temperature of alloys of hypoeutectic composition at least (7.4 at.% As) is more than a degree higher than that of lead (8.4 instead of 7.2°K) and remains constant over the range 3-7.4 at.% As [84]. At the same time, alloys of lead with antimony, which in contrast to arsenic constitutes a superconductor itself at temperatures below 2.6°K, there is a decrease in the transition temperature as compared with that of lead. Although superconductivity appears in this case for a large proportion of the second component, in the alloy of eutectic composition (17.5 at.% Sb) T_c decreases to 6.6°K.

On alloying cadmium with thallium, the transition temperatures of the eutectic alloy (27 at.% Cd) and an alloy of similar composition (28 at.% Cd) differ very little (2.3 and 2.5°K) [85, 86]. The same occurs for Cd–Sn alloys; for the eutectic alloy (33.5 at.% Cd) T_c equals 3.6°K, while for alloys with 0–1% Cd it equals 3.7°K [80, 87]. In the Bi–Sn system the transition temperature of the eutectic alloy (43% Bi) is 3.8°K, and in alloy containing up to 1% Bi it equals at most 3.7°K; the maximum T_c in this system is about 4.18°K [88].

In the Nb–Th system superconductivity has only been studied for the 5%-Nb alloy [89]. In the cast state this alloy exhibits no superconductivity at 4.2°K, although a study of the microstructure reveals discontinuous inclusions of the eutectic and niobium. Only after working and the formation of a network of superconducting inclusions does superconductivity (indicated by the vanishing of the electrical resistance) appear at 4.2°K. The critical magnetic field increases sharply at the same time, this being at least forty times

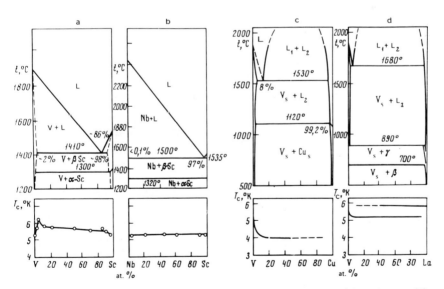

Fig. 119. Phase diagram and T_c of the V–Sc (a), Nb–Sc (b), V–Cu (c), and V–La (d) systems.

greater than that of pure niobium and at least ten times greater than that of worked niobium of commercial purity. However, Nb-Th alloys have not justified the great hopes placed in them at one time as materials with high superconducting characteristics, owing to their instability and the wide scatter in their superconducting properties.

In the simple eutectic V-Sc system (Fig. 119a) there is a rise in the T_c of vanadium to 6.2°K as scandium dissolves in it [89a, 89b]. In the two-phase region the T_c of the V-base solid solution varies little. In the Nb-Sb system (Fig. 119b) solubility of the components hardly exists at all [89a, 89b]. The T_c of niobium remains constant in eutectic alloys with scandium.

There are also two binary systems of a monotectic type in which one of the components is a superconducting transition metal (niobium or vanadium). Such systems are formed on alloying these metals with copper. The addition of up to 2.7% copper to vanadium reduces the transition temperature from 5.3 to 3.9°K [21]. According to our own data, the T_c of a vanadium solid solution completely saturated with copper (5% Cu [90]) is 3.9°K (Fig. 119c) [89b]. In two-phase systems the T_c of the V solid solution remains practically constant as long as the proportion of the superconducting vanadium phase is sufficient to reveal superconductivity; the actual limit depends on the measuring technique and the sensitivity of the measuring apparatus. The same argument applies to Nb-Cu alloys, the structure of these being analogous to that of the V-rich samples in question [18].

An analogous picture appears in the monotectic V-La system (Fig. 119d) [89b]. The T_c of vanadium falls slightly as La dissolves in it. In two-phase alloys the T_c of V and La remain constant.

In alloys belonging to superconducting systems of the eutectic type superconductivity is observed when alloying either two superconductors or else a superconductor and a "normal" element. In the latter case superconductivity may not appear in alloys with a predominance of the nonsuperconducting component, particularly if the superconducting phase is precipitated in the form of individual inclusions and T_c is measured by the resistance method. This may also apply to superconducting systems of the monotectic type. The transition temperatures of alloys belonging to a eutectic system usually lie below that of the superconducting component with the larger T_c; it varies little with changing chemical composition in alloy mixtures. However, certain results indicate a rise in the critical magnetic field under these conditions. It is important that further reliable experimental data should be secured in relation to the superconducting characteristics of alloys belonging to eutectic systems over the whole range of concentrations.

Let us consider some more-complicated superconducting systems in which various intermediate phases with various ranges of homogeneity are formed in addition to eutectic and peritectic mixtures and limited solid solutions based on the original components.

Complex intermetallic systems of this kind may be considered as consisting of a series of simpler systems. The superconducting properties of the complex systems should vary in accordance with the laws of the simple system in particular regions.

The Pb–Bi system (Fig. 120) is formed by two nontransition metals; in addition to the eutectic mixture an intermediate ω phase (22-32 at.%) is formed over a wide concentration range (40-99 at.% Bi).

We measured the superconducting and electrical properties of the Pb–Bi and In–Pb systems on samples of variable composition (Fig. 120b). The transition temperature of lead increases with increasing Bi content up to 7.8°K at the boundary of the solid solution (the T_c of lead equals 7.2°K). Within the limits of the homogeneous ω phase an increase in Bi content leads to an increase in T_c from 8.2 to 8.6°K. On alloying lead with bismuth the critical magnetic field increases from 600 to 2800 Oe; the H_{c2} of the eutectic alloy consisting of a mixture of the ω phase and a Bi–base solid solution reaches 15 kOe [56, 82, 91]. The addition of antimony to lead–bismuth alloys (1:1:1) increases T_c by another 0.1°K (to 8.9°K).

The temperature of the superconducting transition of amorphous phases arising from the quenching of various Pb–Bi condensates was studied in [91a].

Examples of superconducting systems of the peritectic type having intermediate phases present include Pb–In, Pb–Tl, and Pb–Hg. In the Pb–In system (Fig. 121) there are two peritectic horizontals and an intermediate phase of variable composition, separated by two two-phase regions from the ranges of Pb– and In–base solid solutions.

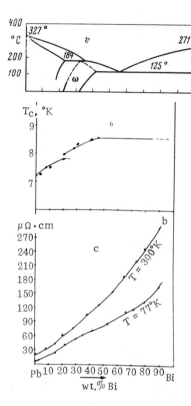

Fig. 120. Phase diagram (a), T_c (b), resistivity (c), and critical magnetic field (d) of the Pb-Bi system.

In the Pb–In system, in the region of the solid solution of In in Pb, T_c decreases to 6.6°K at the saturation point of the solid solution (Fig. 121b) (the measurements were made on a sample of variable composition). In the range of homogeneity of the compound (α_1 phase), any increase in In content lowers T_c, from 6.05°K (55 wt.% In) to 4.65°K (79 wt.% In). No maximum T_c occurs for any composition of this variable-composition phase. In the two-phase region containing the α_1 phase and the Pb-base β solid solution there are two transitions: at 6.6°K (β) and at 6.05°K (α_1).

Solid solutions of lead in bismuth and indium are not superconducting at 4.2°K.

In plotting the composition–ρ diagrams for variable-composition samples of the Pb–In (Fig. 120c) and Pb–Bi (Fig. 121c, d) systems, slight bends appear on passing from the one- to the two-phase regions. The critical magnetic field measured in alloys containing up to 30 at.% In, i.e., within the range of the Pb-base solid solution, increases by a factor of six (to 3750 Oe) as compared with unalloyed lead. In the In-base alloys H_{c2} is low; for 4-6 at.% Pb it never exceeds 100-200 Oe [91, 85].

The same authors studied the superconducting properties of Pb-Tl alloys over a wide concentration range. This system includes a wide range of Pb-base solid solutions, and there is a break in solubility between this region and the region of the Tl-base solid

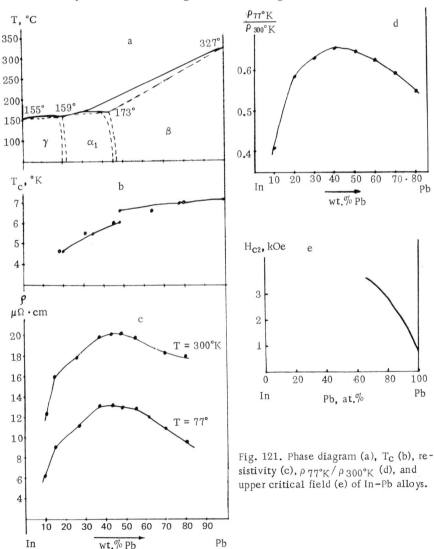

Fig. 121. Phase diagram (a), T_c (b), resistivity (c), $\rho_{77°K}/\rho_{300°K}$ (d), and upper critical field (e) of In-Pb alloys.

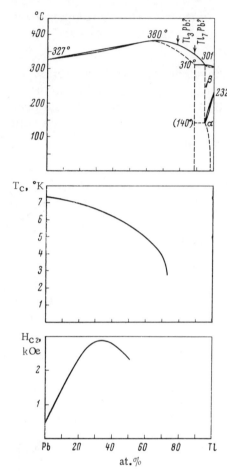

Fig. 122. Phase diagram, superconducting transition temperature, and critical magnetic field of Pb-Tl alloys.

solutions, due to the presence of isomorphism in the thallium. The possibility of ordering processes taking place is suggested by the existence of a maximum on the liquidus curve and the formation of the compounds Tl_3Pb and Tl_7Pb. The boundaries of the phase regions on the thallium side have not been exactly established (Fig. 122). The transition temperature of the alloys decreases from 7.2°K (for lead) to 2.5°K on increasing the thallium content to 75 at.%. Measurements of the critical magnetic field of the alloys containing up to 50 at.% Tl showed that the H_{c2} curve had a maximum (2800 Oe) within the range of the solid solution at about 30 at.% Tl. The dependence of T_c on the composition of Bi-Tl alloys was studied in [92a] in addition to the T_c of the Pb-Tl system.

Alloys of lead with mercury have been studied less; this system is of the peritectic type and also contains a phase of variable composition ($HgPb_2$); on alloying lead with mercury T_c decreases (to 5°K for 30% Hg) while H_{c2} rises sharply (to 4300 Oe for 10% Hg) [82, 85].

The effect of a change in Sn content in the intermediate phase $HgSn_6$ (87-94 at.% Sn) on the transition temperature was studied in [91b]: The T_c of this phase rises from 4.8 to 5.1°K on changing the ratio of the number of valence atoms per atom from 3.74 to 3.88.

The superconducting transition temperature of Cd–Hg alloys, which form an analogous system, has been studied in fair detail [92]. In addition to a determination of T_c over the whole range of concentrations, the effect of ordering on the transition point was studied. The phase boundaries of the system at low temperatures were determined more accurately than before by analyzing the superconducting characteristics. The authors found that on passing from one phase range to another there was a change in the form of the T_c–composition curve, although there was still a gen-

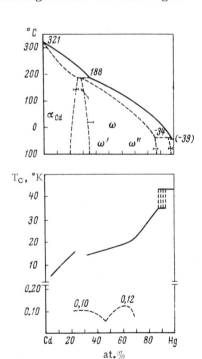

Fig. 123. Phase diagram and superconducting transition temperature of Cd–Hg alloys.

eral tendency for T_c to decrease with increasing cadmium content. In the regions of the solid solutions based on the original components and in the two-phase region T_c varies linearly. In the wide region of the ω phase (with a tetragonal lattice and a variable composition) a bend appears as a result of ordering for a 1:2 ratio of the components (Cd_2Hg and $CdHg_2$). Whereas before ordering the transition temperature varies linearly within the region of the ω phase (37-88 at.% Hg), after ordering a bend appears and there is a decrease in T_c. Figure 123 presents the phase diagram of the Cd-Hg system, indicating the refinement to the boundaries of the phase regions derived from an analysis of T_c. The solubility of cadmium in mercury, the boundaries of the two-phase range ($\omega + \beta$-Hg), the one-phase range ω, and the two-phase range (α-Cd+ω) are all refined in this way. The decrease in the T_c of the alloys after ordering (as compared with the disordered state) constituted the first observation of such an effect. The ω phase is evidently of the same general nature as the σ phases, although its transformation mechanism requires further study. Claeson et al. [92] suggested that the decrease in T_c was due either to a reduction in the electron–electron interaction or else to a change in the density of states on the Fermi level. The electrical resistance in the normal state diminished as a result of ordering. In this case the change in the T_c of alloys of nontransition metals fails to obey the Matthias rule, in contrast to many alloys of transition metals.

In the Tl-Hg system the addition of thallium to mercury also leads to a decrease in T_c. Existing data [93] indicate bends in the T_c curve on passing from one region of the diagram to another, the curve remaining linear within a particular region.

It is interesting to note that Pb-Na alloys have high critical magnetic fields [82]. In an alloy of lead with 5 at.% Na H_{c2} equals 5000 Oe; alloys with higher sodium contents have never been studied. The system contains a number of compounds, while the alloy under consideration consists of a eutectic mixture of the Pb-base solid solution and the compound NaPb [94]. There are few data relating to transition temperature of alloys belonging to this system. It is only known [95] that the hypereutectic alloy of lead with 28% Na has a transition temperature similar to that of lead (7.2°K).

A considerable increase in transition temperature occurs on alloying tin and indium, which have similar T_c values. In an alloy

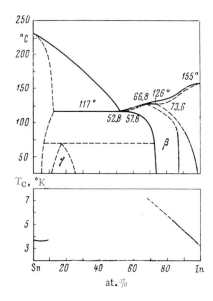

Fig. 124. Phase diagram and superconducting transition temperature of Sn−In alloys.

containing 30% Sn and consisting of a eutectic mixture of γ and β phases it reaches 7.3°K. Figure 124 illustrates existing published data relating to the superconducting properties of this system [93, 96].

It is interesting to note the capacity of bismuth to form superconducting compounds and even superconducting alloys with nonsuperconducting elements, although in the ordinary state, without the application of high pressures, or on preparing samples in film form at low temperatures, no superconductivity is observed in bismuth itself. In addition to existing superconducting bismuth compounds with nonsuperconducting elements (Au_2Bi, BiLi, BiNa, etc.), the critical temperatures of which are fairly low (below 3°K), superconducting alloys have also been obtained. Thus superconductivity has been observed in a Bi−Cu alloy at temperatures under 2.2°K, although no intermediate phases appear in this simple system.

Existing data relating to the superconducting properties of systems formed by nontransition metals lead to the conclusion that the most favorable conditions for the creation of superconducting alloys are realized in eutectic mixtures in the presence of intermediate phases (compounds entering into the composition of the eutectic). Peritectic reactions fail to increase the transition temperatures of the alloys, and the rise in critical field accompanying such reactions may be expected to be smaller than in the case of eutectic interaction.

The highest superconducting characteristics of all known alloys of the nontransition metals are those of binary Pb–Bi alloys, in which the transition temperature reaches 8.8°K, while the critical magnetic field is 15 kOe (or, according to some, even 25 kOe). Further alloying alloys of this system may raise T_c a little more. However, the superconducting properties of even this system of alloys have still not been studied completely; the effect of various phase compositions on T_c and H_{c2} has not really been studied, and there is very little information as to the critical currents achieved. A number of systems have been studied still less.

Nevertheless, despite the general low level of their properties by comparison with the alloys of transition metals, superconducting alloys of the nontransition metals may yet find a wider outlet in science and technology in view of the fact that the general level of their properties and the stability of these may well be improved after systematic research into their structure and the plotting of complete composition–superconducting property diagrams, particularly in a number of the most promising binary and ternary systems.

4. Systems Involving the Formation of Intermediate Phases

On interacting with each other or with other nonsuperconducting elements, superconducting transition metals form a large number of superconducting systems incorporating various intermediate phases. We shall now consider the variations in the superconducting characteristics of alloys belonging to a number of these systems.

Superconductivity has been observed in a large number of alloys of the rare-earth metals, particularly those involving lanthanum as base or additive.

Thus the La–Y system forms the compound LaY and also contains wide ranges of solid solutions based on the original components (Fig. 125) [98]. According to certain data [99] superconductivity occurs in alloys containing up to 50 at.% Y. The transi-

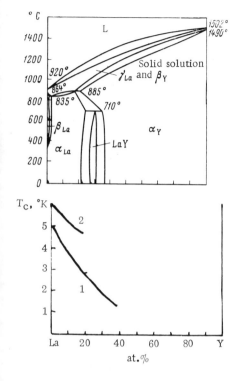

Fig. 125. Phase diagram and superconducting transition temperature of La-Y alloys: 1) hexagonal structure; 2) cubic structure.

tion temperature of lanthanum decreases on alloying with yttrium, and the T_c in the region of the compound LaY formed by a peritectic reaction is equal to 1.7°K. The difference in the transition temperatures of the two forms of lanthanum (the high-temperature form β-La with $T_c = 5.95°K$ and the low-temperature form α-La with $T_c = 4.9°K$) suggests that in the range of La-base solid solutions there may well be alloys of different crystal structure and different T_c (in Fig. 125 the T_c of these alloys is indicated by curves 1 and 2). The low-temperature specific heat of La-Y alloys was studied in [99b]. The addition of Y lowered the T_c of lanthanum while raising θ_D and reducing γ.

Superconductivity also occurs in lanthanum alloys with other rare-earth metals (Lu, Nd, Pr, Sm, Te, and Yb). In all cases except that of the Nd alloys T_c decreases below that of lanthanum; in the La-Nd system T_c increases to 6.3°K [100].

The transition temperature of alloys of lanthanum with various noble metals (Rh, Ir, Pd, Pt) has also been studied [101]. On

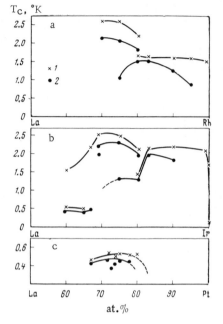

Fig. 126. Superconducting transition temperature of La–Rh (a), La–Ir (b), and La–Pt (c) alloys: 1) cast sample; 2) metalloceramic.

examining the Pd-rich alloys no superconductivity was found. Figure 126 illustrates the T_c of alloys with Rh, Ir, and Pt measured on cast samples and samples prepared by metalloceramic technology. The transition temperature of these alloys is fairly low (never over 2.5°K). In all cases the T_c of the metalloceramic powdered material is lower. Owing to the small number of alloys containing no more than 40-50 at.% La studied, no final conclusion can be drawn as to their superconducting properties.

The same authors studied several alloys of another rare-earth metal (cerium) with iridium (on the iridium side). In this case (Fig. 127) despite the lower transition temperature of iridi-

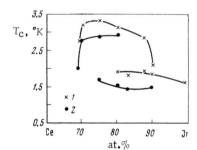

Fig. 127. Superconducting transition temperature of Ce–Ir alloys (notation as in Fig. 126).

um the T_c of the alloys was even a little higher (3.4°K) for 70-80% Ir. It is well known that on alloying rare-earth metals with noble metals [101] compounds with a Laves-phase structure of MeX_2 are formed (X = noble metal). In the present case the maximum T_c appears at approximately the composition $CeIr_2$.

Superconductivity is also found in alloys of lutecium with iridium and rhodium. Alloys of the Lu–Ir system are superconductors even for Lu contents of under 50%, although at very low temperatures (about 0.5°K). The maximum T_c (about 3°K) occurs in alloys of this system with 25% Lu [102].

Yttrium–rhodium alloys constitute an example of the development of superconductivity on alloying two nonsuperconducting elements which form compounds. Superconductivity appears (above 0.3°K) in the majority of alloys of this system. The phase diagram of the system has never been plotted, but it is certainly known that compounds of the YRh_2 and YRh type are formed, the first with the $MgCu_2$ and the second with a cubic structure [101, 103]. There are no other data regarding the form of the Y–Rh phase diagram. The maximum T_c occurs in the alloy with 40 at.% Rh (Fig. 128a); it is rather low (1.5°K). At the compositions of the foregoing compounds a minimum value of T_c appears; there is also a minimum for the alloy containing 30% Rh. It was later concluded [104] that rhodium might itself constitute a superconductor at about 0.025°K; yttrium should have a T_c value very close to this.

Higher transition temperatures occur for Yt–Ir alloys. The addition of yttrium to iridium raises T_c from 0.14 to 3.8-3.5°K (up to 30% Y). The compound YIr_2 formed in this system [101] has a lower transition temperature (about 2.18°K) than the iridium-base alloys (Fig. 128b).

There are certain data relating to the existence of superconductivity in the systems formed by yttrium with other noble metals (Ru, Pd, Os, Pt); however, the superconducting properties have only been measured for a few compounds formed in these systems.

The properties of alloys based on titanium, a transition metal of the following IVA group, were studied elsewhere [104].

In addition to having a continuous series of solid solutions at high temperatures and a eutectoid reaction on reducing the temperature the Ti–Cr system forms a compound $TiCr_2$ (Fig. 129) [56].

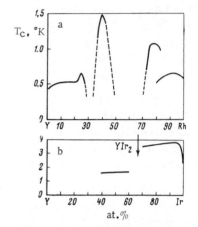

Fig. 128. Superconducting transition temperature of Y–Rh (a) and Y–Ir (b) alloys.

The transition temperature was determined on alloys quenched from the β region containing up to 30 at.% Cr. On studying the structure of this system it was found that the β phase was preserved, on quenching, in alloys containing at least 6.5 at.% Cr. Alloys containing a higher proportion of titanium underwent a transformation of the martensite type $\beta \rightarrow \alpha'$). The change in the structure of the alloys (α' instead of the β phase) disrupted the continuous nature of the T_c-composition curve characterizing the alloys falling within the solid-solution range (all having the same phase structure). In both cases the T_c curve passes through a maximum, this reach-

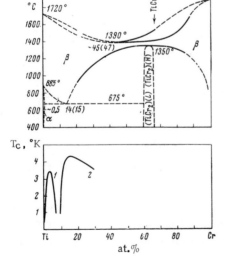

Fig. 129. Phase diagram and superconducting transition temperature of Ti–Cr alloys: 1) with the α' structure; 2) with the β structure.

ing 4.3°K in the case of the alloys with the structure of the β phase. The transition temperature of titanium increases on alloying with chromium (for all cases studied), even on adding comparatively small proportions (under 3 at.%) of this nonsuperconducting metal.

Various authors indicate the formation of one to three compounds in the Ti–Mn system: TiMn, TiMn$_2$, TiMn$_3$. On alloying these two polymorphic metals, a range of solid solutions is also

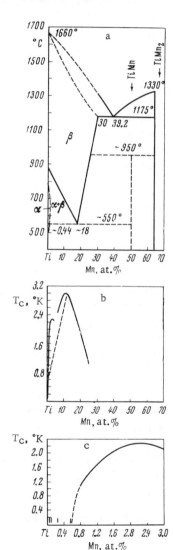

Fig. 130. Phase diagram of the Ti–Mn system (a) and superconducting transition temperature of titanium alloys containing up to 25% Mn (b) and up to 3.5% Mn (c).

formed on the titanium side at high temperatures, although it only extends to 30 at.% Mn (1175°), while on reducing the temperature there is a eutectic decomposition (Fig. 130a) [56]. Measurement of the transition temperature and an examination of the phase composition of quenched alloys containing up to 25 at.% Mn showed that the high-temperature β solid solution was fixed for a manganese content of not less than 7 at.% [104]. Alloys richer in manganese have the structure of the α phase. At the boundary of the α and β phases, the T_c—composition curves exhibit a bend; maxima occur within the ranges of solid solutions having the β and α structures (Fig. 130b). The electrical properties of two-phase Ti–Mn alloys were studied elsewhere [106] at a temperature close to the T_c of these alloys. Below 450° all the alloys containing 1-10% Mn had a two-phase structure and consisted of the α solid solution and the compound $TiMn_2$. The transition temperatures determined in this investigation agreed with those given earlier. On introducing manganese into titanium (2 and 6%) T_c rose to 1.92 and 2.23°K respectively [106]. These values are close to those indicated in Fig. 130c.

In a discussion relating to further work [107] the effect of traces of manganese on the superconducting properties of titanium was described. It is interesting to note that, whereas for manganese contents of over 0.7% the values of T_c obtained in this particular case agreed fairly closely with those presented earlier, smaller proportions of manganese reduced the transition temperature of titanium. Alloys containing under 0.7% Mn failed to exhibit superconductivity. This was apparently because on dissolving in titanium the manganese acquired a magnetic moment. The subsequent rise in T_c with increasing manganese content was attributed to the formation of the β phase, i.e., a change in the phase composition of the alloys.

The broken line in Fig. 130b indicates the effect of rhenium on the superconducting properties of titanium [108]. The solubility of rhenium in titanium extends to 50 at.% [109]. All the quenched alloys studied (up to 10 at.% Re) had the structure of β-titanium with a bcc lattice, and in this range of concentrations T_c rose linearly as the rhenium content increased. However, for a content of 10% Re the value of T_c failed to exceed 2.7°K.

The transition temperature of Ti–Fe alloys was only studied for Ti-rich samples quenched from the region of the β solid solution with the bcc lattice [105]. The structure of the corresponding diagram was similar to that of the Ti–Mn phase diagram in the Ti-rich part. The manner in which the T_c of the samples varied was also very similar (Fig. 131). The main difference lay in the fact that the boundary of the martensite transformation which took place on quenching was slightly displaced in the high-Ti direction, the form of the T_c-composition curve agreeing with this. The transition temperature of titanium rises more sharply on adding traces of iron than on adding manganese and rhenium, and its maximum value for solid solutions with the structure of the β phase is also higher (3.7°K). It should be noted that the foregoing Ti–Re alloys have not been studied over the whole range of titanium solid solutions, and it might well be that the transition temperature would rise further for greater rhenium contents. The same might apply to Ti–Ru alloys, for which only two compositions have been studied (broken line in Fig. 131). It is quite possible that the maximum on the T_c-composition curve observed in many solid solutions based on polymorphic metals might also occur in alloys containing a higher proportion of ruthenium.

The Ti–Co system also has a structure similar to Ti–Mn and Ti–Fe on the titanium side [110], but the solubility of cobalt in the β-titanium is much lower (14.5 at.% Co at 1020°). On analyzing the transition temperatures of these alloys [105] for cobalt contents of up to 20 at.% we find a change in the transition temperature approximately at the boundary between the β bcc solid solution and the two-phase region (β + Ti_2Co), while there is no change in T_c in the two-phase region itself (Fig. 132). Here also there is a rise in the T_c of titanium with increasing content of the alloying component, bends on the T_c curves as the structure of the alloys changes as a result of the martensitic decomposition, and maxima

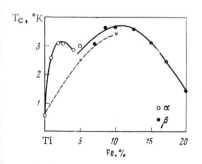

Fig. 131. Superconducting transition temperature of titanium alloys with iron (continuous curve) and ruthenium (broken curve).

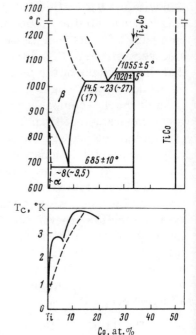

Fig. 132. Phase diagram and superconducting transition temperature of Ti-Co alloys.

on the curves in the regions of the β and α solid solutions. In the regions of the solid solutions of these alloys the transition temperature is rather higher than in Ti-Fe alloys (maximum 3.8°K). The broken line in Fig. 132 shows the effect of alloying titanium with rhodium on the T_c values. Alloys have been studied for Rh contents of up to 15%, and over this range of concentrations T_c rises almost linearly. However, certain other authors obtained quite different results for the same alloys [165]. In cast (rapidly-cooled) alloys containing up to 1 at.% Rh the critical temperature was low, while annealing increased it to 4°K. The introduction of rhodium up to 2-10 at.% slightly increased T_c, but later a reduction set in. The increase in T_c on adding up to 5% Rh was attributed to an increase in electron concentration, which reaches the optimum value at this composition (4.2 electrons/atom), while the subsequent decrease in T_c on further alloying was ascribed to the rise in the mass of the atoms (in accordance with the relation $M^{1/2}T = \text{const}$), which in rhodium is large. The difference in the transition temperature of the Ti-rich alloys as between the cast and annealed states was explained by the coagulation of the high-temperature β phase (which had a higher T_c than the α phase) as a result of annealing. The difference in the results of different authors is doubtless due to the different states of the alloys examined.

Matthias et al. [105] noted that a rise in the transition temperature of titanium solid solutions occurred both on alloying the titanium with transition elements free from ferromagnetism and also on alloying with ferromagnetics. In none of the cases under consideration, however, did the transition temperature of alloys within the range of the Ti-base solid solution exceed 4-4.2°K. The transition temperature of titanium alloys with the hexagonal structure of the α phase is under 3.2-3.6°K.

A number of superconducting systems involving zirconium and forming intermediate phases have also been studied. Thus, for example, in the Zr–Au system several such compounds occur, of which at least one ($AuZr_3$) is superconducting below 1.7°K; in addition to this, superconductivity at 2.75-1.65°K occurs in zirconium-base alloys containing up to 10% Au [112, 113]. There is insufficient information regarding the structure of the Zr–Au phase diagram, particularly on the zirconium side, and the small number of alloys studied prevents us from discussing the nature of the changes in T_c. There is still less information regarding the superconductivity of alloys of the Bi–Zr system; it is only known that in some of the alloys measured in this system T_c lies between 2.8 and 3.2°K.

Vanadium forms a system with zirconium in which a compound ZrV_2 with a hexagonal lattice of the $MgZn_2$ type is formed by a peritectic reaction [58]. The T_c of this compound equals 8.8°K [108], while the alloy containing 25% V and 75% Zr has no superconductivity at 1.2°K [104]. Matveeva [114] refined the structure of the V–Zr phase diagram and measured the transition temperature for alloys containing 53-70 at.% V in the cast state and also after homogenization at 1000°C followed by quenching. The highest transition temperature (8.6°K) occurred for cast alloys containing 60-62 at.% V. Single-phase alloys having the structure of the compound ZrV_2 (63.5-66.5% V) passed into the normal state at a lower temperature: at 7.8-7.7°K in the cast state and 7.1-7.3°K after quenching from 1000°C. Annealing and quenching from 1200°C raised the T_c of ZrV_2 to 9.64°K. On comparing the T_c values of the

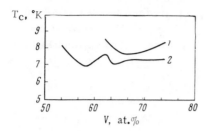

Fig. 133. Superconducting transition temperature of V–Zr alloys: 1) cast; 2) quenched from 1000°C.

two-phase alloys and the alloys in the region of the compound ZrV_2 we notice a minimum corresponding to the latter case (Fig. 133). The transition temperature and critical current of zirconium alloys containing 6 and 9 at.% vanadium were studied in [114a] after various periods of aging at 500°C.

In the Zr–W system superconductivity is only found in an alloy with the composition of the compound ZrW_2. Doubtless other alloys with superconductivity will be found in these systems. Little study has been given to the Zr–Re system, in which several compounds are formed [115]. In several Zr–Re alloys a transition temperature of about 6°K has been obtained [116]; in rhenium–zirconium compounds the T_c is higher (up to 7.5°K).

The superconductivity of a large number of alloys constituting zirconium solid solutions with group VIII elements has been studied, the alloying elements including ferromagnetic metals (Fe, Co, Ni) and noble elements (Ru, Rh, Pd, Os, Ir, Pt) [113, 117]. As in the case of alloying titanium with Fe, Co, Ni, and noble metals, the transition temperature of zirconium thereupon increases. The highest transition temperature (much higher than in the case of titanium) is in alloys of zirconium with rhodium, reaching 9.6°K (Fig. 134). Figure 134 shows the dependence of T_c on the composition of alloys of zirconium with ruthenium and cobalt; in these alloys T_c is lower. The figure also illustrates the results of individual measurements on alloys with palladium, ruthenium, and platinum, which also substantially raise the T_c of zirconium. In the alloy of zirconium with 10% Pd the critical temperature equals 7.5°K. Comparison of the changes in T_c with the phase regions in the phase diagram is hardly possible due to the small amount of data relating to the structure of the alloys.

Superconductivity in hafnium alloys has been little studied; it was recently observed in hafnium itself. We have already considered the properties of Nb–Hf and Ta–Hf alloys. In the Hf–Mo system 10 at.% Mo raises the transition temperature of Hf from 0.165 to 2.5°K [118]. This transition temperature is obtained in an alloy consisting of a mixture of two solid solutions with bcc and fcc lattices. Further increasing the molybdenum content to 35 at.% had no effect on the T_c of alloys which had been quenched from the region of the β solid solution and had a single-phase structure. In the Hf–Re system, apart from the superconductivity of compounds (HfRe$_2$ has a T_c of 5.61°K) superconductivity appears at 7.3°K in an alloy containing 2.5% Hf. A hafnium-base alloy (12.5% Re) quenched from the β solid solution had a T_c of about 1.7°K [118]. Rhodium also raises the transition temperature of hafnium (to 1.51°K) for a content of 4 at.%. The superconducting properties of Hf–V alloys have not been studied at all. The structure of this system has been studied in [118a]. It forms a compound HfV$_2$ with the MgCu$_2$-type structure [56]. There are no data relating to the superconducting properties of either compounds or other alloys.

Extensive investigations into the superconducting properties of alloys of group IVA elements with group VIII elements have shown that in all cases there is a rise in the T_c of Ti-, Zr-, and Hf-base alloys. These experimental results lead to the conclusion [119] that the positive influence of ferromagnetic materials on the T_c of group IVA elements may be explained by the development of a magnetic electron–electron interaction promoting their superconductivity. At the same time, on alloying titanium, zirconium, and hafnium with noble metals (metals of the same group VIII) the effect may be ascribed to the fact that the electron concentration

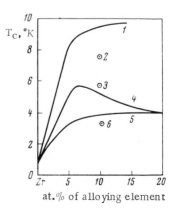

Fig. 134. Superconducting transition temperature of zirconium alloys in relation to rhodium (1), palladium (2), ruthenium (3), osmium (4), cobalt (5), and platinum (6) content.

is thereby raised to the optimum value. We may nevertheless note that, even in this case, in alloys based on group IVA elements, a change in phase composition also changes the course of the transition temperature curve. Hence the empirical rules presented cannot be used to obtain reliable data without comparing the superconducting characteristics with the phase composition and the structure of the alloys in accordance with the phase diagram.

More information is available regarding the superconductivity of alloys formed by VA metals (V, Nb, Ta). Apart from the systems with continuous solubility already described and further data relating to the superconductivity of various compounds of these metals, there are also a great number of other superconducting alloys of these metals with elements belonging to various groups of the periodic table.

Among alloys of vanadium with group I metals, superconductivity occurs in the V–Au system, which forms at least three compounds [120]. Superconductivity is found in only one of these (V_3Au) below 0.74°K, while in the alloy with 5 at.% Au T_c equals 2.25°K [121]. The decrease in the critical temperature of vanadium on

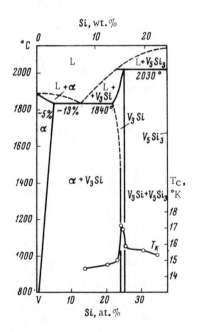

Fig. 135. Phase diagram and T_c of vanadium rich alloys in the V–Si system.

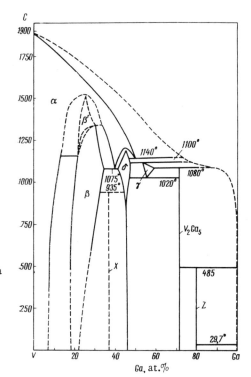

Fig. 136. Phase diagram of the V−Ga system according to Savitskii et al. [123].

alloying with gold agrees with the Matthias rule, but contradicts the proposition [9] that elements with a large atomic radius should increase the T_c of vanadium and niobium.

Of great interest is the more detailed study of superconducting properties in alloys of the V−Si and V−Ga systems, since in them compounds and compound-based alloys with high superconducting characteristics are formed.

The phase diagram of the V−Si system has been studied repeatedly. The results in general are in good agreement [56]. In this system comprising a superconductor (vanadium) and a nonsuperconductor (silicon), only alloys containing a V-base solid solution and a compound V_3Si with the Cr_3Si type of structure are superconducting. The compounds VSi_2 and V_5Si_3 are not superconductors at temperatures above 0.30 and 1.20°K, respectively [64, 122]. The compound V_3Si is homogeneous over a narrow range of concentrations (Fig. 135) [123]. The lattice constant of the compound falls in this range from $a = 4.727$ Å (for 24 at.% Si) to $a = 4.725 \pm 0.001$ Å

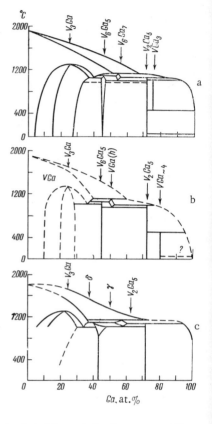

Fig. 137. Versions of the V–Ga phase diagram according to various other authors [133] (a), [134] (b), and [135] (c).

(for 25 at.% Si). For an almost exact stoichiometrical composition the T_c of the compound is 17.0°K [64, 124-126]. Within the range of homogeneity of the compound there is a slight change in the superconducting transition temperature of the alloys [127]. The maximum T_c, 17.2°K, occurs in the alloy with 24.0 at.% Si, almost at the boundary of the homogeneous range. The transition to the two-phase region leads to a sharp decrease in T_c (to 14.5°K on reducing the silicon and to 15.8°K on increasing the silicon in the alloy) as compared with the stoichiometrical composition of the compound V_3Si [127]. Increasing the degree of equilibrium of the vanadium–silicon alloys helps in securing a sharper transition into the superconducting state. The ΔT for cast alloys is 0.25-0.5°K and for heat-treated material 0.15-0.4°K. Attempts at increasing the T_c of vanadium silicide were mentioned in the previous chapter.

The V−Ga system attracts the attention of research workers in view of the high superconducting characteristics of the compound V_3Ga ($T_c = 16.5°K$ [124, 129], $H_c \approx 350$ kOe [130]). An approximate phase diagram of this system was first published in 1964 by Savitskii, Kripyakevich, Baron, and Efimov. An improved version of the phase diagram was published by the authors in 1967 (Fig. 136) [132]. There have been three other versions of the V−Ga phase diagram (Fig. 137) [133-135]; there have also been a number of papers concerned with individual phases in this system [136-139]. Published data are to a certain extent contradictory; they provide different treatments for the effect of composition on the superconducting transition temperature of V_3Ga [135, 139]. These contradictions are primarily associated with the experimental difficulties encountered in alloying and heat-treating alloys of the extremely chemically-active elements concerned (gallium and vanadium), and also with the large number of intermediate phases formed.

Measurements of the superconducting transition temperature of V−Ga alloys showed that at temperatures above 4.2°K only the β phase with the cubic Cr_3Si structure and a low-alloy V-base solid solution were superconducting. It may well be that other V−Ga phases pass into the superconducting state at temperatures below 4.2°K [132]. The superconducting transition temperature of these

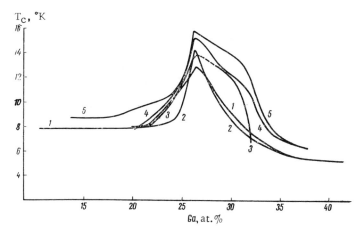

Fig. 138. Superconducting transition temperature of the β phase (Cr_3Si structure) of V−Ga alloys in relation to composition and heat treatment. Annealing temperature: 1) No annealing (cast); 2) 800; 3) 900; 4) 1050; 5) 1150°C.

alloys is shown in relation to the chemical composition and heat treatment in Fig. 138. The critical temperature of the β phase depends on its chemical composition [132, 146]. In the two-phase regions adjacent to the range of homogeneity of the β phase the critical temperature varies very little. In equilibrium with the V-base solid solution, the critical temperature of the β phase equals 7.9-8.8°K [132] (10.1°K [146]). The very similar values of the critical temperature of alloys belonging to this region after various kinds of heat treatment are due to the very slight change in gallium content in the compound V_3Ga with changing temperature (Fig. 136). On passing into the range of homogeneity the critical temperature starts rising, reaching a maximum at 26 at.% Ga [132, 135]. The displacement of the critical-temperature maximum from the compound of stoichiometric composition in the direction of higher gallium content agrees with the analogous displacement in the composition of complete ordering of the β phase [139]. On further increasing the gallium content there is a decrease in the critical temperature (slow within the range of homogeneity and sharp on passing into the two-phase region). In the two-phase region adjacent to the range of homogeneity of the β phase on the gallium side, the β phase has a lower superconducting transition temperature (5.5-6.5°K [132], 7.6°K [146]) by comparison with alloys in the two-phase region on the vanadium side.

On dissolving Ga the critical temperature of vanadium decreases and at about 3 at.% Ga it is lower than 4.2°K [132].

The V-Sn system contains a superconducting compound V_3Sn ($T_c = 6°K$). There is no information as to the superconducting properties of other alloys of this system in the literature; since it is formed by two superconductors and contains two-phase regions of alloys consisting of a mixture of the compound V_3Sn with solid solutions of vanadium and tin [147], we may assert with some confidence that all the V-Sn alloys will possess superconductivity, and that the critical temperature of the majority of the two-phase alloys will be close to the T_c of the compound V_3Sn. The compound V_2Sn_3 with an orthorhombic structure is formed on low-temperature annealing (600°C) [147a].

It would be extremely interesting to follow the transition temperature of V-Ta alloys as a function of their chemical and phase compositions. In this system the components form a con-

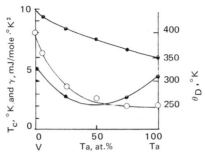

Fig. 139. Transition temperature T_c, specific heat (γ), and Debye temperature (θ) as functions of composition in the V-Ta system.

tinuous series of solid solutions above 1300°C. Figure 139 shows the dependence of T_c, γ, and θ_D in the region of solid solutions [9a, 9b]; T_c has a minimum (~ 2.5°K) on the curve for a composition of 50 at.% Ta [147, 89b], and θ_D falls additively from V to Ta. On the basis of electrical-resistance measurements (ρ_{100} and ρ_{23}) of V-Ta alloys [9a, 9b] the authors found that the electron–phonon interaction diminished with increasing amount of vanadium in the alloys; this explains the minimum on the T_c-composition curve of V-Ta alloys. On reducing the temperature, the compound V_2Ta precipitates from the solid solution; this has a range of homogeneity between 32 and 39 at.% Ta [148, 149].

The transformations in the solid state were determined more accurately in [89b]. The compound TaV_2 with the $MgZn_2$ type of structure is formed at 1420° and ~ 33 at.% Ta (Fig. 140). At 1125°C and 29 at.% Ta this phase decomposes eutectoidally into a solid solution with a bcc lattice and a phase with the $MgCu_2$ type of structure. The latter is also formed by a peritectoid reaction at 1280° and 37 at.% tantalum; it is homogeneous at 800° over the range 32-39.5 at.% Ta; the T_c of the high-temperature phase with the $MgZn_2$ hexagonal type of structure reaches 10°K.

A more detailed investigation into the properties of V-Re alloys is also a matter of interest. This system has ranges of solid solutions based on both components and forms a σ phase by a peritectic reaction [150]. Alloys in the range of the Re-base solid solutions (10 at.% V) pass into the superconducting state at 9.4°K, while in the σ phase the transition occurs at lower temper-

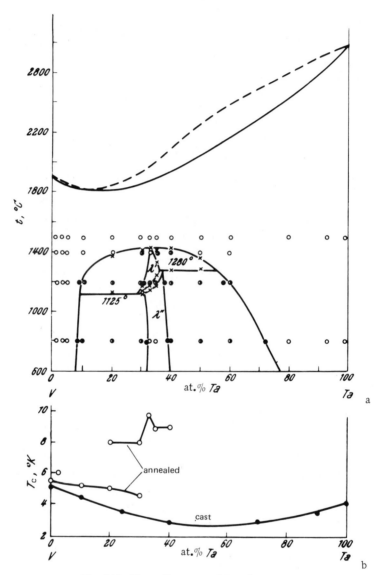

Fig. 140. Phase diagram and T_c of V–Ta alloys.

atures (4.52°K) [64]. There are no further data relating to the superconducting properties of V−Re alloys in the literature.

The solubility of technetium in vanadium reaches ∼ 60 at.% and that of vanadium in technetium around 10 at.% [64a, 64b]. In the range 60-90 at.% Tc there are two solid solutions. The phase with the CsCl structure ($a = 3.025$ Å at 700°) is formed at 50 at.% Tc, evidently in the solid state [64a]. The superconducting properties of Tc−25-35-at.%−V alloys were studied in [64c]. At high temperatures these alloys were single-phased with a bcc structure; below 1150°C there were two phases: an hcp solid solution and a cubic phase of the CsCl type. The structural changes taking place on annealing were compared with the changes in superconducting properties. Thus T_c increased from 6.6°K in the quenched state to 8.8°K on annealing; the critical magnetic field and the current density at 4.2°K varied correspondingly [64c].

The transition temperature of vanadium alloys with Mn, Fe, Co, and Ni was studied in detail by Matthias et al. [105] (in addition to the V−Ti and V−Cr systems, which we described earlier). It was found that in all cases the addition of the elements in question reduced the transition temperature of vanadium. The greatest reduction in T_c occurred on adding cobalt; an alloy of vanadium with 3% Co had a T_c of only 1.5°K.

In the Ru−V system superconductivity has been studied for alloys containing up to 30 at.% V. Figure 141 [121] shows the almost linear increase in the T_c of the alloys with increasing vanadium content. According to the approximate phase diagram [151], alloys of these compositions constitute solid solutions of the substitutional type with an hcp lattice. The solubility of vanadium in

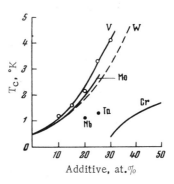

Fig. 141. Critical superconducting temperature of ruthenium alloys in relation to alloying with other metals.

ruthenium is considerable and at 1500°C reaches 31%. The system forms several intermediate phases, the first of which (having the greatest vanadium content) has a lattice of the CsCl type. No superconductivity has been found in the compounds. The transition temperature of an alloy close to the boundary of the Ru-base solid solution reaches 4.1°K as opposed to 0.5°K for ruthenium itself. Figure 141 also shows the effect of molybdenum and tungsten on the T_c of ruthenium; although the transition temperature of these metals into the superconducting state is considerably lower than that of vanadium, their influence on the ruthenium alloys is closely analogous to that of vanadium.

Alloys of palladium containing up to 40 at.% V and having the lattice of the fcc solid solution at temperatures up to 1.2°K proved to be nonsuperconducting [121].

An approximate V-Ir phase diagram was plotted later on the basis of microstructural and x-ray data [152]. In addition to the earlier-known phases V_3Ir and VIr_3, the system contained the phases α-VIr and β-$(V_{1-x}Ir_x)Ir_2$. The latter phase is homogeneous in the range 52-59.5 at.% Ir and has a tetragonal lattice of the AuCu type. The compound V_3Ir with a Cr_3Cs-type structure is formed from the melt at 1930°C. On the basis of this compound a range of solid solutions occurs between 25 and 39 at.% Ir. All the remaining phases in the system crystallize by a peritectic reaction. No superconductivity appears in V_3Ir above 0.35°K, but the alloy containing 26.7% V passes into the superconducting state at 1.39°K. There is no further information in the literature.

Niobium forms systems incorporating a number of superconducting compounds with a series of elements belonging to the III and IVB subgroups.

Niobium forms two compounds with aluminum, having high superconducting transition temperatures [153]. The phase diagram of the Nb-Al system was plotted by Savitskii and Baron (Fig. 142) [154]. One other version [155] differs essentially from the former. A slightly different type of diagram was also obtained by Svechnikov, Pan, and Latysheva [156]. There are three compounds in the system: Nb_3Al, Nb_2Al, and $NbAl_3$. The compound Nb_3Al with the Cr_3Si structure ($a = 5.187$ Å) [129] and the σ phase Nb_2Al ($a = 9.943$, $c = 5.186$ [157]) are formed at 2120 and 1890°C respectively by a peritectic reaction [154, 155]. According to Svechnikov et al. [156]

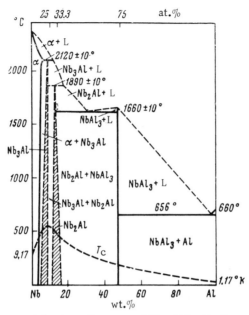

Fig. 142. Phase diagram and T_c of Nb–Al alloys.

Nb$_3$Al is formed by a peritectoid reaction at 1730°C. The compound NbAl$_3$ (tetragonal lattice, $a = 5.427$, $c = 8.584$, $c/a = 1.582$ [158]) crystallizes from the melt at 1660°C [154, 155]. There are narrow ranges of solid solutions based on Nb$_3$Al and Nb$_2$Al [154]. The lattice constant of Nb$_3$Al varies comparatively little [159]. According to some data [155] the compound NbAl$_3$ also has a range of homogeneity. The existence of two other compounds as well as Nb$_3$Al, Nb$_2$Al, and NbAl$_3$ stable above 1000°C [160] is improbable. The limiting solubility of aluminum in niobium is about 6 wt.% at 2120°C and falls to 4.5% at room temperature [154]. Up to 0.08 wt.% Nb dissolves in aluminum at room temperature; with increasing temperature this increases to a limit of 0.22% [161]. According to the same investigation [161] an aluminum-base solid solution is formed by a peritectic reaction at 668.5°C.

All alloys of the Nb–Al system are superconducting. The maximum critical temperature occurs for Nb$_3$Al. Within the range of homogeneity the T_c of this compound varies from 15.2 to 17.6°K [162, 163], however the exact composition dependence of the T_c of Nb$_3$Al has never been established. Certain data [128] suggest that the maximum critical temperature of Nb$_3$Al after annealing the cast alloy at 800°C for 1 h equals 18.2°K. The critical temperature of the σ phase varies with composition (within the range of homogeneity) between 7 and 12°K [164]. The effect of niobium on the T_c of aluminum and of aluminum on the T_c of niobium is also unknown; we may suppose that in accordance with the Matthias rule [64] aluminum reduces the critical temperature of niobium; the T_c of the Al-base solid solution is no greater than that of pure aluminum (1.196°K [165]). The compound $NbAl_3$ has never been studied in relation to superconductivity; remembering the type of crystal lattice of this compound, we can hardly expect it to have a high transition temperature. The character of the possible variation in the transition temperature of Nb–Al alloys is indicated in Fig. 142.

In the Nb–Ga system the compound Nb_3Ga with the Cr_3Si structure ($a = 5.171 \text{Å}$) has $T_c = 14.5°K$ [129]. The phase diagram of the system (Fig. 143) has been plotted fairly recently [166]. The system contains a limited solid solution based on niobium. The solubility of gallium in niobium at 800°C is approximately 8-10 wt.%;

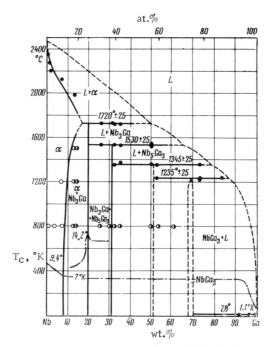

Fig. 143. Phase diagram and T_C of Nb–Ca alloys.

on raising the temperature the solubility increases and probably reaches 16% at the temperature of the peritectic reaction. Apart from the compound Nb$_3$Ga (20 wt.% Ga), the system forms three other compounds: Nb$_5$Ga$_3$ (31.08% Ga), Na$_2$Ga$_3$ (about 51% Ga) and NbGa$_3$ (69.29% Ga). All the compounds are formed by peritectic reactions taking place at 1720, 1530, 1350, and 1235°C [166]. The compound Nb$_5$Ga$_3$ has a tetragonal structure of the W$_5$Si$_3$ type (a = 10.32 ± 0.01 Å, c = 5.06 ± 0.01 Å). The compound NbGa$_3$ also has a tetragonal structure of the TiAl$_3$ type (a = 5.378 ± 0.002 Å, c = 8.73 ± 0.002 Å).

As we might expect from physicochemical considerations [1], all the alloys of the Nb−Ga system are superconducting. With increasing gallium content in the Nb-base solid solution the value of T_c diminishes (Fig. 143), and for the alloy with 5.4 wt.% Ga it equals 7°K. An analogous picture was obtained on alloying niobium with tin or vanadium with gallium. The highest T_c in the Nb−Ga system is that of Nb$_3$Ga at stoichiometric composition (14.5°K [129], 14.2°K [166]). Any deviation from stoichiometric composition reduces T_c [166]. In the two-phase region a measurement of the transition temperature confirms the existence of two superconducting phases: There are two steps on the transition curves (Fig. 143), 7.7°K (solid solution of gallium in niobium) and 9.05°K (compound Nb$_3$Ga). Another two-phase region consists of superconducting phases with a transition temperature of 13.5°K (Nb$_3$Ga) and 7.5°K (Nb$_5$Ga$_3$) [166]. From the changes in the T_c of Nb$_3$Ga in neighboring two-phase regions (9.05 and 13.5°K) we may deduce the existence of a narrow range of homogeneity of this compound; on the T_c curve the compound corresponds to a singular point, a maximum of the transition temperature. The critical temperature of the NbGa$_3$ phase (7.5°K) is constant within the limits of experimental error [166]. The effect of niobium on the T_c of gallium has never been studied. In the range 7-32 wt.% Ga the critical magnetic fields of the alloys considerably exceed 28 kOe. The variation in the H_c of the alloys at 4.2°K with larger and smaller gallium contents is indicated in Fig. 144 [167]. Alloys containing 7-12 wt.% Ga are the most promising for the manufacture of thin wire.

The phase diagram of the Nb−Ge system (Fig. 145) was studied by Pan et al. [171]. The system contains several compounds: Nb$_3$Ge with a cubic Cr$_3$Si structure [172, 173], Nb$_5$Ge$_3$ with a tetragonal W$_5$Si$_3$ structure (a = 10.14 Å, c = 5.15 Å) [174], Nb$_3$Ge$_2$ [171],

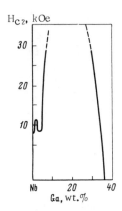

Fig. 144. Values of the critical field in relation to the composition of Nb–Ga alloys.

and NbGe$_2$ with a hexagonal structure of the CrSi$_2$ type ($a = 4.967$, $c = 6.784$ Å) [175]. The compound Nb$_3$Ge is formed by a peritectic reaction at 1970°C [171, 172]; the range of homogeneity of this compound is displaced in the niobium direction from the stoichiometric composition (16-21 at.% Ge at 1850°C, 15-19 at.% at 1450°) [159, 171, 172]; the lattice constant of the phase decreases from 5.177 to 5.167 Å on increasing the germanium content [172, 173]. The compound Nb$_5$Ge$_3$ (34-36 at.% Ge) melts congruently at 2150° [171]; the authors in question suggested that this phase underwent a polymorphic transformation at 1870°C. A eutectic is formed between the

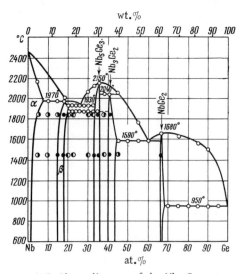

Fig. 145. Phase diagram of the Nb–Ge system.

compounds Nb_3Ge and Nb_5Ge_3 (1930° and 26.27 at.% Ge). The compound Nb_3Ge_2, melting incongruently at 2040°C, is single-phased over the range 40-43 at.% Ge [171]; its crystal structure has never been solved. The compound $NbGe_2$ melting congruently participates in two eutectic reactions: with Nb_3Ge_2 at 1596°C and with germanium at 950°C. The limiting solubility of germanium in niobium is about 7 at.% at 1970°C. The solubility of niobium in germanium is very low indeed.

Until recently the values of T_c for Nb_3Ge in the Nb—Ge system determined by different authors [64, 122, 168, 169] fluctuated from 4.9 to 6.9°K. Recently Matthias et al. [170] found that after quenching at a cooling rate of over 10^6 deg/sec in the presence of an excess of germanium the T_c of this compound equalled 17.0°K. The substantial rise in the T_c of the compound is associated with its approach to stoichiometric ratio (25 at.% Ge). For the compound at the boundary of the two-phase region on the niobium side T_c is 4.9°K; on passing into the region of homogeneity T_c increases to 6.9°K, and on further increasing the germanium content to stoichiometric composition it increases to 17°K [170]. Ordinary germanium is not superconducting down to 0.05°K [176]. The compounds Nb_5Ge_3, $NbGe_2$, and evidently Nb_3Ge_2 are also not superconductors [159].

In the Nb—Sn system the compound Nb_3Sn with the Cr_3Si structure and containing 29.87 wt.% Sn was first observed in 1954 [153]; in 1955 its lattice constant was determined ($a = 5.289$ Å) [177]. It was established in these papers that of all known compounds Nb_3Sn had the highest superconducting transition temperature (18.05°K).

The first phase diagram of the Nb—Sn system was published in 1959 by Savitskii, Baron, and Agafonova [178]. Using vacuum-melted samples it was found that the maximum solubility of tin in niobium reached 14 at.% and decreased very little on reducing the temperature (9 at.% below 1400°C). The compound Nb_3Sn is formed by a peritectic reaction at about 2000°C. No other phases appeared in the system after annealing at 1100 and 1220°C. On the tin side a pseudoeutectic equilibrium was established (232°). The publication of this investigation evoked more than 50 further papers in various countries. The part of the phase diagram close to Nb_3Sn (important for the study of superconducting materials) determined by the authors in question remained almost without any alteration in subsequent work. Certain research workers found some new compounds in the part of the diagram adjacent to tin, particularly metastable phases; these were attributed to the use of metalloceramic methods for preparing the samples [179].

Recently a fair number of modified versions of the Nb–Sn phase diagram have appeared [180-186]. In all these diagrams differences appear in the number of compounds formed, their composition, the extent of the ranges of homogeneity, the ranges of stability, and the solubilities of the components. The substantial variations in the phase diagrams of this system may be attributed to the complexity involved in the preparation of Nb–Sn alloys in view of the great difference in the melting and boiling points of the components, the high vapor tension of tin above 1000°C, the low velocity of the diffusion processes at low temperatures, and so on.

The results of various authors regarding the number, composition, crystal structure, and temperature and concentration ranges of existence of Nb–Sn compounds are presented in Table 30. The same table contains the results of measuring the superconducting transition temperatures of the compounds. The majority of authors consider that niobium forms three compounds with tin. Disagreement mainly concerns the composition of the compounds and the temperature ranges of their stability.

On correlating all the results in question, we may assert that the compound Nb_3Sn is formed by a peritectic reaction at 2040-2130°C and is stable down to room temperature. In those investigations in which the compound Nb_3Sn failed to appear at low temperatures, the samples were prepared by the metalloceramic or diffusion methods. As a result of the comparatively low-temperature and brief sintering or annealing, the compound Nb_3Sn was unable to form in these samples because of the low velocity of the diffusion processes. In all alloys obtained by arc melting the compound Nb_3Sn was stable down to room temperature.

In a number of papers the range of homogeneity of Nb_3Sn was regarded as narrow and was expressed as a vertical line on the diagram [178, 180-182, 197]. In some cases [182-184, 200, 201] it was mentioned that the composition might differ from stoichiometric in the niobium direction. In one case [202] the lattice constant of Nb_3Sn was regarded as constant ($a = 5.291 \pm 0.002$ Å) for alloy com-

TABLE 30. Niobium–Tin Compounds According to Various Authors

Method of preparing the samples	Observed compounds (at.% Nb in brackets)	Crystal structure	Temperature ranges of existence, °C	Critical temperature, °K
Arc melting [178]	Nb_3Sn	Cubic type Cr_3Si	0—2000	—
Sintering of powders [194]	Nb_3Sn (80) (75) (50)	Ditto — — —	— — — —	18.2 18.0 17.6
Sintering of powders	Nb_4Sn	—	0—2040	—
Diffusion layers [181]	Nb_3Sn Nb_2Sn Nb_2Sn_3	— — —	0—730 0—690 0—893	— — —
Ditto [193]	Phase 1 Phase 2 Phase 3 Phase 4	—	$T_m > 1000$ $T_m \sim 900$—950 $T_m \sim 850$—900 $T_m < 800$	— — — —
Sintering of powders [182, 192]	Nb_3Sn Nb_3Sn Nb_3Sn_2	— — Tetragonal ($a = 6.901$; $c = 9.533$Å)	— 860—2000 770—900	18.5 — 16.6
	Nb_2Sn_3	Ditto ($a = 11.304$; $c = 5.013$Å)	0—860	Does not superconduct down to 4.2°K
Sintering of pressed foil of composition Nb_4Sn [191]	$NbSn_2$	Orthorhombic $a = 5.645$; $b = 9.852$; $c = 19.126$Å	800	2.6 ($H_0 = 620$ Oe)
Sintering of powders [190]	Nb_3Sn (75) Nb_5Sn_4 (56) Nb_2Sn_3 (38)	— — —	— — —	— — —
Sintering of powders, diffusion layers, arc melting [183]	Nb_3Sn (75) Nb_3Sn_2 (56) Nb_2Sn_3 (39)	— — Triclinic	775—2000 600—1000 0—1000	— 17.7 3.8
Sintering of powders	(35—40)	Orthorhombic ($a = 5.64$; $b = 9.86$; $c = 19.13$Å)	600 and 700	2.4—3.4 ($H_{1.95°K} = 200$ Oe)
Sintering of powders, diffusion layers, arc melting [189]	Nb_3Sn (75) (79) Nb_3Sn_2	Cubic type Cr_3Si $a = 5.289$Å $a = 5.280$Å Orthorhombic ($a = 5.65$; $b = 9.21$; $c = 16.84$Å)	775—2000 600—925	— —
	$NbSn_2$ (35—40)	Ditto ($a = 5.64$; $b = 9.86$; $c = 19.13$Å)	0—850	—

TABLE 30 (Continued)

Method of preparing the samples	Observed compounds (at.% Nb in brackets)	Crystal structure	Temperature ranges of existence, °C	Critical temperature, °K
Heat treatment of Nb$_3$Sn "wire" [188]	Nb$_3$Sn Nb$_3$Sn$_2$ Nb$_2$Sn$_3$	— — —	>875 875—925 <875	— — —
Single crystals separated from supersaturated liquid [195]	Nb$_3$Sn (78,5) Nb$_3$Sn$_2$ (59) NbSn$_2$ (33)	— — Orthorhombic ($a=5.655$; $b=9.860$; $c=19.152$Å)	— Peritectic at 915 ± 10 Peritectic at 840 ± 15	— — —
Diffusion layers, sintering of niobium powder in tin [180, 195]	Nb$_3$Sn NbSn	Cubic type Cr$_3$Si ($a=5.29$Å)	0—2000 0—850	1.8 2.7 ($H_0=840$ Oe)
Arc melting, sintering of powder, diffusion layers [187]	Nb$_3$Sn (75±1) Nb$_3$Sn$_2$ (58) NbSn$_2$ (33,5)	Ditto Orthorhombic ($a=5.65$; $b=9.85$; $c=19.2$Å) —	Stable to room temperature Peritectic at 910 ± 10° (stable at 820-810°) Peritectic at 840 ± 10° (stable to room temperature)	— — —
Arc melting, sintering of Nb$_3$Sn powder with tin [186, 187, 196]	Nb$_3$Sn (68—86 at 900°) Nb$_6$Sn$_5$ NbSn$_2$ Nb$_3$Sn Nb$_3$Sn$_2$ Nb$_2$Sn$_3$	Cubic type Cr$_3$Si (stable to room temperature) Orthorhombic type β-Ti$_6$Sn$_5$ Orthorhombic type CuMg$_2$ — — —	Peritectic at 2130° Peritectic at 920° (stable to 815°) Peritectic at 860° (stable to room temperature) Ditto, at 430° Ditto, at 915° Ditto, at 820°	— — — — — —
Sintering of powders (750°, 300 h)	Nb$_3$Sn Nb$_3$Sn$_2$	— —	805—2000 805—925	— —
Diffusion layers [197]	NbSn$_2$	Orthorhombic ($a=5.63$; $b=9.85$; $c=18.96$Å)	20—950	—

TABLE 30 (Continued)

Method of preparing the samples	Observed compounds (at.% Nb in brackets)	Crystal structure	Temperature ranges of existence, °C	Critical temperature, °K
Arc melting [135, 136, 198]	Nb_3Sn Nb_6Sn_5 (44 ± 2)	— Orthorhombic ($a = 5.6549$; $b = 9.2037$; $c = 16.814$Å)	— —	— —
	(33.3)	Ditto ($a = 5.6477$; $b = 9.860$; $c = 19.127$Å)		
Arc melting (data obtained by authors of this monograph)	Nb_3Sn	—	Peritectic at 2000° (stable to room temperature)	18.1
	Nb_6Sn_5	—	Peritectic at 950° (stable at 850-950°)	Does not superconduct above 4.2°K
	$NbSn_2$	—	Peritectic at 850° (stable to room temperature)	Ditto

positions from 50 to 90 at.% Nb. On the other hand it was indicated elsewhere [183, 184] that the lattice constant of Nb_3Sn varied from 5.280 to 5.289 Å over the concentration range 75-79% Nb at high temperatures (1300-1400°C); other reported variations were from 5.2803 to 5.291 Å over the range 81.5-75% Nb [200], 5.282 to 5.2887 Å over the range 82.3-75.1% Nb [201], 5.2826 to 5.2887 Å over the range 78.5-72.8% Nb [203], and 5.282 to 5.292 Å over the range 23-26 at.% Sn [187]. In the very latest investigations [186, 187, 187a] the existence of a range of homogeneity in the compound Nb_3Sn was confirmed by metallographic and x-ray methods. However, the various results are in poor agreement. It was indicated in one case [187] that Nb_3Sn was homogeneous at 1500°C from 23.1 to 26 at.% Sn, while on reducing the temperature the range of homogeneity contracted. Svechnikov, Pan, and Beletskii [186] established a wider range of homogeneity, extending from 14 to 32 at.% Sn at 900°C. According to these last data, on raising or lowering the temperature the range of homogeneity of Nb_3Sn contracted, the solubility curve of niobium in Nb_3Sn varying monotonically, while

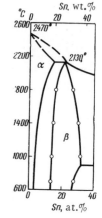

Fig. 146. Range of homogeneity of the compound Nb_3Sn [186].

the solubility curve of tin had a bend on intersecting the peritectic horizontal at about 900°C (Fig. 146). It was indicated in one paper [187a] that the range of homogeneity of the compound Nb_3Sn remained the same (27-37 at.% Sn) over quite a wide temperature range (up to 1600°C). Apart from the Nb_3Sn, two other compounds are formed in this system: Nb_6Sn_5 and $NbSn_2$.

The compound Nb_4Sn observed in a number of the earlier papers [181, 182, 192] in the metalloceramic or diffusion samples was not subsequently confirmed; its observation was evidently due to the methods of producing the alloys and to the existence of a range of homogeneity in the compound Nb_3Sn.

For the compound richest in tin we may take the generally-accepted formula $NbSn_2$ and an orthorhombic structure of the $CuMg_2$ type (a = 5.6450, b = 9.8576, c = 19.121 Å [204]). The compound $NbSn_2$ is formed by a peritectic reaction at 840-860°C; it has a constant composition and is stable to room temperature.

The compound lying between Nb_3Sn and $NbSn_2$ is given the formula Nb_3Sn_2 by the majority of authors. In one case [205] the formula Nb_9Sn_7 was proposed for a compound which, according to the results of this particular investigation, contained 43-44 at.% Sn and had a tetragonal face-centered cell. In later investigations [185, 186, 189, 198, 199, 206], however, it was found that the crystal structure of this compound was orthorhombic of the β-Ti_6Sn_5 type with lattice parameters a = 5.5649, b = 9.2057, c = 16.814 Å [204]. On considering the structural type (associated with the formula

A_6B_5) and also remembering that in two cases [183, 189] the chemical composition of the compound was regarded as only approximate and in fact corresponded to the formula Nb_6Sn_5, it would appear reasonable to accept the latter formula [135, 136, 186, 204, 206]. The compound Nb_6Sn_5 is formed by a peritectic reaction at 910-920°C and has a constant composition. This intermediate phase is only stable over a narrow temperature range, which is confirmed by the results of the majority of investigations [182, 183, 187-189, 192, 197]. The observation of this compound in certain investigations [181, 199] in alloys cooled to room temperature is once again attributable to the methods of producing the samples. At the present time we may regard it as fully established that the compound Nb_6Sn_5 decomposes eutectoidally at 805-820°C [186, 187, 197].

The existence of compounds of other compositions (Nb_2Sn [181], NbSn [180, 195]) is refuted by the majority of research workers.

In almost all research work on this subject the solubility of tin in niobium has been studied; however, there is considerable disagreement as to its value. The maximum solubility of tin in niobium evidently reaches 12 at.% at 2000°C [178, 189] or about 14 at.% at 2130°C [186, 204]. On reducing the temperature the solubility of tin decreases fairly quickly; below 1000-1400°C it decreases less rapidly [178] and at 700-1000°C equals about 6 at.% [189]; at lower temperatures still, approximately 2.5 at.% Sn dissolve [186, 187, 199, 204]. The lattice constant of the Nb-base solid solution changes very little (from 3.12 to 3.329 Å) [189].

The existence of a monotectic equilibrium in the system at 730°C in the range between 25 and almost 100 at.% Sn was suggested by Agafonova [178]. Svechnikov [186] also obtained data relating to the presence of a monotectic equilibrium in the Sn-rich alloys at 960°C. The region of phase separation was very small (width about 2 at.%, height 30°), and was substantially displaced in the tin direction. The monotectic point occurred in the region of 97 at.% Sn. Other authors, however, categorically refuted the existence of a monotectic equilibrium by drawing attention to the slight solubility of niobium in molten tin and the experimentally-determined form of the liquidus. It is possible that a monotectic equilibrium is a characteristic of the metastable phase diagrams

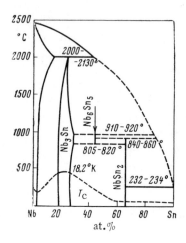

Fig. 147. Phase diagram and T_c of Nb–Sn alloys.

of the Nb–Sn system. The metastability of this equilibrium provides for a possible shift in the composition and temperature of the monotectic point and a change in the temperature and concentration ranges of the phase-separation region. This may explain the difference in the monotectic parameters obtained by different authors [178, 186]. Both these diagrams are evidently metastable in the Sn-rich range of compositions.

Evidently there is no monotectic reaction in the equilibrium phase diagram. On the tin side there is a pseudoeutectic (232°C) [178, 186, 195, etc.] or a peritectic (234°C) [187].

The phase diagram of the Nb–Sn system plotted on the basis of existing data is given in Fig. 147. The system contains five superconducting phases: a Nb-base solid solution (bcc lattice), the compound Nb_3Sn (Cr_3Si structure), the compound Nb_6Sn_5 (orthorhombic structure of the β-Ti_6Sn_5 type), the compound $NbSn_2$ (orthorhombic structure of the $CuMg_2$ type), and an Sn-base solid solution. The highest transition temperature is that of Nb_3Sn (18.2°K). The Nb-base solid solution also has a high transition temperature. The transition temperature of niobium purified by electron-beam melting is 9.3°K; the addition of tin reduces the transition temperature of niobium; at the boundary between the solid solution and the two-phase region this reaches 5.6°K [207]. This change in T_c obeys the empirical Matthias rule [64] to the effect that alloying a transition with a nontransition element reduces the critical temperature of the alloy. In the two-phase region (Nb-base solid solution and compound Nb_3Sn) the transition temperature in the alloys with a fair proportion of Nb_3Sn is close to that of the pure single-phase compound (see Fig. 147).

The superconducting properties of the compound Nb_3Sn depend considerably on the method of its production as well as its chemical composition and mode of heat treatment. It was found in two cases [182, 202] that the temperature of the superconducting transition rose slightly on increasing the niobium content and reached 18.5°K [182] or 18.1°K [202] for 80 at.% Nb. In other cases [169, 200, 201, 208, 209] it was noted, on the other hand, that the highest critical current density and critical temperature occurred for samples with a high tin content, while samples of composition close to 80 at.% Nb had a low current density and a T_c value of 5.6°K [169], 6.2°K [209], 7°K [200, 201], 9.0°K [208]. However, an analysis of these results [169, 200, 201, 208] fails to provide us with a clear relationship between the critical temperature and composition of Nb_3Sn. The values exhibit a wide scatter due, no doubt, to the inhomogeneity of the samples and the inaccurate determination of composition. Svechnikov et al. [186] obtained an almost linear dependence of T_c on the composition of Nb_3Sn. The critical superconducting temperature of Nb_3Sn rises linearly with increasing tin content of the alloys. According to Savitskii, Baron, and Bychkova, the T_c of niobium alloys with 20-30 at.% Sn annealed at 900°C equals 18.1°K, while for annealing temperatures of over 900°C it begins a gradual decline. This change in the critical temperature (and other superconducting parameters) of Nb_3Sn wire is attributed to the nonmonotonic variation in the solubility of tin in Nb_3Sn with varying temperature [186]. On raising the annealing temperature to 850-860°C [180], 900° [210, 211], 930-950° [188], or 1000°C [212] the critical temperature and critical current density increase sharply; on further raising the annealing temperature these parameters diminish.

The other phases in the system (Nb_6Sn_5, $NbSn_2$, and the tin-base solid solution) have low transition temperatures. In $NbSn_2$ the transition is no higher than 3.4°K. It is well known that the transition temperature of this compound may fluctuate from 3.4 to 2.4°K [184], the reasons for this effect not having been established; according to later data [213] the T_c of $NbSn_2$ is 2.68°K. The critical temperature of Nb_6Sn_5 is under 2.8°K [214]; according to other data [213] it is 2.07°K. However, owing to the diffuse transition ($\Delta T = 0.25°K$) it was considered in the latter case that the real critical temperature was a little higher (probably 2.2°K). It was further noted that the T_c of sintered Nb–Sn alloys was lower than that of cast samples. The effect of niobium on the transition temperature of tin (3.74°K) was not established.

The compound $NbSn_2$ has a critical magnetic field H_0 equal to 840 Oe [184, 191, 195] and the compound Nb_6Sn_5 under 600 Oe [214]. Hence the superconductivity of Nb–Sn alloys in fields of over 50 kOe is due solely to the presence of Nb_3Sn ($H_c = 220$ kOe) if only in very small quantities.

In contrast to the Nb–Al, Nb–Ga, Nb–Ge, and Nb–Sn systems, no compounds with the Cr_3Si structure are formed in the Nb–Si system. Earlier [215, 216] it was considered that a compound Nb_4Si was formed in the Nb–Si system. Later it was found that a compound Nb_3Si was formed by a peritectic reaction at 1920°C (tetragonal structure of the Ti_3P type, $a = 10.23$, $c = 5.189$ Å, $c/a = 0.507$), this only being stable at 1800-1920°C [216a]. Prolonged annealing at lower temperatures leads to the decomposition of this compound into an α solid solution of silicon and another phase. At 1880°C the eutectic $\alpha + Nb_3Si$ crystallizes. The eutectic point lies at 18 at.% Si. The maximum solubility of silicon in niobium is ~ 5 at.% at 1880°C. In none of the small number of investigations into this system of alloys has superconductivity been observed above 1.2°K [59], but it is reasonable to suppose that more careful research using materials of greater purity will find superconductors among the Nb-rich alloys of this system.

In the Nb–Cr system there are ranges of solid solutions based on each of the components (the solubility of chromium in niobium at 1600°C is 20 at.% and at 1200°C 11 at.%). A compound $NbCr_2$ with an fcc lattice is formed by a peritectic reaction at about 1670°C. The system contains ranges of eutectic mixtures $NbCr_2 + \alpha_1$ and $NbCr_2 + \alpha_2$ (Fig. 148) [217]. The transition temperature has been determined in Nb-base cast alloys (up to 20 at.% Cr), the lowest temperature (corresponding to the maximum Cr content) being 1.8°K [8]. We see from Fig. 148 that on adding chromium to niobium T_c decreases linearly, reaching 1.8° (as against 9.2°K for Nb) at 20% Cr.

In studying this system [8] the effect of the homogeneity of the sample on the form of the transition curve was determined.

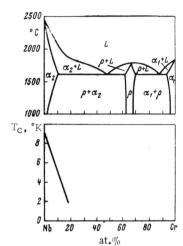

Fig. 148. Phase diagram and T_c of Nb–Cr alloys.

Figure 149 illustrates two curves for an alloy containing 10% Cr plotted for the cast and homogenized states respectively (annealing at 2200°C for 100 h in a vacuum of $3 \cdot 10^{-6}$ mm Hg). The transition temperature was determined by the induction method, using a ballistic galvanometer. The vertical axis gives the values of the so-called effective permeability μ (ratio of the galvanometer deflection at the measuring temperature to the deflection at liquid-nitrogen temperature), and the horizontal axis gives the temperature in °K. We see clearly from this figure that the width of the transition for samples in the cast state is several degrees. After homogenization in the manner indicated the same sample has a narrow transition range. There is also a slight decrease (1°K) in the transition temperature after annealing; this was attributed [8] to the composition homogenization of the sample, leading to the

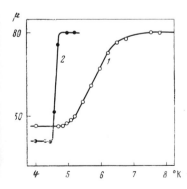

Fig. 149. Superconducting transition temperature of an Nb–10%-Cr alloy in the cast (1) and annealed (2) states.

dissolution of the Nb-rich regions with the highest T_c values, as may readily be seen from Fig. 148. Microstructural and gas analysis of the samples before and after the homogenizing anneal confirmed this explanation. This example demonstrates the importance of bringing the test samples into an equilibrium state, even those not undergoing phase decompositions, in order to secure reliable data regarding the level of their superconducting characteristics.

The phase diagram of the Nb–Re system was studied by Savitskii et al. [218]. This revealed a compound $Re_{22}Nb_7$ with the structure of α-Mn; a σ phase was also apparently formed. There were wide ranges of Nb-base solid solutions (about 42% Re at 800°C) and narrow Re-base regions (Fig. 150). The superconducting properties of this system were not studied in any systematic manner. Over the wide range of Nb-base solid solutions T_c was hardly measured at all. For the alloys in the middle of the diagram, consisting of a mixture of the solid solution and the compound $Re_{22}Nb_7$ (50-60% Re), $T_c = 2.5$-$3.8°K$. The transition temperature of the Re-rich alloys was considerably higher, and in the region of the compound $Re_{22}Nb_7$ (82% Re) – 9.7°K [219-220].

On considering the influence of group VIII elements (Fe, Ru, Os) on the superconducting properties of niobium [17], it was noted that for up to 10% of these elements the transition temperature fell linearly in each case (Fig. 151). This effect does not support the Matthias rule as to the optimum mean number of valence electrons

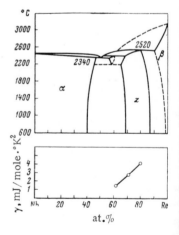

Fig. 150. Phase diagram and electron specific heat of Nb–Re alloys.

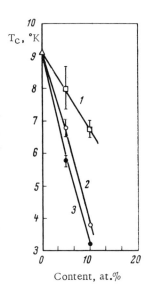

Fig. 151. Superconducting transition temperature of alloys of niobium with iron (1), ruthenium (2), and osmium (3).

per atom, while the correction elsewhere [9] proposed for this rule in order to allow for the difference in the atomic volumes of the basic metal and the alloying component makes the disagreement still worse.

In the Nb−Fe system the solubility of iron in niobium is quite low (no more than 1.96 at.% at 1000°C [220]); the alloys studied in one particular investigation [221] were two-phased. The phases with higher iron contents were nonsuperconducting at the temperatures studied.

In contrast to this, in the Nb−Ru and Nb−Os systems wide ranges of Nb-base solid solutions are formed. Thus at approximately room temperature the solubility of ruthenium in niobium reaches 45 at.% [222]. The authors of this investigation studied the structure of the alloys in the range of niobium solid solutions (without giving the full phase diagram) and also their critical superconducting temperature, electron specific heat, and Debye temperature. In addition to the results presented in Fig. 151 (T_c of niobium alloys containing up to 10 at.% Ru) it was found that on further increasing the ruthenium content T_c decreased still further, while alloys containing 20 and 30 at.% Ru failed to superconduct at 1°K; increasing the ruthenium content to 40% produced an increase in T_c to 1.2°K; in alloys with 60% Ru T_c reached 2.5°K.

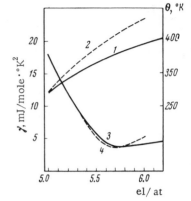

Fig. 152. Debye temperature θ_D and electron specific heat γ as functions of the mean number of electrons per atom [222]: 1) θ (Nb-Mo); 2) ditto (Nb-Ru); 3) γ (Nb-Mo); 4) ditto (Nb-Ru).

It is interesting to compare the T_c curve with the electron specific heat and the Debye temperature of alloys of the Nb-Ru and Nb-Mo systems. The γ-composition and T_c-composition curves pass through a flat minimum at the same mean number of electrons per atom (5.3) calculated on the weighted-mean principle. In both cases the Debye temperature (θ_D) of niobium increased with increasing proportion of the second component. Figure 152 shows the dependence of γ and θ on the mean number of electrons per atom of the alloys of these two systems. Careful x-ray analysis and a microstructural study of niobium-ruthenium alloys in the region of the minimum of T_c and γ did not reveal the formation of any second phase. After analyzing the lattice-constant measurements and the x-ray analysis it was suggested that an ordering process might occur in the alloys containing 40% Ru, thus leading to a rise in T_c. In the Nb-Os system in the region of the niobium solid solutions the T_c curve also passes through a minimum. Despite the formation of a compound with the Cr_3Si structure in this system, the transition temperature of the alloy of this composition is quite low (1.05°K). In all the remaining alloys measured (those containing more osmium) the T_c was considerably lower than in the niobium-base alloys (Fig. 151).

The structure of the Nb-Ir phase diagram has still not been fully established. It is well known that this system contains a compound with the Cr_3Si structure. There are also some indications as to the existence of other intermediate phases and a limited solubility in the system [223]. The maximum transition temperature is that of the Nb-40%-Ir alloys. In the compound Nb_3Ir the transition temperature is only 1.7°K [125].

The compound Nb_3Pt with the Cr_3Si structure has the highest superconducting transition temperature in the Nb−Pt system (9.2°K). The alloy containing about 62% Nb becomes superconducting at a lower temperature, about 4°K [219]. There is no other information as to the superconducting properties of Nb−Pt alloys.

In the Nb−Pd system only a single alloy has been studied, the composition of this approximately corresponding to that of a compound with the structure of a σ phase (60 at.% Nb). The transition temperature of this alloy is quite low, various authors giving values between 1.7 and 2.47°K [219, 224]. No other alloys have been studied. In this system in addition to the σ phase a compound $NdPd_3$ is formed in the solid state; this may be a compound of the Kurnakov type [225].

In systems involving tantalum and containing various intermediate phases the superconducting properties of a limited number of alloys have been studied. Thus Re−Ta alloys have only been studied in the region of the λ phase [219]. For alloys in the region of the compound Re_3Ta (75 at.% Re), which is formed by a peritectic reaction and has the structure of α-Mn [226], T_c equals 6.78°K; in alloys adjacent to this region T_c is much lower: for 65% Re it is 1.58 and for 62% 1.4°K [219]. We may therefore consider that T_c passes through a maximum in the region of solid solutions based on the χ phase.

In a Ta−50-at.%-Si alloy the superconducting transition temperature is about 4.3°K [122]; the alloy consists of a mixture of the compounds $TaSi_2$ and Ta_5Si_3 [56]. Since no superconductivity appears in $TaSi_2$ on cooling to 1.2°K, we may consider that the second compound entering into the composition of this alloy possesses superconductivity. Superconductivity will clearly also occur in the Ta-rich alloys consisting of a mixture of the Ta-base solid solution and the compound $Ta_{4.5}Si$. Nothing is known as to the superconducting properties of this compound, which is the richest in tantalum.

On alloying molybdenum with rhenium the transition temperature rises sharply. The maximum T_c occurs for alloys lying at the boundary of the solid solutions of rhenium in molybdenum (~ 12°K). The transition temperature of the σ and χ phases formed in this system is much lower. Data as to the structure and superconducting properties of Mo−Re alloys [8, 118, 226] are presented in Fig. 153. The transition temperature was determined for cast

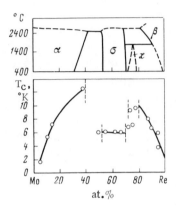

Fig. 153. Phase diagram and superconducting transition temperature of Mo–Re alloys.

samples annealed at 1500-2500°C for 50-100 h in a vacuum of $3 \cdot 10^{-6}$ mm Hg. The minimum measuring temperature was 1°K. In the region of molybdenum solid solutions, the transition temperature rises from molybdenum to the boundaries of the solid solution (around 40 at.% at 2000°C). The properties of alloys containing 25 and 38 at.% Re were studied in the form of single crystals [9, 118]. The transition temperature of the single-crystal alloys equalled 9.8°K (25 at.% Re) and 12.25°K (38 at.% Re). In both cases the superconducting transition of the samples was very sharp, indicating their high degree of equilibrium and homogeneity. The critical temperature falls sharply on passing from the two-phase region ($\alpha + \sigma$) into the region of solid solutions based on the σ phase (to 6°K), remaining constant over this region. The χ phase has a higher critical temperature than the σ phase, and in the range of homogeneity $T_c \approx 9$°K. In alloys with the structure of the Re-base solid solution, obtained by rapid cooling after high-temperature annealing, T_c falls as the rhenium content increases. It is interesting to note the effect of reaching equilibrium on the transition temperature of alloys with the composition of the χ phase ($Mo_{23}Re_{77}$). According to certain data [9], on quenching this alloy from the melt there is a transition into the superconducting state over a wide temperature range and T_{av} equals 10.25°K; annealing the alloy at 1800°C (100 h) reduces T_c to 9.25°K, and the transition becomes sharp.

The example of the Mo–Re alloys shows the sensitivity of T_c to the phase composition of the alloys; at the boundaries of the phase regions in the phase diagram it changes sharply.

According to [228] the critical magnetic field H_{c2} in alloys of this system reaches 27 kOe (50% Re).

The highest transition temperature of binary alloys of molybdenum is found in the Mo–Tc system [228]. The transition temperature of alloys containing 5-25 at.% Mo rises from 10.8 to 15.8°K; alloys containing 30-35% Mo have a T_c of 12-13.3°K, while increasing the molybdenum content to 45% raises T_c to 14°; there is also a high transition temperature in the alloy with 50%, but at 60% Mo T_c decreases below 12.8°K. It is well known [229] that the Mo–Tc system contains regions of solid solutions based on both components, σ and β phases (the latter contains some 55 at.% Tc and has a structure of the Cr_3Si type). The range of the σ phase extends from 65 to 74 at.% Tc at 1800°C. If we compare the results obtained by measuring the T_c of the alloys with the boundaries of the phase regions in this system, we notice that the observed changes in T_c are associated with the formation of the σ phase, the β phase, and the molybdenum-base α solid solution.

The transition temperature of molybdenum also rises on forming alloys with Ru, Rh, Pd, Os, Ir, and Pt, i.e., with all the noble metals of the eighth group. Figure 154 shows the effect of rhodium on the T_c of molybdenum [3]. The transition temperature rises to 8.2°K for a content of 20% Rh. In alloys with ruthenium T_c is higher; for 50% Mo T_c reaches 10.5°K [230]. It is hard to compare existing data relating to the superconducting properties of alloys of molybdenum with the noble metals, as the relevant information is sparse. The phase diagrams of the molybdenum with these metals have also been little studied.

Figure 155 shows the phase diagram of the W–Re system and the corresponding T_c values [118]. The alloys were studied

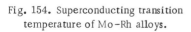
Fig. 154. Superconducting transition temperature of Mo–Rh alloys.

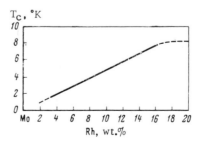

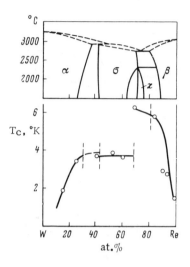

Fig. 155. Phase diagram and superconducting transition temperature of W-Re alloys.

after annealing at 1500-2500°C. A more precise phase diagram was obtained elsewhere [226]. The amounts of σ and χ phases and the general appearance of the system are the same in the two cases; the difference lies mainly in the temperature of formation of the χ phase and the boundaries of the solid solutions based on the components. On the basis of the general composition dependence of T_c and the structure of the alloys it would appear that the later phase diagram [226] is the more accurate, since T_c suffers no change on passing into the range of homogeneity of the χ phase. According to the investigation in question, the χ phase is formed at 1050°C, and therefore in the case of alloys quenched from higher temperatures there should be no change in the form of the T_c-composition curve in this range of concentrations. The transition temperature of rhenium also increases sharply with increasing tungsten content close to the boundary of σ phase formation (about 9°K). The transition temperature of the σ phase is lower (5.12°K) and alloys in the range of solid solutions based on this phase have approximately the same transition temperature. In the range of solid solutions based on tungsten the T_c is never greater than 4.5°K (close to the solubility limit); it decreases sharply on approaching pure tungsten.

The transition temperature of W–Os alloys has also been studied [118, 227]. This system (illustrated in Fig. 156) forms a σ phase (by virtue of a peritectic reaction) in the range 20-37 at.%

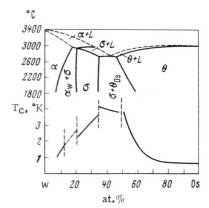

Fig. 156. Phase diagram and superconducting transtion temperature of W-Os alloys.

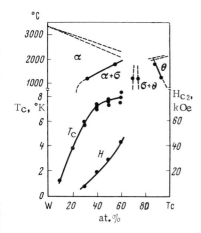

Fig. 157. Structure, superconducting transition temperature, and critical magnetic field of W-Tc alloys.

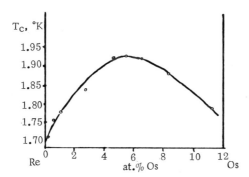

Fig. 158. Transition temperature as a function of composition in the Re-Os system.

Os. The transition temperature changes linearly in the range of solid solutions based on the σ phase, rising from 2.3 to 3.8°K with increasing osmium content. The maximum T_c (about 4°K) is found in the alloy lying at the boundary between the σ and σ + θ regions (θ being the Os-base solid solution, see Fig. 156). On further increasing the osmium content T_c decreases again. In this system also we notice a change in the form of the T_c curve on passing from one part of the diagram to another.

The critical temperature of tungsten also rises on alloying it with other group VIII metals. Thus platinum increases the T_c of tungsten to 2.15°K and iridium to 1.39°K [231].

The structure and superconducting properties of tungsten–technetium alloys have also been studied [232]. As in the case of molybdenum, alloying with technetium substantially increases the transition temperature of tungsten and also its critical magnetic field. The alloys studied (up to 60 at.% Tc) were melted in an electric arc furnace after previously sintering in hydrogen at 1000°C. After melting, the alloys contained a considerable amount of oxygen (0.26 wt.%) but less nitrogen and carbon (0.02 and 0.039 wt.%). Microstructural and x-ray investigations established the boundary of the W-base solid solutions (about 48 at.% Tc at 2000°C). Alloys containing 50 and 60% Tc (at.) had a two-phase structure consisting of an α solid solution and a σ phase with a tetragonal crystal lattice (Fig. 157). The accuracy of measuring the transition temperature was 0.5°K and the critical magnetic field 0.5 Oe. In the region of the α solid solution the transition temperature rose linearly, but in the two-phase region the curve had a flat appearance. The maximum transition temperature (in an alloy containing 60 at.% Tc) equalled 7.88°K, while the H_{c2} of this alloy amounted to 43.5 kOe at 4.2°K.

The effect of osmium on the T_c of rhenium was studied in [232a] in the range of solid solution (Fig. 158). The T_c of rhenium first increases with increasing osmium content and reaches a maximum (1.93°K) at ∼ 5.5 at.% Os, then it falls again. The change in T_c with the composition of the alloys does not agree with the change in the density of states, which rises continuously in this region.

TERNARY AND MORE COMPLEX SUPERCONDUCTING SYSTEMS

1. Ternary Systems

At the beginning of this chapter we considered the criteria of alloy formation in binary metal systems. At the present time there are no such criteria for ternary systems; however, the general laws of physicochemical analysis, the principles of continuity and correspondence, are also applicable to ternary and more complicated systems [233]. The theory of ternary systems has been discussed earlier [138, 140].

Niobium−Zirconium−Titanium. Of all ternary superconducting systems the one which has been studied most fully is the Nb−Zr−Ti system. The very widespread binary superconducting alloys of the Nb−Zr and Nb−Ti systems, in spite of the high critical superconducting parameters which have given them a leading place among other solid-solution alloys, also have certain disadvantages. Alloys of the Nb−Zr system have the highest critical currents and superconducting transition temperatures by comparison with alloys of the Nb−Ti system, which have higher critical magnetic fields and better technological properties, but at the same time comparatively low transition temperatures and a lower critical current in the cold-worked state.

The high-temperature part of the Nb−Zr−Ti system was studied by Mikheev et al. [234], and the solidus surface was constructed by a method based on optical pyrometry (Fig. 159). The results of these and later investigations [236-240] into the structure of alloys belonging to this particular system confirmed the existence of continuous series of solid solutions with bcc lattices above the temperatures of the polymorphic transformations in zirconium and titanium. Figure 160 shows the parameters of Nb−Zr−Ti alloys quenched from the region of the solid solutions [235, 237]. In the region denoted by the broken line the isoparametric lines indicate an anomalously large compression of the lattice of the β solid solution. According to certain data, limiting solubility exists in the system up to 975°C.

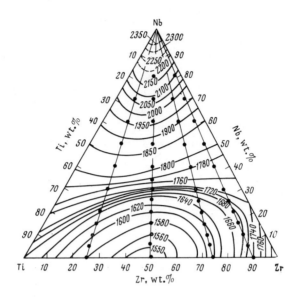

Fig. 159. Solidus lines of the Nb–Zr–Ti system.

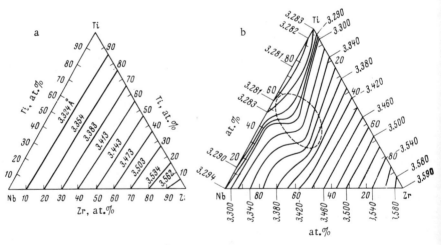

Fig. 160. Lattice parameters (Å) of Nb–Zr–Ti alloys quenched from 1100°C (a) and 1050°C (b).

For the majority of the alloys the β solid solution is a nonequilibrium phase at room temperature, and annealing the alloys at various temperatures causes a change in structure. Phase transformations in the ternary system below 1100°C [238] are due to transformations occurring in the binary systems: in Nb−Zr (monotectoid transformation at 610°C together with the polymorphic transformation of zirconium), Nb−Ti (two-phase $\alpha + \beta$ region below the temperature of the polymorphic transformation of titanium), and Zr−Ti (unlimited solubility of the α and β forms). On the side adjacent to the Nb−Zr system there is a region characterized by a two-phase state, the alloys being formed from two ternary solid solutions based on niobium and zirconium [235-238]. This region expands with decreasing temperature (Fig. 161), and below the temperatures of the polymorphic transformations of titanium (882°C) and zirconium (862°C) two single-phase regions appear on the isothermal cross sections (these consisting of ternary solid solutions based on α_{Zr} and α_{Ti}), together with the corresponding two-phase regions $\alpha_{Ti} + \beta_{Ti}$ and $\alpha_{Zr} + \beta_{Zr}$ (Fig. 161b) [235]. Figure 161c shows an isothermal section at 600°C [237]; in addition to regions associated with the polymorphic behavior of titanium and zirconium and the region of immiscibility of the solid solutions $\beta_{Nb} + \beta_{Zr}$, this section also contains a region of three-phase monotectoid equilibrium $\beta_{Zr} \rightleftarrows \beta_{Nb} + \alpha_{Zr}$. At 500°C almost the whole phase field of the triangle is occupied by the two-phase region $\alpha_{Ti\text{-}Zr}$ (Fig. 161d) [237]. According to certain data [238] titanium reduced the temperature of the monotectoid transformation from 610°C in the binary system to 580-600°C in the ternary system, the monotectoid reaction in the ternary system proceeding over a wide range of temperatures; this contradicts earlier data [236] according to which the ternary region ceased to exist at 500°C. It may well be that in this latter case [236] the alloys under examination were not in an equilibrium state (cast samples were annealed at 350-600°C for 20-120 h). Figure 161e shows the distribution of the phase regions in the Nb−Zr−Ti system at room temperature [238]. Close to the Ti−Zr side is a narrow band of ternary α solid solution starting from the binary Nb−Zr system (up to 2% Nb) and ending in the binary Nb−Ti system (4% Nb). In the Nb-rich region the ternary system includes a region of solid solution starting from the binary Nb−Ti system (37% Ti) and extending to the Nb−Zr system (6% Zr).

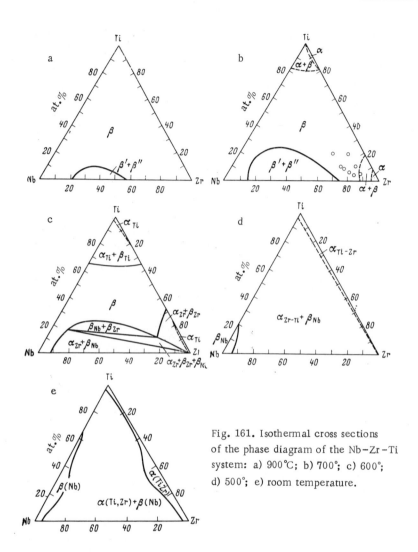

Fig. 161. Isothermal cross sections of the phase diagram of the Nb–Zr–Ti system: a) 900°C; b) 700°; c) 600°; d) 500°; e) room temperature.

Our own x-ray structural investigations into the Nb–Zr–Ti system agree with earlier data [235, 237]. We studied alloys with a constant niobium content (about 50%) and also alloys of the $NbTi_3$–$NbZr_3$ section. The lattice constant of the alloys increases on introducing zirconium into Nb–Ti and decreases on introducing titanium into Nb–Zr alloys. In all these alloys we measured the superconducting characteristics: the superconducting transition temperature (Table 31) and the critical current. The superconducting properties of the alloys were also studied in a number of other papers [237, 239-242, 244, 245]. Using our own data and the results of certain others [237, 240], we plotted curves of equal superconducting-transition temperature for quenched alloys of the Nb–Zr–Ti system on the concentration triangle (Fig. 162). The values of T_c for binary Nb–Zr and Nb–Ti alloys were incorporated in the plot [8]. The maximum T_c occurred for alloys close to the Nb–Zr side.

As already indicated, in the majority of alloys of the system under consideration the bcc solid solution decomposes under annealing for a content of over 10 at.% Zr. Measurements of T_c carried out on alloys annealed at 600°C or lower showed [240] that as a result of the decomposition of the β solid solution the transition temperature of the alloy rose from 8.5 to 9.2°K; this may be attributed to the presence of a β_{Nb} phase (precipitated as a result of the decomposition), with a niobium content greater than that of the original β phase. If there is no decomposition of the solid solution, no changes occur in T_c as a result of annealing [240]; this may of course also be the case if T_c is the same in the original β phase and the precipitating β phase.

TABLE 31. Composition and Superconducting Transition Temperature of Nb–Zr–Ti Alloys

Composition of the alloy, at.%						Phase composition and lattice constant a, Å	T_c, °K
original mixture			chemical composition (analysis)				
Nb	Ti	Zr	Nb	Ti	Zr		
25	75	—	24.7	75	—	β; 3.28	7.2
25	74	1	25.3	74.0	0.65	—	7.4
25	70	5	25.5	69.6	4.78	—	7.5
25	65	10	23.6	66.1	10.3	β; 3.30	7.7
25	50	25	24.05	48.5	24.9	—	7.2
25	40	35	—	—	—	β; 3.38	7.7
25	30	45	28.4	29.5	42.0	—	7.7
25	20	55	33.4	19.3	47.3	β; 3.42	7.5
25	10	65	—	—	—	—	7.5
25	0	75	25.5	—	74.4	β; 3.515	8.3
50	—	50	49.5	—	50.0	β; 3.450	10.2
50	5	45	49.5	4.42	46.08	β; 3.423	10.2
50	10	40	49.6	8.64	41.7	β; 3.410	9.8
50	25	25	47.1	22.3	30.6	β; 3.354	9.7

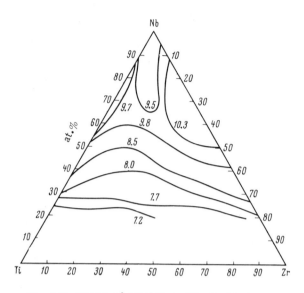

Fig. 162. Curves of equal T_c in Nb–Zr–Ti alloys.

In addition to measuring the T_c of the alloys, the critical magnetic field and critical current density were also determined [240, 241, 241a]. The critical magnetic fields H_{c1} were determined from measurements of magnetic induction and magnetic moment in relation to the external magnetic field (the measurements were carried out at 4.2°K in magnetic fields up to 40 kOe) [240, 241]. The critical magnetic field H_{c2} was measured at 4.2°K in pulsed magnetic fields by observing the restoration of electrical resistance on passing from the superconducting to the normal state. Figure 163 gives the results of these measurements. The H_{c2} surface is concave with a fairly wide plateau in the center of the concentration triangle, where it equals 70-80 kOe [240].

Lazarev et al. [240, 241] note that the low-temperature annealing of Nb–Zr–Ti alloys (600°C and under) has no effect on the value of H_{c2}. Only after severe plastic deformation does the H_{c2} of these alloys rise slightly (by a few percent) [240]. The H_{c1} and H_{c2} values were measured in particular for an Nb–5Zr–48Ti (at.%) alloy [241]. For a sample in the cast state, H_{c1} = 600 Oe, and in the worked state 850 Oe, while on annealing after working the value rose to 1500 Oe; the upper critical magnetic field of all the samples was nevertheless exactly the same (78 kOe) [241].

For ternary superconducting alloys in the cast state, the value of H_{c1} depends little on the composition, and over a wide range of concentrations from Nb–50Zr–15Ti to Nb–5Zr–48Ti (at.%) H_{c1} varies only between 500 and 600 Oe [240].

The critical magnetic fields of niobium alloys containing 30, 50, and 70 wt.% (Ti + Zr) were also measured [242]. The critical magnetic fields were measured in wire samples (diameter 0.3 mm, annealed at 550°C for 1-3 h), using a pulse method [243]. Figure 164 gives the critical magnetic fields of these alloys in relation to their composition. We see from Fig. 164 that in the alloys of section 1 (constant Nb content of 70 wt.%) the magnetic fields decrease on adding zirconium in place of titanium, from 113.6 kOe (Nb–45.4 at.% Ti) to 95.5 kOe (Nb–30 at.% Zr); the initial critical magnetic field decreases by only 4-6 kOe (for 12-14 at.% Zr and 27-25 at.% Ti). A sharper decrease in magnetic field occurs on further increasing the zirconium content, when the alloys exist in the two-phase region of $\alpha_{Zr} + \beta_{Nb}$ [242]. In alloys of other sections (e.g., sections 2 and 3, with respective constant niobium contents of 50 and 30 wt.%), the critical magnetic field is also sensitive to the phase composition of the alloys. However, in section 2 the maximum magnetic field corresponds to the region of the β solid solution, while on increasing the zirconium content and passing into the $\alpha_{Zr} + \beta_{Nb}$ region the field diminishes; in section 3 the region of the β solid solution corresponds to the minimum field. Doi

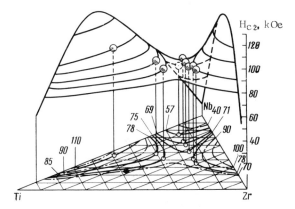

Fig. 163. Spatial representation of the surfaces of the critical magnetic fields H_{c2} at 4.2°K.

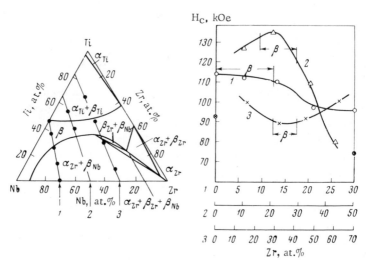

Fig. 164. Isothermal cross section of the Nb–Ti–Zr system at 550°C and values of the critical magnetic fields for alloys lying along sections 1, 2, and 3 annealed at 550°C for 1-3 h.

et al. [244] gave the magnetic field for cold-worked Nb–45.4-at.%-Ti and Nb–30-at.%-Zr alloys (93 and 74 kOe). The annealed alloys of the same compositions studied by Alekseevski et al. [242] had higher values of the critical magnetic fields.

Niobium – Zirconium – Hafnium. On considering the structure of the binary Nb–Zr, Nb–Hf, and Zr–Hf systems, it is reasonable to expect that the ternary Nb–Zr–Hf system will also have continuous series of solid solutions above the temperatures of the polymorphic transformation. Alloys obtained from zirconium of 99.8% purity, niobium (99.9%), and hafnium (99.2%) quenched from 1700 and 1000°C after annealing for 100 and 1000 h respectively have been studied by microstructural and x-ray structural analyses [246]. The x-ray analysis of alloys quenched from 1700°C showed that only alloys with a high niobium content (over 60 at.%) were single-phased; alloys containing 20-60 at.% Nb consisted of two solid solutions (with bcc and hexagonal structures); the x-ray diffraction patterns of alloys containing less than 20 at.% Nb only exhibited the lines corresponding to the hexagonal structure. In the opinion of the authors, the solid solutions with the bcc structure were only fixed by quenching for a high Nb content; for a lower content they decomposed. According to certain other results [247]

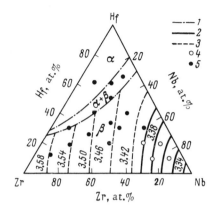

Fig. 165. Proposed phase-equilibrium diagram of the Nb-Zr-Hf system at 1000°C: 1) boundaries of the phase fields: 2) isoparametric lines established; 3) ditto, hypothetical; 4) single-phase alloys; 5) alloys with cubic structure not fixed by quenching.

the temperature of the polymorphic transformation of hafnium equals 1760°C, and it may therefore well be that the quench temperature chosen [246] was insufficient to produce a single-phase solid solution. Figure 165 shows the proposed phase-equilibrium diagram of the Zr–Nb–Hf system at 1000°C.

We have studied arc-melted alloys of the same system (Table 32). The original materials were electron-beam niobium (99.8%), zirconium refined by the iodide method (99.5%), and pure hafnium. The superconducting characteristics T_c and J_c were studied on cold-worked wire samples and the results were correlated with x-ray analysis.

All the alloys were single-phase in the cold-worked state and had the structure of the β solid solution; the bcc lattice constant increased with increasing hafnium content; the value of T_c remained constant within the limits of measuring error (0.2°K) for solid solutions of the compositions indicated.

TABLE 32. Composition and Superconducting Transition Temperature of Nb–Zr–Hf Alloys

Composition of alloys in original mixture						Composition of alloys according to chemical analysis, wt.%		Phase composition and lattice constant a, Å	T_c, °K	Concentration of valence electrons per atom, el./at.
at.%			wt.%							
Nb	Zr	Hf	Nb	Zr	Hf	Nb	Zr+Hf			
50	45	5	48.39	42.75	8.86	47.3	52.6	β; 3.435	10.0	4.5
50	40	10	46.05	36.25	17.7	46.67	52.92	β; 3.438	10.1	4.5
50	25	25	40.75	20.05	39.2	40.54	59.16	β; 3.451	10.1	4.5

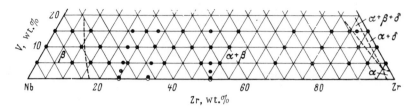

Fig. 166. Chemical and phase compositions of the alloys of the Nb–Zr–V system studied.

Niobium–Zirconium–Vanadium. The structure and superconducting properties of various alloys belonging to this system were studied by Savitskii et al. [248]. Alloys close to the Nb–Zr side (with vanadium contents of 5, 10, and 15 wt.%) were melted in an arc furnace with a nonconsumable tungsten electrode. The original materials were iodide-type zirconium (99.9%), electron-beam-melted niobium (99.99%), and carbothermal vanadium (99.75%). In order to determine the structure of the alloys, these were subjected to microstructural analysis in the cast, annealed, and cold-worked states, and also to x-ray phase analysis. Figure 166 shows the chemical and phase compositions of the alloys studied. Most of the cast alloys were single-phase (a solid solution with a bcc lattice), this being due to their having been quenched in a water-cooled copper mold attached to a vacuum arc furnace.

TABLE 33. X-Ray Analysis of Niobium–Zirconium Alloys Containing 5% Vanadium

Zr, wt.%	Treatment*	Phase composition	Lattice constants		Zr, wt.%	Treatment*	Phase composition	Lattice constants	
			a	c				a	c
0	W	β	3.289	—	45	W + A	β	3.275	—
0	W + A	β	3.289	—			α	3.241	5.183
10	W + A	β	3.306	—	47.5	W	β	3.422	—
20	W + A	β	3.314	—	47.5	W + A	β	3.288	—
24	W + A	β	3.285				α	3.247	5.183
		α	3.241	5.187	50	W + A	β	3.295	—
24	W	β	3.379	—			α	3.246	5.184
25	W + A	β	3.310	—					
		α	—	—					
35	W + A	β	3.288	—			β	3.303	—
		α	—	—	65	W + A	α	3.245	5.182

* W = 99% cold-worked; A = annealed at 900°C for 1 h.

After annealing, only the Nb-base alloys were single-phase, while the majority were two-phased and consisted of solid solutions based on niobium (bcc lattice) and zirconium (hexagonal lattice). In the alloy containing 3% Nb and 15% V a compound V_2Zr appeared in addition to the α and β solid solutions. X-ray data relating to the cold-worked and annealed Nb–Zr alloys containing 5% V are presented in Table 33.

The superconducting transition temperature was measured for all these alloys. The transition temperature of the ternary Nb–Zr–V alloys was lower than the T_c of the binary Nb–Zr alloys with the same Nb:Zr ratio (Fig. 167); with increasing vanadium content the value of T_c diminished.

<u>Niobium–Zirconium–Tantalum</u>. There is almost no information at all relating to the structure of the Nb–Zr–Ta phase diagram, except for some confined to the zirconium corner of the diagram [249]. On the basis of the structure of the Nb–Zr, Nb–Ta, and Zr–Ta binary systems we may well expect that in the ternary Nb–Zr–Ta system all the alloys will crystallize below the melting point in the form of ternary solid solutions with a bcc structure. On reducing the temperature, a two-phase region of solid solutions due to the monotectoid decomposition in the Nb–Zr and Zr–Ta system appears, and below the temperature of the monotectoid equilibrium yet another two-phase region of a ternary bcc solid solution and a hexagonal phase appear as a result of the polymorphic transformation of zirconium.

The superconducting properties of particular alloys belonging to the system in question have been studied by a number of authors [250, 251].

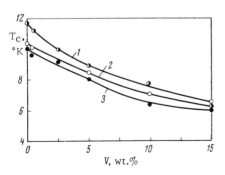

Fig. 167. Superconducting transition temperature of alloys of niobium with 25 (1), 33 (2), and 50 wt.% Zr (3) in relation to their vanadium content.

TABLE 34. Composition and Superconducting Transition Temperature of Nb-Zr-Ta Alloys

Alloy composition, wt.%						Phase composition and lattice constant a, Å	T_c, °K
original mixture			by chemical analysis				
Nb	Zr	Ta	Nb	Zr	Ta		
48.15	42.5	9.35	48.77	40.2	11	β; 3.422	10.4
46	36.1	17.9	50.02	36.48	13	β; 3.385	10.3
40.6	19.9	39.5	38.5	20.7	40.9	β; 3.350	9.9

We have studied alloys belonging to this system for a constant niobium content of about 50 at.% (Table 34). In the cast (quenched) state all the alloys had a bcc structure. The critical superconducting temperature and critical current density of cold-worked alloys were measured in transverse magnetic fields up to 26 kOe, using wire samples 0.25 mm in diameter. The transition temperature of alloys of the compositions indicated was measured by the magnetic method (by noting the change in the magnetic permeability of the sample). We see from Table 34 that an increase in the tantalum content of these alloys from 11 to 40.9 wt.% has hardly any effect on the critical superconducting temperature.

The I_c values of alloys subjected to electron-beam melting, homogenized, and cold-worked were measured by Rose et al. [250] in fields up to 90 kOe.

Shmulevich et al. [251, 252] studied the possibility of increasing the superconducting characteristics in the presence of microstructural inhomogeneities. In an Nb-35Zr-15Ta (wt.%) alloy obtained by electron-beam melting with various degrees of inhomogeneity (the melting being carried out under different thermal conditions), these authors studied the I(H) relationship, and discovered an effect in which I_c rose in the presence of a high degree of microinhomogeneity [251]. According to x-ray structural analysis the alloys constituted a single-phase solid solution with a bcc lattice ($a_{\text{homog}} = 3.398$ Å, $a_{\text{liquat}} = 3.390$ Å).

Figure 168 shows the dependence of the T_c of Nb-Zr-Ta alloys on the composition and various degrees of inhomogeneity [252] created by zone melting. We see from Fig. 168 that T_c decreases with diminishing degree of liquation (from 10.6°K in the cast state to 10.17°K in the homogenized state). The authors considered that the tendency for the superconducting transition temperature to decrease with diminishing microstructural inhomogeneity might be explained not simply on the basis of fluctuations in composition but also on that of the segregation of various impurities contained in the alloy in the course of liquation. Plotting the transition temperature against the composition of the alloys showed that, among all the alloys with a 50% Nb content, the alloy containing 35% Zr and 15% Ta had the greatest value of T_c. Further increasing the tantalum content reduced the critical superconducting temperature.

Niobium−Zirconium−Molybdenum. The structure of the Nb−Zr−Mo ternary system was studied in a number of papers [253-255]. In the first of these [253] the structure of three polythermal sections of the phase diagram (with constant Mo:Zr ratios of 2:1, 1:1, and 1:3) was studied. The investigation was based on microstructural and x-ray analysis, measurements of microhardness, and the initial and final melting temperatures. The alloys were prepared in an arc furnace with a nonconsumable tungsten electrode using metalloceramic niobium, molybdenum (99.9% purity), and iodide-type zirconium. The resultant samples were subjected to a homogenizing vacuum-anneal for 2-2.5 h at 1900-1500°C. The annealed alloys were hammer-forged (upset) with a 50% degree of deformation. Then the alloys were quenched from 2000°C, 1900, 1800, 1700, 1500, and 1100°C after holding for 5, 7, 10, 15, 20, 25 min and 100 h.

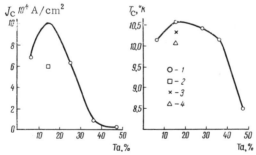

Fig. 168. Variation in the J_c and T_c of Nb−Zr−Ta alloys containing 50 wt.% Nb: 1) alloy with substantial microinhomogeneity; 2) alloy with slight microinhomogeneity; 3) after homogenization at 1200°C (3 h); 4) after homogenization at 1200°C (8 h) + 1500° (4 h).

The range of homogeneity of the β solid solution (Nb-base) was deduced from the microhardness measurements and microstructural analysis. The combined solubility of molybdenum and zirconium taken in a ratio of 2.1:1 was 23-27 wt.% at 1800°C and 11-12% at 1100°C; the solubility decreased sharply with decreasing temperature. All the alloys quenched from the various temperatures had either a single-phase structure (β solid solution of molybdenum with zirconium in niobium) or else a two-phase structure (crystals of the β solid solution and the compound Mo_2Zr).

These results facilitate a systematic construction of the projection of the liquidus of the ternary Nb-Zr-Mo diagram on the plane of the concentration triangle (Fig. 169). The peritectic curve pk passes into the eutectic curve ke, forming a region of crystallization of the compound Mo_2Zr from the liquid (pke). The branch pk corresponds to the composition of the liquid which reacts with the β solid solution in accordance with the reaction $l + \beta \rightleftarrows Mo_2Zr$; the composition of the molybdenum-rich crystals varies along the curve an. The section ke characterizes the liquid solutions from which the binary eutectic precipitates $l \to \beta + Mo_2Zr$; the curve nb characterizes the composition of the β solid solution based on the high-temperature form of zirconium. The region pkna is the surface constituting the locus of temperatures corresponding to the onset of the monovariant peritectic reaction $l + \beta \to Mo_2Zr$; the surface corresponding to the onset of crystallization of the binary

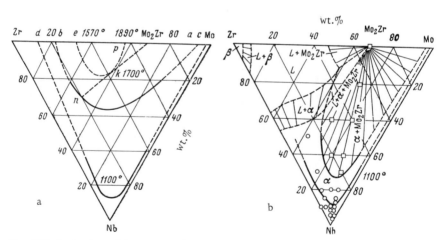

Fig. 169. Projection of the liquidus (a) and isothermal section of the ternary Nb-Zr-Mo system at 1800°C (b).

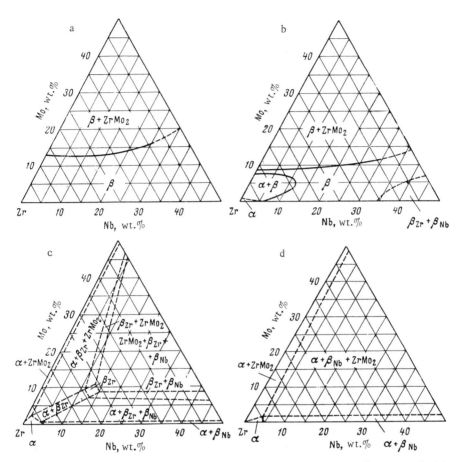

Fig. 170. Isothermal sections of the ternary Nb-Zr-Mo system at 1100°C (a), 800°C (b), 600°C (c), and below the eutectoid temperature (d).

eutectic ($\beta \rightarrow Mo_2Zr$) is the region eknb. The straight line Mo_2Zr–kn corresponds to the moment of the transition of the monovariant peritectic transformation $l + \beta \rightarrow Mo_2Zr$ into the eutectic transformation $l \rightarrow \beta + Mo_2Zr$. Figure 169b gives the isothermal section of the system at 1800°C. The same figure indicates the boundary of the β solid solution (Nb-base) at 1100°C.

Figure 170 presents the isothermal sections of the ternary system under consideration at 1100, 800, 600, and 500°C [254]. The temperature of the peritectic $l + Mo \rightleftarrows ZrMo_2$ at a composition of 32.2% Zr falls gradually with increasing niobium content and coincides with the eutectic $l \rightleftarrows \beta + ZrMo_2$ at a composition of

TABLE 35. Composition and Superconducting Transition Temperature of Nb–Zr–Mo Alloys

Composition of the alloys, wt.%						Phase composition and lattice constant a, Å	T_c, °K
original mixture			by chemical analysis				
Nb	Zr	Mo	Nb	Zr	Mo		
50.9	48.05	1.05	—	52.57	0.94	β; 3.439	9.8
50.35	44.45	5.2	51.2	44.31	3.97	β; 3.356	9.9
50.2	39.4	10.4	49.4	39.3	12.0	β; 3.384	8.8
49.8	24.45	25.75	49.7	27.4	23.8	β; 3.292	5.3

69% Zr. The ternary eutectoid $\beta \rightleftarrows \beta_{Nb} + \alpha + ZrMo_2$ is formed at a temperature of 535°C and a content of 13% Nb and 8% Mo [254]. These data supplement other results [253], in which the ternary alloys were studied at 1100°C and over, but the low-temperature region was omitted. The results of individual investigations disagree in relation to the limits of existence of the ternary β solid solution. Murakami et al. [254] give a wider range of existence of the solid solution at 1000°C than that illustrated in Fig. 169b for the β solid solution; this may be due to different prequench holding periods in the two cases.

Popov et al. [255] also studied the phase diagram of this system; according to these results (obtained at 1440°C) the range of the β solid solution remains intact in alloys containing over 40% Nb.

We have studied the superconducting properties of Nb–Zr–Mo alloys with a constant content of 50 at.% Nb. The method of preparation and studying the alloys was analogous to that employed previously for the Nb–Zr–Hf system. We see from the data presented in Table 35 that all the quenched alloys were single-phase solid solutions with a bcc lattice, in agreement with the earlier data [254, 255] relating to the structure of these alloys.

The critical superconducting temperature measured by reference to the magnetic permeability decreases with increasing molybdenum content (to 5.3°K for 23.8 wt.% Mo), for a constant proportion of niobium.

Niobium–Zirconium–Tungsten. The structure of the phase diagram of this system has been studied on a number of occasions [253, 256]. The alloys were prepared by melting in an arc furnace in an atmosphere of purified helium, using 99.9% pure tungsten, metalloceramic niobium (99.4%), and iodide-type zirconium. The investigation was based on microstructural and x-ray analyses, together with measurements of microhardness and the melting point of the alloys. The alloys were homogenized at 1500-1900°C (depending on composition) in a vacuum of $5 \cdot 10^{-6}$ mm Hg for 2-2.5 h, worked with a 50% degree of deformation, and then quenched from various temperatures. Figure 171 gives the isothermal cross sections at 2000, 1600, and 1100°C and a projection of the liquidus on the plane of the concentration triangle. The

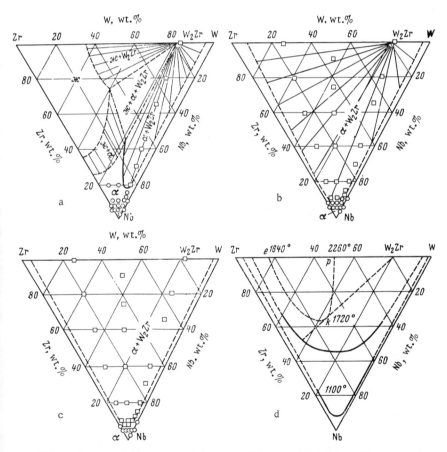

Fig. 171. Isothermal cross sections of the phase diagram of the Nb-Zr-W system: a) at 2000°C; b) 1600°C; c) 1100°C; d) projection of the liquidus on the concentration triangle.

TABLE 36. Composition and Superconducting Transition
Temperatures of Nb–Zr–W Alloys

Composition, wt.%					Phase composition and lattice constant a, Å	Superconducting transition temperature, °K
original mixture			by chemical analysis			
Nb	Zr	W	Zr	W		
49.8	47.2	3.0	49.95	2.92	β; 3.413	10.3
45.8	36.1	18.1	39.30	14.50	β; 3.378	9.1
40.4	19.8	39.8	21.58	39.76	β; 3.235	8.9

combined solubility of tungsten and zirconium in niobium falls from 16-18% at 2000° to 8-9% at 1100°C. The compound W_2–Zr is formed by the peritectic reaction $l + \alpha \rightarrow W_2Zr$ at 1720°C. For temperatures below 1720°C (but above 1640°C) a eutectic is formed $l \rightarrow \alpha + W_2Zr$.

We further studied the superconducting properties of certain alloys belonging to the Nb–Zr–W system, the composition of these being indicated in Table 36. We measured the superconducting transition temperature by observing the change in the magnetic permeability of the sample (measuring accuracy ± 0.2°K). The results show that T_c depends very little on the amount of tungsten in the alloys, decreasing by a total of 1.4°K as the tungsten content is increased from 1.5 to 25 at.%.

Niobium – Zirconium – Copper. The ternary phase diagram of this system has never been fully studied. After considering the binary phase diagrams we may well expect that there will be a certain range of solubility of zirconium and copper in niobium at 1000°C, close to the Nb–Zr side, while for greater copper contents chemical compounds will be formed.

Frolov et al. [257] studied a number of alloys along two sections with constant Zr:Nb ratios of 4:6 and 3:7 (copper content up to 40 and 24 wt.% respectively) and several alloys in the niobium corner (Fig. 172). The alloys were prepared by arc melting from electron-beam-refined niobium (99.6% purity), iodide-type zirconium (99.7%), and very pure copper (99.99%). After homogenization at 1000°C for 300 h the alloys were quenched in water. The alloys were studied by microstructural and x-ray analysis and by measurements of hardness, microhardness, and the superconducting characteristics (the transition temperature and critical current)

The solubility of copper in the alloys close to the Nb–Zr side was less than 0.3 wt.%.

Measurements of the superconducting transition temperature (Fig. 173) showed that on introducing copper into the niobium alloy with 30-50% Zr the T_c decreased particularly sharply for low copper concentrations, while further addition of copper left the transition temperature at a fairly high level (10-10.1°K) [273].

Niobium – Titanium – Hafnium. Bychkova et al. [258] studied the structure and superconducting properties of several such alloys. The phase diagram of the system was not fully constructed; however, on the basis of the structures of the binary Nb–Ti, Nb–Hf, and Hf–Ti systems it is reasonable to expect a continuous series of solid solutions with a bcc lattice above the

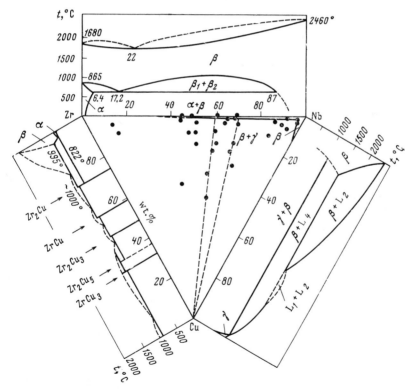

Fig. 172. Composition of the Nb–Zr–Cu alloys studied, expressed in the form of points in the concentration triangle, and range of existence of the solid solution of zirconium and copper in niobium at 1000°C.

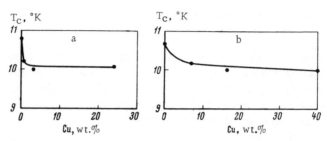

Fig. 173. Variation in the superconducting transition temperature of Nb-Zr-Cu alloys along sections with a Zr:Nb ratio of 3:7 (a) and 4:6 (b).

temperatures of the polymorphic transformations of the constituents.

Bychkova et al. [258] carried pit microstructural and x-ray analyses of alloys containing about 25 at.% Nb and measured the critical superconducting temperature and current. The alloys were prepared by arc melting, the original materials being electron-beam-refined niobium and iodide-refined titanium and hafnium. The resultant samples were homogenized in a vacuum furnace at 1500°C for 10 h, then further annealed for 200 h at 1100°C and water-quenched from this temperature. Cold-worked wire samples (d = 0.25 mm) were subjected to x-ray analysis (the phase composition and lattice constants being determined), the electric resistance was measured at room temperature, and the critical superconducting temperature and current were determined in a transverse magnetic field of up to 26 kOe.

Alloys of the Nb–65Ti–10Hf and Nb–50Ti–25Hf (at.%) type constituted single-phase solid solutions with a bcc lattice. The lattice constant increased after introducing hafnium, from 3.282 Å for Nb–75Ti to 3.35 Å for Nb–50Ti–25Hf. The transition temperature of the alloys changed very little on increasing the hafnium content from 10 to 25 at.% (6.8 and 6.9°K respectively).

Niobium – Titanium – Vanadium. Vanadium, niobium, and β-titanium form solid solutions at all concentrations (Fig. 174) [259-263]. The solidus surface rises continuously from the titanium corner of the system. The presence of flat minima on the solidus curves of the binary V–Ti and V–Nb systems produces indentations on the solidus surface of the ternary system. At temperatures below 885°C there is a range of α-Ti-base solid solutions in the titanium corner of the system, and a two-phase region separating the single-phase α and β regions. At 600°C the surface bounding the α region passes from a composition of 4 wt.% Nb on the Ti–Nb side to 3 wt.% V on the Ti–V side. The $\beta/\alpha + \beta$ surface of separation at 600°C passes through a line connecting the titanium alloy with 53 wt.% Nb to the titanium alloy with 30 wt.% V. No ternary compounds are formed in the system.

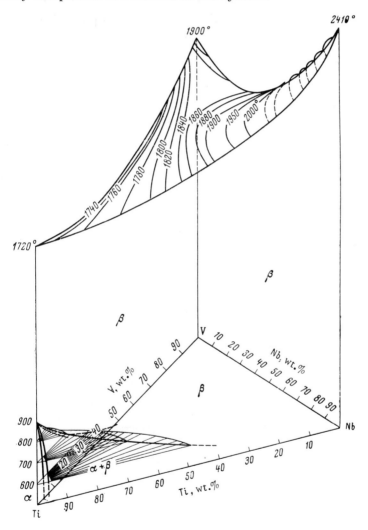

Fig. 174. Spatial representation of the phase diagram of the Ti–Nb–V system.

The structure and superconducting properties of Nb–V–Ti alloys were studied in [258, 262, 262a]. The compositions of the alloys required in order to calculate the composition–property diagram by the Simplex-lattice method [262b, 262c] are presented in the concentration triangle of the composition–T_c diagram of the system in question (Fig. 175). A number of other alloys were also studied; data relating to these appear in Table 37 [258, 262].

We see from Fig. 175 that the T_c surface falls off sharply from the maximum of the Nb–Ti system in the Ti direction and slowly in the V direction. The T_c of the majority of the ternary alloys is higher than that of vanadium or titanium and lower than that of niobium. The ternary alloys with the highest T_c (around 10°K) lie close to the T_c maximum on the Nb–Ti side (up to 5 at.% V). The critical temperature in the sections with constant V (up to 40 at.%) or Nb (up to 30 at.%) content has a maximum at an elec-

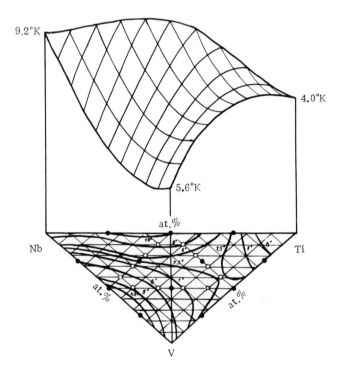

Fig. 175. Spatial composition–T_c diagram of Nb–V–Ti alloys.

TABLE 37. Composition and Superconducting Transition Temperature of Nb-V-Ti Alloys

Composition, at.%			Phase composition and lattice constant a, Å	T_c, °K	Concentration of valence electrons per atom, el/at	Composition, at.%			Phase composition and lattice constant a, Å	T_c, °K	Concentration of valence electrons per atom, el/at
Nb	V	Ti				Nb	V	Ti			
—	50	50	β; 3.175	7.8	4.5	25	—	75	β; 3.28	7.2	4.25
1	50	49	β; 3.166	7.5	4.51	25	5	70	β; 3.27	7.7	4.3
5	50	45	β; 3.155	7.4	4.55	25	10	65	β; 3.25	8.0	4.35
10	50	40	β; 3.149	7.1	4.6	25	25	50	β; 3.215	7.3	4.5

tron concentration of 4.5-4.7 electrons/atom; the maximum moves in the direction of lower electron concentration on moving away from the side of the binary alloys in the concentration triangle.

Niobium−Titanium−Tantalum. The structure and properties of the alloys of this system were studied in [258, 262a, 262d, 262e]. At high temperatures this system contains a continuous series of solid solutions (beta) with a bcc lattice. At temperatures below the polymorphic transformation of titanium there is a region of ternary alpha solid solution with a hexagonal structure and a corresponding ($\alpha + \beta$) region. The composition−T_c diagram plotted with the help of a computer (Simplex-lattice method) on the basis of experimental data obtained for alloys melted in an arc furnace in an atmosphere of purified helium, homogenized at 1400°C (3 h) after 50% deformation, and quenched from this temperature is shown in Fig. 176 [262a]. The T_c of the single-phase ternary alloys with the bcc lattice varies smoothly without any extremal points. The highest T_c correspond to alloys of the binary Nb-Ti system (40-60 at.% Ti). The T_c surface of the ternary system has a saddle-shaped plateau at 40-60 at.% Ti and 5-20 at.% Nb. The T_c of the ternary alloys in all sections with a constant Nb (up to 40 at.%) or tantalum (up to 30 at.%) content has a maximum at 4.7-4.8 electrons/atom. Our own data agree with the result of [262d] in which the T_c surface is plotted by traditional methods using a large number of alloys.

Our own data also agree with earlier studies regarding individual alloys of this system. The results of [258] appear in Table 38.

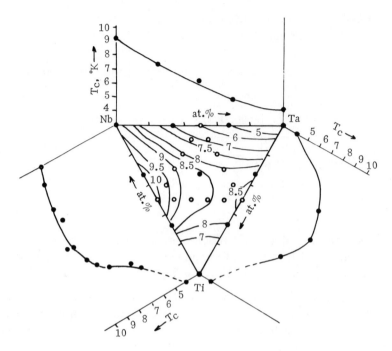

Fig. 176. Composition-T_c diagram of Nb-Ta-Ti alloys.

TABLE 38. Composition and Superconducting Transition Temperature of Nb-Ti-Ta Alloys

Composition in the original mixture, at.%			Composition by chemical analysis, at.%			Phase composition and lattice constant a, Å	T_c, °K
Nb	Ti	Ta	Nb	Ti	Ta		
25	75	—	24.8	75.1	—	β; 3.29	7.2
25	74	1	25.3	72.5	2.1	—	7.55
25	70	5	24.2	69.8	6.0	—	7.85
25	65	10	22.4	66.6	11.0	—	8.4
25	50	25	28.3	43.8	28.8	β; 3.282	8.6
24	75	1	23.6	75.6	0.8	—	7.4
20	75	5	19.7	75.0	5.3	—	7.1
15	75	10	13.7	76.0	10.3	β; 3.274	6.7
0	75	25	—	75.0	25.0	β; 3.281	6.0

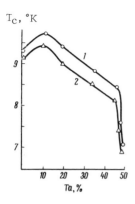

Fig. 177. Dependence of the T_c of Nb–Ti–Ta alloys (with 50 wt.% Nb) on composition: 1) cast state; 2) after homogenization at 1500°C for 4 h.

The dependence of T_c on the composition of alloys belonging to this system was studied elsewhere [252] along a section with a constant niobium content (from Nb–50% Ti to Nb–50% Ta). We see from Fig. 177 that replacing some of the titanium in the alloy by tantalum (up to 10 wt.%) raised the T_c value, while further increasing the tantalum content reduced it (to 6.9°K for Nb–50% Ta). The transition temperature of the alloys after a homogenizing anneal (1500°C, 4 h) was slightly lower than in the cast samples.

The critical magnetic fields (Fig. 178) were measured under pulse conditions at 4.2°K with a current density of 30 A/cm² [262d]. There is a wide range of equal values of H_{c2} (> 120 kG) around the composition 63 at.% Ti–31 at.% Nb–6 at.% Ta.

Niobium – Titanium – Molybdenum. The phase diagram of this system was studied by Kornilov [263]. The alloys were prepared from titanium (99.5% pure), niobium (98.9%), and molybdenum (99.9%) by sintering and melting. The investigation included differential thermal, metallographic, and x-ray analysis, and also measurements of the specific gravity, hardness, and electrical resistance. It was found that, above the temperature of the polymorphic transformation of titanium, there was a continuous series of solid solutions in the Nb–Ti–Mo system. At room temperature the continuity of the solid solution was limited. In the titanium corner there was a narrow range of Ti-base α solid solutions bounding the $\alpha + \beta$ two-phase region, which on increasing the niobium and molybdenum content passed into a region of ternary β solid solution with a bcc lattice (Fig. 179).

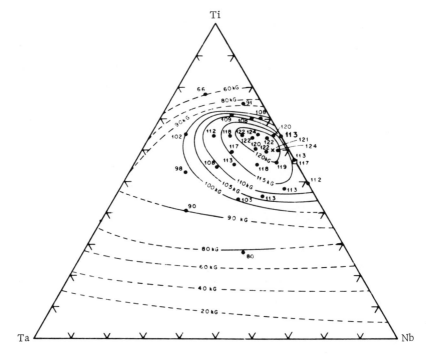

Fig. 178. Composition—upper critical field diagram of Nb–Ta–Ti alloys.

Bychkova et al. [258] studied the structure of several alloys belonging to this system and also measured their superconducting properties. The results of microstructural and x-ray analyses confirmed the existence of a single-phase solid solution with a bcc lattice in alloys of all compositions quenched from 1100°C the lattice constant of the alloys decreasing with increasing molybdenum content (Table 39).

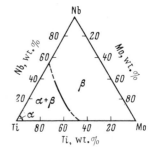

Fig. 179. Phase diagram of the Nb–Ti–Mo system at room temperature.

TABLE 39. Composition and Superconducting Transition
Temperature of Nb–Ti–Mo Alloys

Composition of the alloys, at.%						Phase composition and lattice constant a, Å	T_c, °K
original mixture			by chemical analysis				
Nb	Ti	Mo	Nb	Ti	Mo		
25	74	1	25.4	74.3	0.44	—	7.3
25	70	5	25.4	69.6	5.0	—	7.4
25	65	10	27.77	63.6	8.8	β; 3.24	6.8
25	50	25	27.7	46.3	26.05	β; 3.22	4.3

On increasing the Mo content above 5 at.% the critical temperature falls more sharply than on introducing small quantities of molybdenum into the alloy.

Niobium – Titanium – Tungsten. The ternary Nb–Ti–W phase diagram was studied by Bychkova et al. [264]. Seventy alloys were studied, these lying on six radial sections with different Ti:W ratios (1:19, 1:4, 7:13, 1:1, 6:4, 17:3), chosen in such a way as to lie simultaneously on sections parallel to the Nb–Ti side, i.e., with a constant tungsten content. The alloys were prepared by arc melting and were studied by microstructural, x-ray, and thermal analyses and by measuring the microhardness. It was established as a result of studying alloys annealed at 1000°C that in the Nb–Ti–W system (at 1000°C and over) a wide range of ternary solid solutions (up to 42 wt.% Nb) with a bcc lattice was formed, together with a two-phase region constituting a mixture of two bcc solid solutions; the boundaries of these appear in Fig. 180. Polythermal sections and liquidus isotherms were plotted for this system.

We have studied the superconducting transition temperature of these alloys as a function of composition (Fig. 181). On considering a cross section with a constant tungsten content in this system, it is easy to see that, on introducing titanium into an alloy with 10% W, the critical temperature rises from 6.7 to 7.3°K, while further increasing the titanium content (to 48 wt.%) changes it very little (to 7.5°K). On increasing the tungsten content in ternary alloys with 20 or 30% W beyond 30%, T_c decreases.

Niobium – Titanium – Rhenium. There are no published data regarding the structure of the ternary Nb–Ti–Re sys-

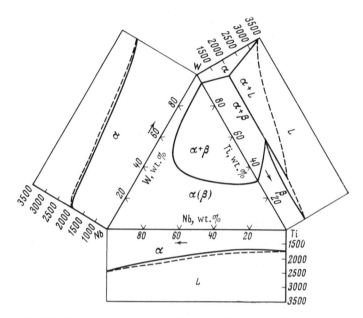

Fig. 180. Isothermal section of the phase diagram of the Nb-Ti-W system at 1000°C.

tem. On the basis of the structure of the binary Nb-Ti, Ti-Re, and Nb-Re diagrams we may expect that, at temperatures above the polymorphic transformation of titanium, there will be a wide range of solid solutions with a bcc structure adjacent to the Nb-Ti side of the concentration triangle. The structure and superconducting properties of several alloys in this system have in fact been studied [258]. The composition of the alloys was chosen in such a way that, for a constant niobium content (about 25 at.%), the titanium and rhenium concentration varied (from 1 to 25 at.% Re), while all the alloys lay within the range of the ternary solid solution. Wire 0.25 mm in diameter was cold-drawn from samples with rhenium contents of 1 and 5%; alloys with higher rhenium contents were not susceptible to cold working. In addition to the lines of the bcc solid solution, the x-ray patterns of all the alloys exhibited two unidentified lines, which may have belonged to a hexagonal phase (no single-phase solid solution was fixed by quenching). With increasing proportion of rhenium in the alloys the transition temperature decreased from 7.2°K (for Nb-74% Ti-1% Re) to 6.85°K (for Nb-70% Ti-5% Re). Further raising the rhenium

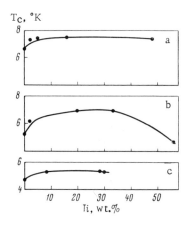

Fig. 181. Change in the superconducting transition temperature of Nb–Ti–W alloys with constant W contents: 10 (a), 20 (b), and 30 wt.% (c).

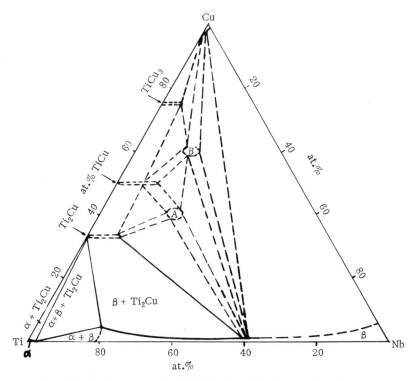

Fig. 182. Isothermal section (700°C) of the Nb–Ti–Cu phase diagram.

content to 10% reduced T_c to 5.4°K. The alloy containing 25 at.% Re was nonsuperconducting at the measuring temperature (4.2°K).

Niobium − Titanium − Copper. The structure of an isothermal section of the Nb−Ti−Cu system was studied at 700°C (Fig. 182). In the β-(Ti, Nb) solid solution existing at 700°C between 20 and 100 at.% Nb, 1-5 at.% of copper are able to dissolve. In alloys richer than this in copper the β solid solution is in equilibrium with α-Ti, the compound Ti_2Cu, proposed ternary compounds of the intermetallic type A and B, and a Cu-base solid solution. The ternary compounds A and B may be copper-stabilized forms of the binary compounds Ti_2Cu_3 (B) and TiCu (A). When studying the superconducting properties of certain alloys of this system, an increase occurred in the critical current density of Nb−45 wt.% Ti and Nb−61 wt.% Ti alloys on introducing 3-8 wt.% of copper into these; the critical superconducting temperature remained high [264a]. For example, in the cold-worked state the critical current density of Nb−43.1 wt.% Ti−8.4 wt.% Cu was $1.1 \cdot 10^5$ A/cm^2 in a transverse magnetic-field of 36.5 kOe at 4.2°K, while in the case of the alloy with the same ratio of Nb and Ti but no Cu and under the same conditions it equalled $8 \cdot 10^3$ A/cm^2. The T_c of this alloy without Cu was 9.1°K and with Cu 7.9°K. The effect of the rise in the critical current density of the Nb−Ti−Cu alloys was attributed to the presence of precipitates of the nonsuperconducting phase Ti_2Cu, forming "pinning" centers for the magnetic flux.

Niobium − Titanium − Aluminum. The phase diagram of the Nb−Ti−Al system was studied in [264b, 264c]. The system contains a fairly wide range of the β solid solution of aluminum in Nb−Ti alloys. We studied the alloys in the part of the β solid solution close to the Nb−Ti side, with a constant Nb:Ti ratio of 45:55 and an Al content of up to 7.2 wt.% in the cast state and after annealing at 1200°C [264d]. According to the results of microstructural and x-ray analysis the cast and annealed samples constituted a single-phase β solid solution of titanium and aluminum in niobium. In all the alloys studied we found a monotonic fall in T_c both in the cast and in the annealed states as aluminum content increased. The smallest values of T_c were 4.4°K in the cast state and 6.5°K in the annealed state in the alloy containing 7.2 wt.% Al. The decrease in T_c may be attributed to the well-known influence of a nontransition metal (Al) on the T_c of transition metals. The in-

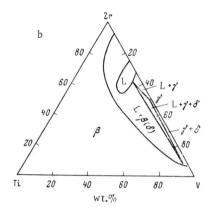
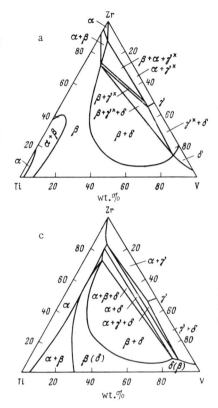

Fig. 183. Isothermal section of the V–Ti–Zr ternary phase diagram at 750°C (a), 1300°C (b), and room temperature (c).

troduction of aluminum into an Nb–45-wt.%-Ti alloy leads to a decrease in the critical current of cold-worked samples 0.2 mm in diameter in magnetic fields up to 40 kOe. In alloys containing 7.2% Al the critical current decreases to a very small value (0.05 A at 6 kOe and 0.03 A at 11 kOe).

Vanadium–Titanium–Zirconium. Nowikow [265] made an experimental study of an isothermal section of the ternary system at 750°C (Fig. 183). The beta solid solution of the binary Ti–Zr and Ti–V systems forms a wide range of ternary solid solutions with a bcc lattice on combining with the δ solid solution of the V–Zr system. The β region is separated from the V–Zr side by a wide two-phase (β + δ) region. The solubility of vanadium in the α solution of titanium with 1.3 at.% Zr at 750°C is 1.2 at.% [266]. The three-phase (β + γ^x + δ) region exists in an extremely wide range of concentrations (from 10 wt.% V with 70% Zr to 80% V with

TABLE 40. Composition and Superconducting Transition Temperature of V–Ti–Zr Alloys

Composition, at.%			Phase composition and lattice constant a, Å	T_c, °K
V	Ti	Zr		
50	50	—	β; 3.170	7.8
50	49	1	β; 3.079	7.7
50	45	5	β; 3.148	7
50	40	10	β; 3.158	5.5

15% Zr); γ^x is a phase based on the compound ZrV_2, but having a different crystal structure [265]. At temperatures exceeding the polymorphic transformation of titanium and zirconium, the β range widens, and at 1300°C it occupies the whole space with titanium contents of over 20 wt.% (Fig. 183). At room temperature, there is a continuous series of solid solutions between α-titanium and α-zirconium containing up to 10% V [265]. The temperature of the eutectic $l \rightleftarrows \beta + \gamma$, peritectic $l + \delta \rightleftarrows \gamma$, and eutectoid $\beta_{Zr} \rightleftarrows \alpha + \gamma'$ reactions of the V–Zr system decreases in the ternary system as the titanium content of the alloys increases.

Efimov et al. [262] studied the properties of several alloys belonging to the system under consideration (Table 40). X-ray

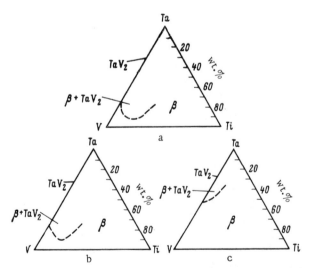

Fig. 184. Isothermal sections of the ternary phase diagram of the V–Ti–Ta system at 800 (a), 1000 (b), and 1200°C (c).

analysis showed that all the alloys in the cast (quenched) state constituted solid solutions with a bcc lattice, which agrees with the structure of the V−Ti−Zr phase diagram. The lattice constants of the ternary alloys increase with zirconium content, while the critical temperature is lowered.

Vanadium − Titanium − Tantalum. Isothermal sections of this partially-studied system are given in Fig. 184 for 800, 1000, and 1200°C [267]. The system has a wide range of β solid solutions with a bcc lattice. The boundary of the two-phase (β + TaV_2) region is indicated in the figure. The solubility of titanium in TaV_2 is unknown. There is a continuous series of solid solutions in the system at high temperatures [267a].

The composition−T_c diagram of the whole V−Ti−Ta system (β phase) (Fig. 185) was recently plotted by our Simplex-lattice computer method [267a]. The T_c surface of this system has a

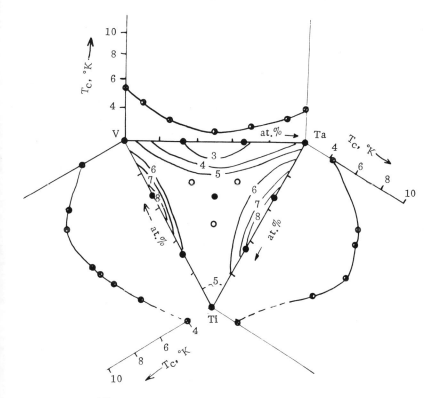

Fig. 185. Composition−T_c diagram of the V−Ta−Ti system.

TABLE 41. Composition and Superconducting
Transition Temperature of V–Ti–Ta Alloys

Composition, at.%			Phase composition and lattice constant a, Å	T_c, °K	Concentration of valence electrons per atom, el/at
V	Ti	Ta			
50	50.0	—	β; 3.175	7.8	4.5
50	49.5	0.5	β; 3.208	7.8	4.505
50	49.0	1.0	β; 3.164	7.0	4.51
50	45.0	5.0	β; 3.152	7.0	4.55
50	40.0	10.0	—	6.7	4.6

saddle-shaped plateau, this being rather insignificantly expressed as a result of the fact that the critical temperatures of the original components are almost identical and very low. The greatest T_c values occur in binary V–Ti and Ta–Ti alloys containing 50-70 at.% of Ti. On adding a third component (3-7 at.%) to these alloys their T_c decreases sharply (Fig. 185, Table 41) [262, 267a]. The critical temperature of the ternary alloys passes through a maximum in sections with a constant Ta or V content (up to 30 at.%) for an electron concentration of 4.5-4.7 electrons/atom.

Vanadium–Titanium–Molybdenum. At high temperatures the system forms continuous β solid solutions with a bcc lattice [268, 269]. The solidus surface falls smoothly from the molybdenum corner toward vanadium and titanium (Fig. 186); in the titanium corner the solidus surface has a minimum at 2-10 wt.% Mo and 8.35% V (1600°C) [268]. The limit of stability of the β solid solutions passes through a line joining the binary alloys of titanium with 15 wt.% V and 11.5% Mo; alloys with a lower content of the alloying elements experience a martensitic transformation on cooling and contain a nonequilibrium α' phase.

At temperatures of under 885°C there is a narrow range of ternary solid solutions based on α-titanium in the titanium corner of the system [268, 269]. The lattice constant of the β solid solution of titanium with 15 wt.% V is $a = 3.24$ Å; as the β phase is enriched with molybdenum and vanadium, the lattice constant a falls smoothly (at 980°C) to 3.10-3.14 Å for Mo-base alloys and to 3.04 Å for those enriched with vanadium [268].

The superconducting properties of several alloys of this system (Table 42) were studied by Efimov et al. [262]. The alloys

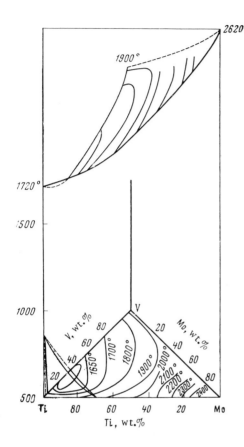

Fig. 186. Spatial representation of the V−Ti−Mo phase diagram.

TABLE 42. Composition and Superconducting Transition Temperature of V−Ti−Mo Alloys

Composition, at.%			Phase composition and lattice constant a, Å	T_c, °K
V	Ti	Mo		
50	50	—	β; 3.175	7.8
50	49	1	β; 3.155	6.9
50	45	5	β; 3.158	6.9
50	40	10	β; 3.138	6.2

TABLE 43

Element	Fermi energy, eV	γ, mJ/mole·°K^2	χ, 10^{-6} °K	θ, °K	a, Å	T_c, °K
Vanadium	6.8	9.26	296	360	3.024	5.3
Niobium	15.4	7.79	214	230	3.3007	9.2
Tantalum	5.2	5.90	154	240	3.3026	4.48

were cold-forged and drawn into wires 0.2 mm in diameter (the degree of deformation was about 99%). In the cast state all the alloys were single-phase solid solutions with a bcc lattice. The T_c values were determined by studying the change in the magnetic permeability of the samples. We see from Table 42 that the transition temperature decreases slightly with increasing molybdenum content. Comparing the transition temperature of V–49% Ti–1% Mo with that of the binary alloy V–50% Ti, we see that the introduction of 1% Mo reduces T_c appreciably (from 7.8 to 6.9°K).

Vanadium–Niobium–Tantalum. Vanadium, niobium, and tantalum have analogous electron structures [$(n-1)d^4ns^2$] and Fermi surfaces [269a], and hence also similar electron properties (Table 43) [269b-269d]. The values of γ and χ are highest in vanadium, which indicates a greater density of states on the Fermi surface as compared with tantalum. The electron characteristics of niobium occupy an intermediate position. However, the Fermi energy calculated for niobium by the free-electron model, with due allowance for the collectivization of the 5d zone, is the highest. Correspondingly the T_c of niobium is approximately double that of vanadium and tantalum. These data, taken together with the established existence of a correlation between T_c and the electron-interaction parameter for the binary alloys of these elements [9a], arouse particular interest in the study of the ternary V–Nb–Ta system.

The phase diagram of this system plotted by our computer (Simplex lattice) method [262b, 262c] is presented in Fig. 187. The critical temperature decreases smoothly from the T_c of niobium in the direction of the V–Ta side. As in the constituent binary systems, the T_c surface of the alloys of the ternary system passes through a minimum. The microhardness and specific electrical resistance of the ternary alloys with the lowest T_c values, determined at 4.2-300°K, are higher than those of other alloys.

Hafnium–Molybdenum–Rhenium. Taylor and others [270] studied the structure of the Hf–Mo–Re phase diagram. Alloys prepared by arc melting from pressed and sintered powders were studied by x-ray and microstructural analyses. Isothermal sections were studied at 1600, 2000, and 2400°C as well as the Mo_2Hf–Re_2Hf section. No new ternary compounds were found in the system. The alloys contain phases corresponding to the binary systems [Fig. 188]: solid solutions based on molybdenum, rhenium, and hafnium, a narrow range of the σ phase along the Mo–Re side with a Hf content of under 1 at.%, a χ phase based on the compounds $MoRe_2$ and $HfRe_9$ existing in the binary systems,

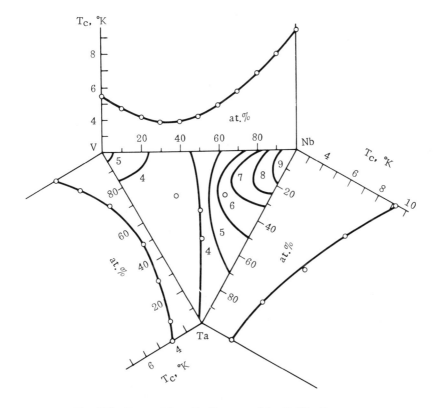

Fig. 187. Composition–T_c diagram of the V–Nb–Ta system.

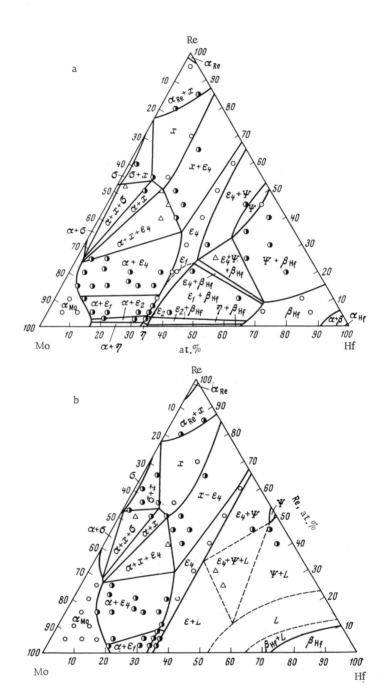

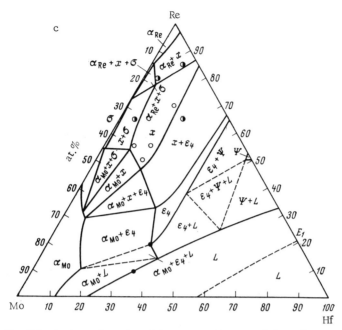

Fig. 188. Isothermal section of the ternary Mo-Re-Hf phase diagram at 1600 (a), 2000 (b), and 2400°C (c).

having a structure of the α-Mn type (at 2162°C the χ phase decomposes peritectoidally with the formation of hexagonal α-Re and a σ phase), a range of Laves η phase based on the compound $MoHf_2$ (which passes through the whole system to the compound Re_2Hf), and a ternary Φ phase.

Taylor et al. [270] also measured the transition temperature of several ternary phases of the Hf-Mo-Re system, finding that the χ phase had a T_c of 5.6°K and the Φ phase one at about 6°K; the Φ phase was nonsuperconducting. The transition temperature of the α-Re solid solution equalled 1.76°K at the boundary of the phase range.

Vanadium-Manganese-Silicon. The V-Mn-Si system is at the moment the only system in which a new ternary phase with the Cr_3Si structure has been found [271, 272]. An isothermal section of the system at 800°C is given in Fig. 189 [272].

At high temperatures the compound V_3Si dissolves up to 49 at.% Mn [272]. At 800°C there is a break in the range of homoge-

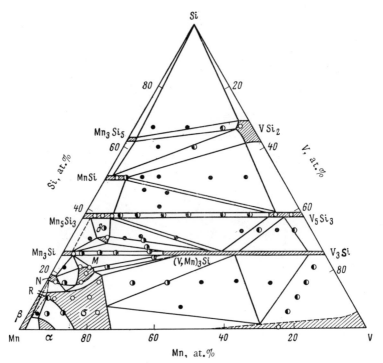

Fig. 189. Isothermal section of the ternary V–Mn–Si phase diagram at 800°C.

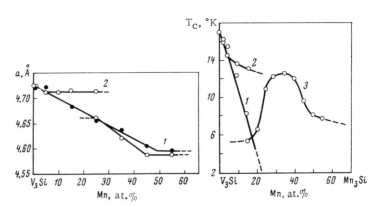

Fig. 190. Changes in the lattice constant and T_c of alloys belonging to the V_3Si–Mn_3Si system: 1) cast alloy; 2) alloy in the annealed state; 3) T_c of the ternary phase.

neity of the $(V, Mn)_3Si$ solid solution in the V_3Si-Mn_3Si system, with the formation of a more limited solid solution of Mn in V_3Si (5 at.% Mn) and an intermediate ternary phase $(V, Mn)_3Si$ in the region of 25-45 at.% Mn with an analogous structure of the Cr_3Si type [271]. The dissolution of manganese in V_3Si causes a reduction in lattice constant (Fig. 190) [271].

Apart from the ternary compound $(V, Mn)_3Si$ [271] two other ternary compounds, the M and δ phases, are formed in this system [272].

The critical temperature of this system was only measured for alloys of the V_3Si–Mn_3Si system. According to our data manganese sharply reduces the T_c of the compound V_3Si. Cast V_3Si alloys containing more than 20 at.% Mn become nonsuperconducting at temperatures exceeding 4.2°K. Within the range of homogeneity of the solid solution based on V_3Si (at 800°C) there is a sharp decrease in the T_c of the alloys, from 17.1°K for the binary compound to 14.6°K for the alloy with 5 at.% Mn. The transition into the two-phase region is accompanied by a bend on the curve relating the T_c of the V_3Si-base solid solution to composition (Fig. 190). The extent to which the critical temperature of this solid solution is reduced within the two-phase region is considerably less than in the single-phase region. The signal representing the superconducting transition of the V_3Si-base solid solution becomes weaker with increasing manganese content within the range of the two-phase region. In practice the superconducting transition of this phase fails to appear at all for alloys containing 15.5-25 at.% Mn. In the alloy with 15 at.% Mn there are two steps corresponding to a transition into the superconducting state: 13°K for the V_3Si-base solid solution, and 5.5°K for the ternary compound $(V, Mn)_3Si$.

The homogeneous ternary phase $(V, Mn)_3Si$ has a fairly high superconducting transition temperature, varying with composition from 9.6 to 12.5°K over the range of homogeneity; the maximum critical temperature is reached for 25 at.% Mn and 25 at.% Si. In the two-phase region (in equilibrium with the Mn_3Si-base solid solution) the T_c of the ternary phase equals 7.6-8.1°K.

The compound Mn_3Si (bcc lattice of the α-Fe type) and the solid solution of vanadium in this compound are not superconducting at temperatures of over 4.2°K.

We shall now present some brief details relating to the ternary phase diagrams in which no superconducting properties have yet been studied [275-296]. Many alloys of these systems should certainly exhibit conductivity, judging from general physicochemical considerations as to the properties of the alloys formed by the superconducting components. The laws governing the variations in the superconducting characteristics of ternary and more complicated systems call for urgent investigation.

Nb − Mo − V. Continuous range of solid solutions is noted over the whole temperature range. The melting point of the alloys decreases with increasing vanadium content [275].

Nb − Mo − W. Unlimited solubility of the components is observed in the liquid and solid states [276].

Nb − Mo − U. At temperatures above 1450°C unlimited solubility holds between γ-U and a solid solution of Nb−Mo. Below 1450°C the γ solid solution decomposes into two phases: γ-U and the solid solution of Nb−Mo; these coexist from 50 to 70 Nb and Mo. Below 700°C the uranium corner of the system exhibits a three-phase eutectoid and a four-phase peritectoid equilibrium associated with the low-temperature form of uranium [277].

Nb − Mo − Cr. At 1000°C there is a considerable range of bcc solid solutions in the molybdenum corner. This region expands with rising temperature. At 1000-1200°C the system contains: three two-phase regions of Cr-solid solution + $NbCr_2$, Nb-solid solution + $NbCr_2$, and Cr-solid solution + Nb-solid solution, a three-phase region of Cr-solid solution + Nb-solid solution + $NbCr_2$; the chemical compound $NbCr_2$ has a certain range of homogeneity and dissolves up to 10 at.% Mo at 1000°C [278, 279].

Nb − Zr − Sn. Isothermal sections have been plotted at 1050, 940, 850, and 725°C, as well as projections of the zirconium corner of the phase diagram of this system [307].

Nb − Zr − V, Nb − Zr − Al, Zr − V − Mo, Zr − Mo − Nb. The construction of the zirconium corners of the phase diagrams of these systems has been discussed [308].

V − Ti − Cr. A continuous series of solid solutions is noted with a bcc lattice at high temperatures. On reducing the temperature narrow regions of solid solutions based on α-Ti and $TiCr_2$ appear [280].

V − Ti − Fe. The V−Ti−TiFe$_2$ region has been studied. Pseudobinary sections divide the system into several secondary ternary systems. A ternary compound (Ti, Fe)$_1$V$_1$ formed by a peritectic reaction has been established. The range of ternary solid solutions is limited even at high temperatures [282-283].

V − Nb − Cr. There are no ternary compounds. The phase equilibria are determined by a fairly wide (particularly in the vanadium corner) range of homogeneity of a ternary solid solution with a bcc lattice and a narrow region of solubility of this solid solution in NbCr$_2$ [284].

V − Mo − W. A continuous series of solid solutions is believed to occur at all temperatures [285].

Mo − Zr − Ti. There is a wide range of combined solubility of molybdenum and zirconium in β-Ti; this diminishes with decreasing temperature. At about 1750°C there is a transition from the peritectic transformation L + β → Mo$_2$Zr into a eutectic transformation L → β + Mo$_2$Zr [286].

W − Zr − Ti. The combined solubility of tungsten and zirconium in β-Ti in the solid state diminishes with decreasing temperature; for example, the solubility of tungsten and zirconium in β-Ti along the Ti−W$_2$Zr section is 43-44 wt.% at 1500°C. Apart from the β solid solution, at 1500 and 1000°C the system contains an α solid solution based on tungsten, two-phase regions (α + W$_2$Zr), (β + W$_2$Zr), + (α + β), and a three-phase region (α + β + W$_2$Zr) [287].

W − Mo − Zr. The combined solubility of tungsten and molybdenum in zirconium (and conversely zirconium and tungsten in molybdenum) diminishes with decreasing temperature. Between the chemical compounds W$_2$Zr and Mo$_2$Zr, continuous series of solid solutions (W, Mo)$_2$Zr are formed. The three-phase peritectic and eutectic transformations may be written as L + α → (W, Mo)$_2$Zr (at 2260-1890°C) and L → β + (W, Mo)$_2$Zr (at 1640-1570°C), where α is the W- or Mo-base solid solution and β is the solid solution based on β-Zr [288].

W − Mo − Ti. At 1500°C the system forms continuous series of solid solutions between tungsten and β-Ti for a molybdenum content of about 20 wt.%, and at 1000°C the same for a molybdenum content of 25 wt.%. At these temperatures a two-phase region of solid solutions α + β with bcc lattices still exists [289].

W – Mo – Re. This system contains a ternary solid solution. The limit of solubility at 1000°C extends from the limit of solubility of rhenium in tungsten (32 wt.% Re) to the limit of solubility of rhenium in molybdenum (46 wt.% Re); with increasing temperature the solubility rises a little. This system forms a ternary σ phase. In the rhenium corner there is a small region of rhenium-base ternary solid solution [290, 291].

Re – Os – Ru. These metals form continuous series of solid solutions. The diagram is the first known example of unlimited mutual solubility in the liquid and solid states between three metals with hexagonal lattices [292].

W – Pd – Re. In the palladium corner there is a wide range of ternary solid solutions (from the alloy W-25% Pd to Pd-20% Re) [293].

Re – Ta – Nb. The limit of solubility of the ternary solid solution based on tantalum and niobium at 1000-1500°C is a straight line lying between the limits of solubility in the binary systems Ta–Re and Nb–Re [294].

Nb – Ta – Cu. The system is characterized by the existence of a wide range of Nb-base ternary solid solutions (at 1900°C and over) and the existence of a region of monotectic equilibrium in the liquid state, characteristic of the binary system Nb–Cu [295].

Nb – W – Mo – Zr. This system contains a single-phase region of a quaternary solid solution of tungsten, molybdenum, and zirconium in niobium; the solubility falls sharply with reducing temperature. In the two-phase regions the excess phase $(Mo, W)_2 Zr$ is in equilibrium with the quaternary solid solution. A three-dimensional projection of this quaternary system has been constructed in the form of a polythermal tetrahedron (Fig. 191) [296].

Changes taking place in the T_c of Bi–Pb–Tl alloys with fcc lattices on varying the number of valence electrons (Tl = 3, Bi = 5, Pb = 4 electrons/atom) were studied in [296a] on the basis of the previously-studied binary systems Pb–Tl, Bi–Tl, and Pb–Bi; using the McMillan theory, the electron–phonon interaction was calculated on the basis of T_c and θ_D measurements.

Recently [309] the critical magnetic fields of highly-ductile, many-component superconducting alloys of the Nb–Ti–Zr, Nb–Ti–

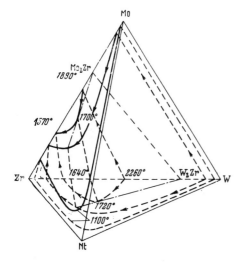

Fig. 191. Polythermal tetrahedron of the quaternary system Nb-W-Mo-Zr.

Hf, Nb-Ti-Ta, Nb-Ti-Mo, Nb-Ti-W, Nb-Ti-Fe, Nb-Zr-Hf, Nb-Ti-Zr-Ta, Nb-Ti-Hf-V, Nb-Ta-Ti-Hf, Nb-Zr-Hf-Ti systems were studied. All the alloys with high H_{c2} values were based on the Nb-Ti system. The maximum critical fields occurred in alloys of the following compositions (at.%): 34Nb-65Ti-1Mo (110 kOe), 32Nb-64Ti-4Hf (116 kOe), 36Nb-60Ti-4Ta (125 kOe), 35Nb-64Ti-1W (114 kOe), 39Nb-53Ti-4Zr-4Ta (120 kOe). On measuring the H_{c2} of cold-worked Nb-Ti-Zr alloys, no rise in H_{c2} relative to the H_{c2max} of the Nb-Ti alloys was observed.

2. Quaternary System

The V-Nb-Ta-Ti system forms a continuous series of solid solutions with a bcc lattice at 1400°C [262d]. The composition-T_c diagram of the V-Nb-Ta-Ti system was plotted on the basis of experimental measurements by our computer technique. Figure 192 shows the T_c isosurfaces. On moving along the normal to the V-Ti-Ta face of the concentration tetrahedron, T_c increases monotonically. The lowest T_c occurs for alloys with a medium concentration of vanadium, niobium, and tantalum, containing up to 8 at.% titanium; these are not superconducting at temperatures above the boiling point of liquid helium. A characteristic feature is the existence of special regions with high T_c values lying close to the V-Ti and Ta-Ti edges. Alloying with a third or fourth component leads to a decrease in the critical temperature.

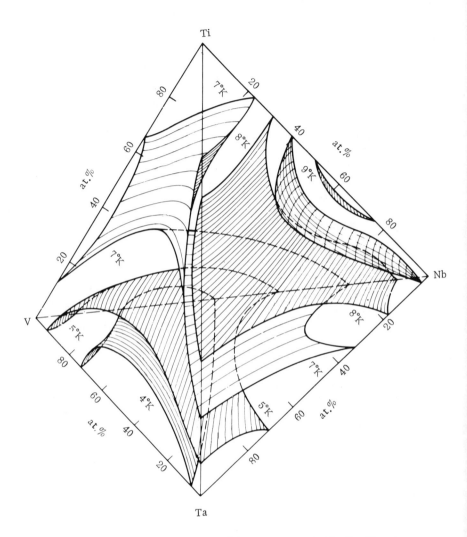

Fig. 192. Composition–T_c diagram of the quaternary V–Nb–Ta–Ti system.

In any cross section of the quaternary system with a constant Ti content, the specific electrical resistance of the alloys at room temperature passes through a maximum slightly displaced in the V–Nb direction. On reducing the measuring temperature to the liquid-helium level, the isosurfaces of the specific electrical resistance of the alloys straighten out, and the sharp maxima in the sections with constant titanium content degenerate into maxima of a less sharply-expressed type. The microhardness of the quaternary alloys in the sections under consideration passes through a maximum. Thus in the quaternary system a rise in the critical temperature is accompanied by a fall in the microhardness and a rise in the specific electrical resistance.

3. Pseudoternary Superconducting Systems

The binary compounds Nb_3Sn, Nb_3Al, and Nb_3Ge with identical crystal structures have very similar lattice constants (5.289, 5,187, and 5.168 Å respectively) and electron concentrations (4.7, 4.5, and 4.7 electrons/atom). On this basis Alekseevskii, Ageev, and Shamrai [297] suggested fairly similar positions of the Fermi surface with respect to the boundaries of the Brillouin zones in these compounds together with the formation of a continuous series of pseudoternary solid solutions with the Cr_3Si structure between them. The validity of this proposition was confirmed experimentally. The range of homogeneity of the solid solution Nb_3 (Sn, Al, Ge) in this system deviated slightly from the isoconcentration characteristic of 75 at.% Nb and occupied a position inclined to this characteristic. On increasing the germanium content the range of homogeneity moved in the direction of greater niobium content (Fig. 193). At 600°C throughout the whole range of concentrations the Nb_3 (Sn, Al, Ge) phase was in equilibrium with the quaternary solid solution of tin, aluminum, and germanium in niobium [297].

In the pseudoternary system the lattice constant of the solid solution varies along smooth curves. For alloys with a large tin

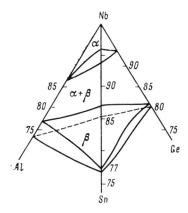

Fig. 193. Position of the range of homogeneity of the Nb_3(Sn, Al,Ge) solid solution with the Cr_3Si structure in the quaternary system Nb-Sn-Al-Ge.

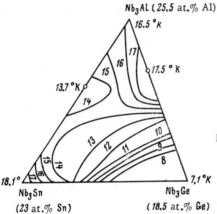

Fig. 194. Critical temperature of the solid solution Nb$_3$(Sn, Al, Ge).

Fig. 195. Critical temperature as a function of the shortest interatomic distance in the crystal lattice of the Nb$_3$(Sn, Al, Ge) solid solution for equal degrees of ordering.

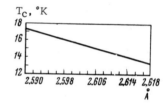

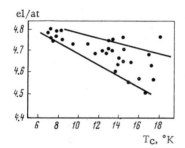

Fig. 196. Dependence of the T_C of the solid solution Nb$_3$(Sn, Al, Ge) on the electron concentration [297].

Fig. 197. Variations in the T_C of the pseudo-ternary solid solution in the Nb$_3$Sn–Ta$_3$Sn–V$_3$Sn.

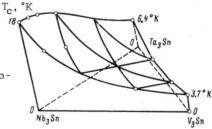

content the changes in the lattice constant obey an additive law. For alloys rich in aluminum and germanium there is some deviation from additivity. The variation in the T_c of the ternary system $Nb_3Sn-Nb_3Al-Nb_3Ge$ is illustrated in Fig. 194. The highest values of the critical temperature occur for alloys rich in aluminum and tin; increasing the germanium content leads to a sharp decrease in the T_c of the alloys. In general the T_c curves in the pseudoternary (T_c as a function of composition) pass through a distinct minimum. The only exceptions are alloys close to the maximum in the binary system Nb_3Al-Nb_3Ge.

With increasing degree of order in the crystal structure, T_c tends to increase. In the ternary system the greatest degree of order occurs for alloys with a large aluminum content, while with increasing germanium content the degree of ordering in the solid solution with the Cr_3Si structure diminishes smoothly. The departure from order leads to a decrease in critical temperature, this evidently resulting from the disruption of the chains of A atoms in the Cr_3Si-type structure. For equal degrees of order the shortest distance between the A atoms controls the critical temperature; as the interatomic distance decreases the critical temperature becomes greater (Fig. 195).

According to published data, there is a direct dependence of T_c on the electron concentration [298-300]. In alloys of the ternary system $Nb_3Sn-Nb_3Al-Nb_3Ge$ there is a tendency for T_c to fall with increasing electron concentration (Fig. 196). However, later investigations showed [301] that a material with the Cr_3Si structure and a high superconducting transition temperature 20.98°K might be created on the basis of the compounds Nb_3Al and Nb_3Ge. In ternary systems the Matthias rule is clearly of limited applicability [297].

Cody et al. [302] noted the formation of a continuous series of solid solutions in pseudobinary and pseudoternary systems of the compounds Nb_3Sn, Ta_3Sn, and V_3Sn. All these compounds have the same crystal lattice and the same electron concentration, but their molecular weights and atomic volumes differ considerably from one another. Identical crystal structures, together with the same B component, result in the formation of continuous series of solid solutions. Differences between the other quantities (atomic volumes etc.) result in different critical temperatures of the original compounds and alloys (Fig. 197) [302]. A formula has been de-

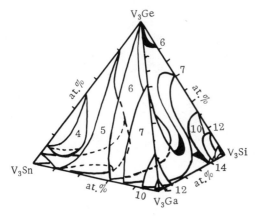

Fig. 198. Composition–T_c diagram of the ternary V_3Si–V_3Ga–V_3Ge system.

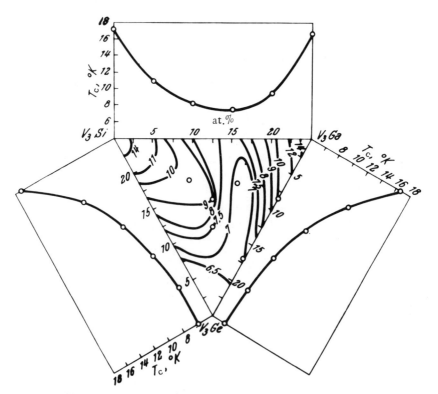

Fig. 199. Composition–T_c diagram of the ternary V_3Si–V_3Ga–V_3Sn system.

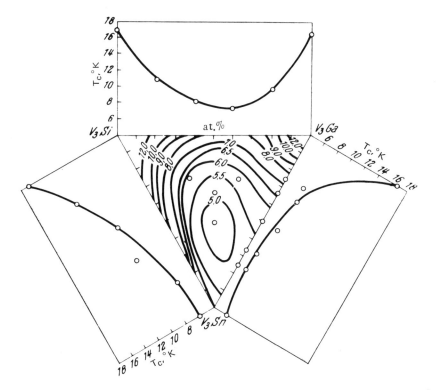

Fig. 200. Composition–T_c diagram of the ternary $V_3Si-V_3Ge-V_3Sn$ system.

rived for calculating the T_c of such alloys, but the accuracy with which the critical temperature may be calculated from this is not very high (40%) [302].

The $V_3Si-V_3Ga-V_3Ge$ system forms a continuous series of solid solutions with the Cr_3Si-type structure at temperatures below that corresponding to the formation of the compound V_3Ga [89b]. The T_c surface of the alloys of this system has a complicated character, incorporating a saddle-shaped plateau (Fig. 198). On alloying the compounds V_3Ga and V_3Si, a decrease in their critical temperature takes place. Alloying the compound V_3Ge leads to some increase in its critical temperature. The pseudoternary alloys of medium concentrations have roughly the same values of T_c (6.5-7.5°K). The lattice constant of the alloys varies smoothly, with a slight negative deviation from the additivity law.

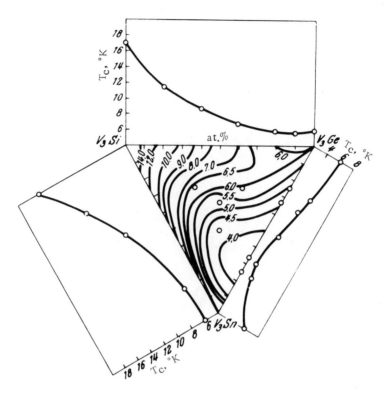

Fig. 201. Composition-T_c diagram of the ternary $V_3Ge-V_3Ga-V_3Sn$ system.

We also studied the $V_3Si-V_3Ga-V_3Sn$, $V_3Si-V_3Ge-V_3Sn$, and $V_3Ga-V_3Ge-V_3Sn$ systems [89b]. In the first of these the T_c surface has a sharp minimum in the range of pseudoternary alloys (Fig. 199). In the other two systems T_c decreases smoothly from the T_c of V_3Si or V_3Ga to the opposite side of the pseudoternary system (Figs. 200 and 201). The T_c of the pseudoternary alloys of all three systems lies below that of the compounds V_3Ga and V_3Si.

4.- Pseudoquaternary System

We established the formation of a continuous series of solid solutions with the Cr_3Si type of structure in the pseudoquaternary system $V_3Si-V_3Ga-V_3Ge-V_3Sn$ (at 800° [89b]). The T_c isosurfaces of this system are shown in Fig. 202; they are very similar to one

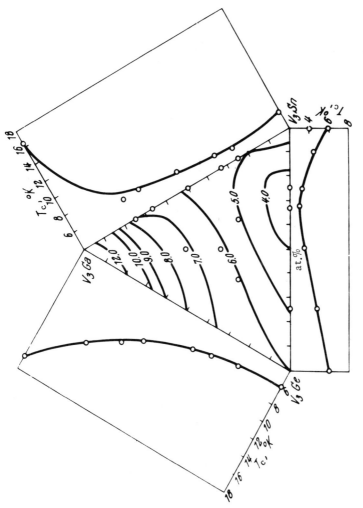

Fig. 202. Composition–T_c diagram of the quaternary V_3Si–V_3Ga–V_3Ge–V_3Sn system.

another and have convexities facing the $V_3Si-V_3Ga-V_3Ge$ side. Hence there are sections with constant tin contents in which the T_c surfaces have maxima, although these remain lower than the T_c of alloys belonging to the pseudobinary systems of the concentration tetrahedron. The T_c of all the pseudoquaternary alloys remain lower than the T_c of the compounds V_3Si and V_3Ga.

According to [309] all the pseudocomponents under consideration may be excellently described on the rigid-band approximation. This means that the special characteristics in the structure of the 3sd band of vanadium are not particularly sensitive to changes in the B components. Since the Fermi level in the V_3X compounds lies within a very high and narrow peak of the density of states, mainly due to the 3d states of the vanadium atoms, and only moves slightly within the concentration tetrahedron under consideration ($\Delta n = 0.25$ electrons/atom), we may suppose that the large changes in T_c (from 17 to 3°K) in the system are due to a change in the phonon spectrum.

In coming to the end of this chapter on superconducting systems, the authors of the monograph wish once again to draw the attention of research workers to the importance and necessity of carrying out physicochemical analyses of superconducting systems, and the promise of fruitful results from such an analysis; it is in particular highly desirable that a systematic investigation should be pursued into the superconducting properties of metallic alloys at various temperatures (especially in the liquid-helium range) in relation to their chemical and phase constitutions.

The application of physicochemical analysis to a study of the structures and properties of alloys based on ferrous and nonferrous, light, refractory, rare, and radioactive metals has been brilliantly justified, and at the present time this serves as a fundamental theoretical basis for the development of alloys in various branches of the popular economy and new technology [303-305]. The use of physicochemical analysis in studying superconductors has only just started, but has already achieved substantial results [1, 1b, 4, 5]. It is important to emphasize the necessity of studying the structure and superconducting properties of alloys only after these have been brought into a thermodynamically-equilibrium state or have a precisely known deviation from the state of equilibrium (this has hardly been done at all at the present time). Essentially, all the work so far carried out on superconducting systems should be repeated once again over the whole range of concentrations with full cooperation between metallurgists and metal physicists.

LITERATURE CITED

1. E. M. Savitskii, Izv. Akad. Nauk SSSR, Metally, No. 10, p. 11 (1967).
1a. E. M. Savitskii, Yu. V. Devingtal', and V. B. Gribulya, Dokl. Akad. Nauk SSSR, Tekh. Fiz., 178(1):79 (1968); 183(5):11 (1968).
1b. E. M. Savitskii, Zh. Neorgan. Khim., 12(7):1721 (1967).
2. B. T. Matthias, in: Superconducting Materials [Russian translation], Izd. Mir, Moscow (1965), pp. 15 and 25.
3. N. V. Ageev, Nature of the Chemical Bond in Metallic Alloys, Izd. AN SSSR, Moscow (1947).
4. E. M. Savitskii and V. V. Baron, Izv. Akad. Nauk SSSR, Metallurgiya i Gornoe Delo, No. 5, p. 1 (1963).
5. E. M. Savitskii, in: Metallography and Metal Physics of Superconductors, Izd. Nauka, Moscow (1965), p. 3.
6. B. W. Roberts, General Electric Research Lab. Rept., R.L.-3552M (1963).
7. H. Wilchelm, O. Carlson, and I. Dickinson, J. Metals, No. 6, p. 195 (1954).
8. J. K. Hulm and R. D. Blaugher, Phys. Rev., 123(5):1569 (1961).
9. W. De Sorbo, Phys. Rev., 130:2177 (1963).
9a. I. M. Corsan and A. I. Cook, Phys. Stat. Sol., 40:657 (1970).
9b. M. Ishikawa and L. E. Toth, Phys. Rev., B., 3(6):1856 (1971).
9c. N. N. Sirota and É. A. Ovseichuk, Dokl. Akad. Nauk SSSR, 174(3):570 (1967).
10. D. Williams and W. Pechin, Trans. ASM, 50:108 (1958).
11. R. R. Hake, T. G. Berlincourt, and D. H. Leslie, Superconductors, New York (1962), p. 53.
12. B. S. Chandrasekhar, J. K. Hulm, and S. K. Jones, Phys. Letts., 5:18 (1963).
12a. E. M. Savitskii, V. V. Baron, and S. D. Gindina, Dokl. Akad. Nauk SSSR, 191(2):338 (1970).
13. I. W. Heaton and A. C. Rose-Innes, Cryogenics, 4:85 (1964).
14. B. T. Ogasawara, I. Kubota, and K. Iasukochi, Phys. Letts., A24(9):463 (1967).
15. I. I. Kornilov and R. S. Polyakova, Zh. Neorgan. Khim., 4:879 (1958).
16. E. M. Savitskii, V. V. Baron, and K. N. Ivanova, Izv. Akad. Nauk SSSR, Otd. Tekh. Nauk, Metallurgiya i Toplivo, No. 4, 143 (1960).
17. R. A. Hein, I. W. Gibson, and R. D. Blaugher, Rev. Mod. Phys., 36(1):149 (1964).
18. G. V. Zakharova, I. A. Popov, L. P. Zhorova, and B. V. Fedin, Niobium and Its Alloys, Metallurgizdat, Moscow (1961), pp. 213 and 216.
19. V. S. Mikheev and D. I. Pevtsov, Zh. Neorgan. Khim., 3(4):861 (1958).
19a. E. M. Savitskii, V. V. Baron, and K. N. Ivanova, Izv. Akad. Nauk SSSR, Otd. Tekh. Nauk, Metallurgiya i Toplivo, 2:119 (1962).
20. E. M. Savitskii, V. V. Baron, and Yu. V. Efimov, Transactions of Inst. Met., No. 4, Izd. AN SSSR, Moscow (1960), p. 230.

21. I. Muller, Helv. Phys. Acta, 32:141 (1959).
22. V. V. Baron, Yu. V. Efimov, and E. M. Savitskii, Izv. Akad. Nauk SSSR, Otd. Tekh. Nauk, No. 4, p. 36 (1958).
23. E. Bucher, G. Buch, and I. Muller, Helv. Phys. Acta, 32:318 (1959).
24. C. H. Schramm, A. R. Gordon, and A. R. Kaufmann, J. Metals, 2(2):80 (1949).
25. R. H. Myers, Metallurgia, 42:3 (1950).
25a. E. M. Savitskii and G. S. Burkhanov, Metallography of Refractory Metals and Alloys, Izd. Nauka, Moscow (1967).
26. B. W. Couser, Industr. and Engng. Chem., 42: 222 (1950).
27. M. Hansen, E. L. Kamen, H. D. Kessler, and D. J. McPherson, Trans. AIME, 191:881 (1951).
28. V. P. Elyutin, M. L. Berenshtein, and Yu. A. Pavlov, Dokl. Akad. Nauk SSSR, 104:546 (1955).
29. P. Duwez, Trans. ASM, 45:934 (1953).
30. K. I. Shakhova and P. B. Budberg, Izv. Akad. Nauk SSSR, Metallurgiya i Toplivo, No. 4, p. 56 (1961).
31. S. Ames and A. McQuillan, Acta Metallurgica, 2:831 (1954).
32. Yu. A. Bagaryatskii, G. I. Nosova, and T. V. Tagunova, Dokl. Akad. Nauk SSSR, 122:4 (1958).
32a. E. M. Savitskii, B. Ya. Sukharevskii, V. V. Baron, A. V. Alanina, M. I. Bychkova, Yu. A. Dushechkina, I. S. Shchetkin, and M. N. Kharchenko, in: Superconducting Alloys and Compounds, Izd. Nauka, Moscow (1971), p. 79.
33. V. V. Baron, M. I. Bychkova, and E. M. Savitskii, in: Metallography and Metal Physics of Superconductors, Izd. Nauka, Moscow (1965), p. 53.
34. V. V. Baron and M. I. Bychkova, in: Metallography, Physical Chemistry, and Metal Physics of Superconductors, Izd. Nauka, Moscow (1967), p. 44.
35. G. Berlincourt and R. R. Hake, Phys. Rev. Lett., 9(7):293 (1962).
36. I. H. Wernick, F. I. Morin, S. L. Hsu, D. Dorsi, I. P. Maita, and J. E. Kunzler, Phys. Rev. Lett., 5:149 (1960).
37. T. G. Berlincourt, Transactions of the Eighth International Conference on Low-Temperature Physics, London (1962).
38. C. Anderson, E. Hayes, A. Robertson, and W. K. Kroll, Rept. Invest. U. S. Bur. Mines, 4658 (1950).
39. F. Litton, Iron Age, 167:95, 112 (1951).
40. C. R. Simcol and W. Mudge, AEC Rept., WARD-38, 21:70 (1951).
41. E. Hodge, AEC Rept., NIID-5061, 31:461 (1952).
42. I. Keeler, J. Metals, 8(2):122 (1956).
43. B. Rogers and D. Atkins, J. Metals, 7(2):1034 (1955).
44. Yu. F. Bychkov, A. N. Rozanov, and D. M. Skorov, At. Énerg., 2(2):146 (1957).
45. C. W. Berghout, Phys. Letts., 1:992 (1962).
46. H. Richter, P. Winciez, K. Anderko, and U. Zwiker, J. Less-Common Metals, 4(3):252 (1962).
47. D. R. Love and M. L. Picklesimer, Trans. AIME, 4:236, 430 (1966).
48. L. F. Myzenkova, V. V. Baron, aand E. M. Savitskii, in: Metallography, Physical Chemistry, and Metal Physics of Superconductors, Izd. Nauka, Moscow (1967), p. 112.

49. V. V. Baron, in: Metallography, Physical Chemistry, and Metal Physics of Superconductors, Izd. Nauka, Moscow (1967), p. 37.
50. Yu. F. Bychkov, I. N. Goncharov, and I. S. Khukhareva, Zh. Éksp. Teor. Fiz., 48(3):817 (1965).
51. S. V. Sudareva, N. N. Buinov, V. A. Vozilkin, E. P. Romanov, and V. G. Rakin, Fiz. Met. Metallov., 21(3):388 (1966).
52. E. M. Savitskii, M. A. Tylkina, and I. A. Tsyganova, Zh. Neorgan. Khim., 9(7):1650 (1954).
53. A. Taylor and N. I. Doyle, J. Less-Common Metals, 7:37 (1964).
54. M. I. Bychkova, V. V. Baron, and E. M. Savitskii, in: Metallography, Physical Chemistry, and Metal Physics of Superconductors, Izd. Nauka, Moscow (1967), p. 68.
55. T. G. Berlincourt and R. R. Hake, Phys. Rev., 131:1 (1963).
56. M. Hansen and K. Anderko, Constitution of Binary Alloys, Vols. 1 and 2 [Russian translation], Metallurgizdat, Moscow (1962).
57. T. G. Berlincourt, Phys. and Chem. Solids, 11:12 (1959).
58. W. Rostoker, Metallurgy of Vanadium [Russian translation], IL, Moscow (1959), pp. 57 and 65.
59. I. Muller, Helv. Phys. Acta, 32(3):141 (1959).
60. C. H. Cheng, K. P. Gupta, E. C. Van Reuth, and P. A. Beck, Phys. Rev., 126:2030 (1962).
61. Yu. V. Efimov, V. V. Baron, and E. M. Savitskii, in: Metallography and Metal Physics of Superconductors, Izd. Nauka, Moscow (1965), p. 59.
61a. Yu. V. Efimov, V. V. Baron, and E. M. Savitskii, in: Metallography, Physical Chemistry, and Metal Physics of Superconductors, Izd. Nauka, Moscow (1967), p. 64.
62. R. R. Hake and D. H. Leslie, Transactions of the Eighth International Conference on Low-Temperature Physics, London (1962).
63. E. Bucher, F. Heiniger, and I. Muller, Low Temperature Physics, LT-9, Plenum Press, New York (1965), p. 482.
64. B. T. Matthias, T. H. Geballe, and V. B. Compton, Rev. Mod. Phys., 35(1):1 (1963).
64a. J. B. Darby, D. J. Lam, L. J. Norton, and J. W. Downey, J. Less-Common Metals, 4(6):558 (1962).
64b. D. O. van Ostenberg, D. J. Lam, H. D. Trapp, and D. E. McLeod, Phys. Rev., 128(4):1550 (1962).
64c. C. C. Koch and G. R. Love, J. Metals, 20(1):56A (1968).
65. D. I. Maykuth, H. R. Ogden, and R. I. Jaffee, Trans. AIME, 197, p. 231 (1953).
66. D. Summers-Smith, J. Inst. Metals, 81:73 (1952, 1953).
67. C. I. Raub and U. Zwicker, Phys. Rev., 137(1A):142 (1965).
68. R. D. Blaugher and W. C. Joiner, Bull. Amer. Phys. Soc., 8:192 (1963).
69. D. A. Colling, K. M. Ralls, and J. Wulff, J. Appl. Phys., 37(13):4750 (1966).
70. S. T. Secula, R. W. Boom, and C. I. Bergerou, Appl. Phys. Letts., 2:102 (1963).
71. R. P. Elliott, Armor Res. Foundat. Chicago, Techn. Rpt. 1, OSR Techn. Note OSR-247 (1954).
72. L. Oden-Laurance, D. K. Deardorf, M. I. Copeland, and H. Kato, Rept. Invest. Bur. Mines, U. S. Dept. Interior, 6521, p. 12 (1964).

73. M. Hausen, E. L. Kamen, H. D. Kessler, and D. I. McPherson, Trans. AIME, 191:881 (1951).
74. R. R. Hake, D. H. Leslie, and T. G. Berlincourt, Phys. and Chem. Solids, 20:177 (1961).
75. R. R. Hake, Phys. Rev., 123:1986 (1961).
76. P. Duwez, J. Inst. Metals, p. 80 (1951-1952).
77. E. Bucher, F. Heiniger, and I. Muller, Rev. Mod. Phys., 36(1):146 (1964).
78. M. A. Tylkina, A. I. Pekarev, and E. M. Savitskii, Zh. Neorgan. Khim., 4(10):2320 (1959).
79. R. A. Hein, Phys. Rev., 102:1511 (1956).
79a. M. A. Jensen and J. P. Maita, Phys. Rev., 149(1):409 (1966).
80. W. Meissner, H. Franz, and H. Westerhoff, Phys. Rev., 13:505 (1932).
81. I. F. Allen, Phil. Mag., 16:1005 (1933).
82. I. Livingston, Phys. Rev., 129:1943 (1963).
83. I. Livingston, Rev. Mod. Phys., 36(1):54 (1964).
84. I. C. McLennan, I. F. Allen, and I. O. Wilchelm, Trans. Roy. Soc. Canada, 24:53 (1930).
85. W. Meissner, H. Franz, and H. Westerhoff, Ann. Phys., 17:593 (1933).
86. K. Clusiuk, Z. Electrochem., 38:312 (1932).
87. D. P. Seraphim, C. Chiou, and D. I. Quinn, Acta Metallurgica, 9:861 (1961).
88. W. Love, Phys. Rev., 92:238 (1951).
89. H. E. Cline, R. M. Rose, and J. Wulff, J. Appl. Phys., 34(6):1771 (1963).
89a. E. M. Savitskii, V. V. Baron, O. P. Naumkin, and Yu. V. Efimov, in: Problems of Superconducting Materials, Izd. Nauka, Moscow (1970), p. 178.
89b. E. M. Savitskii and Yu. V. Efimov, Mh. Chem., No. 10 (1971).
90. E. M. Savitskii, V. V. Baron, U. K. Duisemaliev, and Yu. V. Efimov, Vestn. Akad. Nauk Kazakh SSR, No. 7, p. 39 (1964).
91. W. Meissner, Ergebn. Exakt. Naturwiss., No. 11, p. 219 (1932).
91a. I. Haase and I. Seiberth, Z. Phys., 213(1):79 (1968).
91b. T. Claeson, Phys. Stat. Sol., 25(2):K95 (1968).
92. T. Claeson, H. L. Luo, and M. F. Merriam, Phys. Rev., 141(1):412 (1966).
92a. T. Claeson, Phys. Rev., 147:340 (1966).
93. M. F. Merriam and M. Van Herren, Progress in Cryogenics, London (1964), p. 159.
94. H. Klailer, Z. Elektrochem., 42:258 (1936).
95. I. C. McLennan, I. F. Allen, and I. O. Wilchelm, Trans. Roy. Soc. Canada, 23:278 (1929).
96. I. H. Wernick and B. T. Matthias, J. Chem. Phys., 34:2194 (1961).
97. N. E. Alekseevskii, Zh. Éksp. Teor. Fiz., 38:1 (1960).
98. A. F. Spedding, R. M. Valetta, and A. N. Doane, Trans. ASM, 55:22 (1963).
99. G. S. Anderson, S. Legvold, and F. H. Spelding, Phys. Rev., 109:243 (1958).
99a. T. Sugawara and H. Educhi, J. Phys. Soc. Japan, 23(1):9 (1967).
100. B. T. Matthias, H. Suhl, and E. Corenzwit, Phys. and Chem. Solids, 13:156 (1960).
101. V. B. Compton and B. T. Matthias, Acta Crystallogr., 12(9):651 (1959).
102. T. H. Geballe, B. T. Matthias, V. B. Compton, E. Corenzwit, and G. W. Hull, Phys. Rev., 137(19):4 (1965).

103. K. A. Gschneidern, Alloys of the Rare-Earth Metals, Mir, Moscow (1965).
104. B. T. Matthias, T. H. Geballe, V. B. Compton, E. Corenzwit, and G. W. Hull, Rev. Mod. Phys., 36(1):155 (1964).
105. B. T. Matthias, V. B. Compton, H. Suhl, and E. Corenzwit, Phys. Rev., 115(6): 1597 (1959).
106. L. N. Fedotov, A. I. Rad'kov, and S. M. Khromov, Trans. of the Central Scientific-Research Institute of Ferrous Metallurgy, No. 51, Izd. Metallurgiya, Moscow (1967), p. 34.
107. R. R. Coles, Phys. Rev., 36(1):139 (1964).
108. B. T. Matthias, V. B. Compton, and E. I. Corenzwit, Phys. and Chem. Solids, 19:130 (1961).
109. E. M. Savitskii, M. A. Tylkina, and Yu. A. Zot'ev, Zh. Neorgan. Khim., 4(3):702 (1959).
110. F. L. Orrell and M. G. Foutcma, Trans. ASM, 47:554 (1955).
111. W. Buckel, G. Dummer, and W. Gey, Z. Angew. Phys., 14(12):703 (1962).
112. K. Schubert, M. Balk, S. Bhan, H. Breimer, P. Esslinger, and E. Stolz, Naturwissenschaften, 46(23):647 (1959).
113. B. T. Matthias and E. Corenzwit, Phys. Rev., 100:626 (1955).
114. N. M. Matveeva and T. O. Malakhova, in: Physical Chemistry, Metallography, and Metal Physics of Superconductors, Izd. Nauka, Moscow (1967), p. 145.
114a. T. A. Pollock, R. Shull, and H. C. Gatos, Phys. Stat. Sol., 2(2):251 (1970).
115. E. M. Savitskii, M. A. Tylkina, and I. A. Tsyganova, At. Énerg., 7(3):231 (1959).
116. G. K. White and S B. Woods, Canad. J. Phys., 35:899 (1957).
117. B. T. Matthias, in: Superconducting Materials (Russian translation], Izd. Mir, Moscow (1965), p. 15.
118. R. D. Blaugher, A. Taylor, and J. K. Hulm, IBM J. Res. Devel., 6:116 (1962).
118a. E. Rudy and St. Windisch, J. Less-Common Metals, 15:13 (1968).
119. E. A. Linton, Superconductivity [Russian translation], Izd. Mir, Moscow (1964), p. 160.
120. E. Stolz and K. Schubert, Z. Metallkunde, 55(4):195 (1964).
121. R. Christoph, Z. Metallkunde, 55(4):195 (1964).
122. G. F. Hardy and J. K. Hulm, Phys. Rev., 93:1004 (1954).
123. Yu. V. Efimov, Zh. Neorgan. Khim., 8(6):1522 (1963).
124. G. F. Hardy and J. K. Hulm, Phys. Rev., 89:884 (1953).
125. B. T. Matthias, E. A. Wood, E. Corenzwit, and V. B. Bala, Phys. and Chem. Solids, No. 1, p. 188 (1956).
126. T. V. Samsonov, Silicides and Their Use in Technology, Izd. AN Ukr. SSR, Kiev (1959), p. 21.
127. Yu. V. Efimov, Izv. Akad. Nauk SSSR, Neorgan. Mat., 1(6):875 (1965).
128. W. Kunz and E. Saur, Z. Phys., 189(4):401 (1966).
129. E. A. Wood, V. B. Compton, B. T. Matthias, and E. Corenzwit, Acta Crystallogr., 11:604 (1958).
130. J. H. Wernick, F. J. Morin, F. S. L. Hsu, D. Dorsi, J. R. Maita, and J. E. Kunzler, High Magnetic Fields, Technol. Press, New York–London–Cambridge (1962), p. 609.

131. E. M. Savitskii, P. I. Kripyakevich, V. V. Baron, and Yu. V. Efimov, Zh. Neorgan. Khim., 9(5):1155 (1964).
132. E. M. Savitskii, P. I. Kripyakevich, V. V. Baron, and Yu. V. Efimov, Izv. Akad. Nauk SSSR, Neorgan. Mat., 3(1):45 (1967).
133. V. M. Pan, in: Structure of Metallic Alloys, Izd. Naukova Dumka, Kiev (1966), p. 56.
134. H. G. Meissner and K. Schubert, Z. Metallkunde, 56(7):475 (1965).
135. J. H. N. van Vucht, H. A. C. M. Bruning, H. C. Donkersloot, and A. H. Gomes de Mesquita, Philips Res. Rept., 19(5):407 (1964).
136. J. H. N. van Vucht, H. A. C. M. Bruning, and H. C. Donkersloot, Phys. Letts., 47(5):297 (1963).
137. W. Jeitsenko, H. Nowotny, and F. Benesovsky, Monatsh. Chem., 95(1):156 (1964).
138. D. A. Petrov, Ternary Systems, Izd. AN SSSR, Moscow (1953).
139. H. J. Levinstein, J. H. Wernick, and C. D. Capio, Phys. and Chem. Solids, 26(7):1111 (1965).
140. A. M. Zakharov, Phase Diagrams of Binary and Ternary Systems, Izd. Metallurgiya, Moscow (1964).
141. K. Schubert, H. G. Meissner, W. Rossteutscher, and E. Stolz, Naturwissenschaften, 49:357 (1962).
142. A. M. Clogston and V. Jaccarino, Phys. Rev., 121:1357 (1961).
143. F. Y. Morin and J. P. Maita, Phys. Rev., 120:1115 (1963).
144. H. von Philipsborn, Mischsysteme von Verbindungen des Cr_3Si Typs under deren Polimorphia-Erscheinungen, Juris Verlag, Zurich (1964).
145. C. J. Raub, W. H. Zachariasen, T. H. Geballe, and B. T. Matthias, Phys. and Chem. Solids, 24:1093 (1963).
146. W. E. Blumberg, J. Jaccarino, V. Jaccarino, and B. T. Matthias, Phys. Rev. Lett., 5:149 (1960).
147. W. Koster and K. Haug, Z. Metallkunde, 48:327 (1957).
147a. O. Rapp, Solid-State Commun., 9:1 (1971).
147b. F. Jonault and P. Lecocq, Colloq. Int. CNRS (1967), p. 157.
148. R. P. Elliott, Armor. Res. Found. Chicago Techn. Rept., p. 1, OSR-TN-247 (1954).
149. A. P. Nefedov, E. M. Sokolovskaya, A. T. Grigor'ev, I. G. Sokolova, and N. A. Nedumov, Zh. Neorgan. Khim., 9(4):883 (1964).
149a. A. Raman, Proc. Indian Sci., A65(4):256 (1967).
150. M. A. Tylkina, K. B. Povarova, and E. M. Savitskii, Dokl. Akad. Nauk SSSR, 131(2):332 (1960).
151. E. Raub and W. Fritzsche, Z. Metallkunde, 54(1):21 (1963).
152. B. C. Giessen, P. N. Daugel, and N. I. Grant, J. Less-Common Metals, 13(1):12 (1967).
153. B. T. Matthias, T. H. Geballe, S. Geller, and E. Corenzwit, Phys. Rev., 95:1435
154. V. V. Baron and E. M. Savitskii, Zh. Neorgan. Khim., No. 1, p. 182 (1961).
155. N. A. Nedumov and V. I. Rabezova, Izv. Akad. Nauk SSSR, Otd. Tekh. Nauk, No. 4, p. 68 (1961).
156. V. N. Svechnikov, V. M. Pan, and V. I. Latysheva, Metal Physics, Resp. Mezhvedomstv. Sborn. No. 25 (1968), p. 54.
157. C. R. McKingsey and G. M. Fauling, Acta Crystallogr., 12:701 (1959).
158. G. Brauer, Z. Anorgan. Chem., 242(1):548 (1939).

159. N. E. Alekseevskii, N. V. Ageev, and V. F. Shamrai, Izv. Akad. Nauk SSSR, Neorgan. Met., 2(12):2150 (1966).
160. G. Richards, Mem. Scient. Rev. Metallurgie, 61:265 (1964).
161. V. M. Glazov, V. N. Vigdorovich, and A. M. Korol'kov, Zh. Neorgan. Khim., 4:7 (1959).
162. K. Rdetz and E. Saur, Physik, 169:315 (1962).
163. P. S. Swartz, Phys. Rev. Lett., 9:448 (1962).
164. E. Corenzwit, Phys. and Chem. Solids, 9:93 (1959).
165. B. W. Roberts, in: New Materials and Methods of Studying Metals and Alloys [Russian translation], Izd. Metallurgiya, Moscow (1966), p. 9.
166. V. V. Baron, L. F. Myzenkova, E. M. Savitskii, and E. I. Gladyshevskii, in: Metallography and Metal Physics of Superconductors, Izd. Nauka, Moscow (1965), p. 86.
167. Z. A. Guts, N. I. Krivko, V. K. Morozova, T. A. Sidorova, and L. A. Fogel', Zh. Tekh. Fiz., 35(9):1675 (1965).
168. C. T. Thompson, and J. F. Gerber, Solid-State Electronics, 2:259 (1961).
169. T. B. Reed, H. C. Gatos, M. J. La Fleur, and J. T. Roddy, Metallurgy of Advanced Electronic Materials, New York (1963), B, p. 71.
170. B. T. Matthias, T. H. Geballe, R. H. Willens, E. Corenzwit, and G. W. Hull, Phys. Rev., 139(5A):1501 (1965).
171. V. M. Pan, V. I. Latyesheva, and E. A. Shishkin, in: Metallography, Physical Chemistry, and Metal Physics of Superconductors, Izd. Nauka, Moscow (1967), p. 157.
172. J. H. Carpenter and A. W. Searcy, J. Amer. Chem. Soc., 78:2079 (1956).
173. J. H. Carpenter, J. Phys. Chem., 67:2141 (1963).
174. H. Nowotny, A. W. Searcy, and J. E. Orr, J. Phys. Chem., 60:677 (1956).
175. J. H. Carpenter and A. W. Searcy, J. Phys. Chem., 67:2144 (1963).
176. N. Kurti and F. Simon, Proc. Roy. Soc. Lond., A151:610 (1935).
177. S. Geller, J. Amer. Chem. Soc., 77:1502 (1955).
178. M. I. Agafonova, V. V. Baron, and E. M. Savitskii, Izv. Akad. Nauk SSSR, Otd. Tekh. Nauk, No. 5, p. 138 (1958).
179. E. M. Savitskii, Zh. Neorgan. Khim., 12(7):1721 (1967).
180. V. S. Kogan, A. I. Krivko, B. G. Lazarev, L. S. Lazareva, A. A. Matsakova, and O. N. Ovcharenko, Fiz. Met. Metallov., No. 15, p. 143 (1963).
181. L. L. Wyman, J. R. Cuthill, G. A. Moorse, J. J. Park, and H. Jakowitz, J. Res. Nat. Bur. Standards, 66A:351 (1962).
182. T. B. Reed, H. C. Gatos, M. J. La Fleur, and J. T. Roddy, Superconductors, New York (1962), p. 143.
183. R. Enstrom, T. Courtney, G. Pearsall, and J. Wulff, Metallurgy of Advanced Electronic Materials, Vol. 1, Interscience, New York, (1962), p. 121.
184. R. E. Enstrom, G. W. Pearsall, and J. Wulff, Appl. Phys. Lett., 3:81 (1963).
185. H. W. Schaeder and H. S. Rosenbaum, J. Metals, 16:97 (1964).
186. V. N. Svechnikov, V. M. Pan, and Yu. I. Beletskii, in: Metallography, Physical Chemistry, and Metal Physics of Superconductors, Izd. Nauka, Moscow (1967), p. 152.
187. V. G. Kuznetsova, V. A. Kovaleva, and A. V. Beznosikova, in: Metallography, Physical Chemistry, and Metal Physics of Superconductors, Izd. Nauka, Moscow (1967), p. 146.

187a. J. P. Charlesworth, T. Macphail, and P. E. Madsen, J. Materials Science, 5(7): 580 (1970).
188. C. L. Kolbe and C. H. Rosner, Metallurgy of Advanced Electronic Materials, Vol. 19, Interscience, New York (1962), p. 17.
189. R. E. Enstrom, N. Y. Pearsall, G. W. Pearsall, and J. Wulff, J. Metals, 16:97 (1964).
190. R. E. Enstrom, G. W. Pearsall, and J. Wulff, Bull. Amer. Phys. Soc., 7:323 (1962).
191. D. J. van Ooijen, J. H. van Vucht, and W. R. Oruynesteyn, Phys. Lett., 3:128 (1963).
192. T. B. Reed and H. C. Gatos, J. Appl. Phys., 33:2657 (1962).
193. M. L. Picklesimer, Appl. Phys. Lett., 1:64 (1962).
194. H. G. Jansen and E. J. Saur, Transactions of the Seventh International Conference on Low-Temperature Physics (Canada, 1960), Univ. Toronto Press (1961), p. 185.
195. V. S. Kogan, A. I. Krivko, B. G. Lazarev, L. S. Lazareva, A. A. Matsakova, and O. N. Ovcharenko, in: Metallography and Metal Physics of Superconductors, Izd. Nauka, Moscow (1965), p. 76.
196. V. N. Svechnikov, V. M. Pan, and Yu. I. Beletskii, Dokl. Akad. Nauk SSSR, 6:1328 (1966).
197. H. J. Levinstein and E. Buchler, Trans. Metallurg. Soc. AIME, 230(6):1314 (1964).
198. A. H. Gomes de Mesquita, C. Langereis, and J. I. Leenhonts, Philips Res. Repts., 18(5):377 (1963).
199. T. G. Ellis and H. A. Wilhelm, J. Less-Common Metals, 7(1):67 (1964).
200. L. J. Vieland, RCA Review, 25(3):366 (1964).
201. J. J. Hanak, K. Strater, and G. W. Cullen, RCA Review, 25(3):342 (1964).
202. H. G. Janen, Z. Physik, 162(3):275 (1961).
203. R. G. Maier and G. Wilhelm, Z. Naturforsch., 19A(3):399 (1964).
204. J. H. N. van Vucht, D. J. van Ooijen, and H. A. C. M. Brunning, Philips Res. Repts., 20:136 (1965).
205. T. R. Anantharaman, Nucl. Sci. Abstracts, 19(16):30730 (1965).
206. J. R. Ogren, T. D. Ellis, and J. F. Smith, Acta Crystallogr., 18(5):968 (1965).
207. G. E. Telentyuk, V. V. Baron, and E. M. Savitskii, in: Metallography and Metal Physics of Superconductors, Izd. Nauka, Moscow (1965), p. 83.
208. J. J. Hanak, G. D. Cody, P. R. Aron, and H. C. Hitchcock, High Magnetic Fields, Technol. Press, New York-London-Cambridge (1962), p. 592.
209. T. H. Courtney, G. W. Pearsall, and J. Wulff, Trans. Metallurg. Soc. AIME, 233(1):212 (1965).
210. E. Saur and P. Schuit, Z. Physik, 167:170 (1962).
211. F. Lange, Monatsber. Dtsch. Akad. Wiss. Berlin, 1:408 (1959).
212. E. Buchler, J. H. Wernick, K. M. Olsen, F. S. L. Hsu, and J. E. Kunzler, in: Metallurgy of Advanced Electronic Materials, Vol. 19, Interscience, New York (1963), p. 105.
213. J. P. Charlesworth, Phys. Letts., 21(5):501 (1966).
214. R. E. Enstrom, J. Appl. Phys., 37(13):4880 (1966).
215. R. Kieffer, F. Benesovsky, H. Nowotny, and H. Schachrer, Z. Metallkunde, 47:247 (1956).

216. G. V. Samsonov, V. A. Ermakova, and V. S. Neshpor, Zh. Neorgan. Khim., 3:868 (1958).
216a. V. M. Pan, V. V. Pet'kov, and O. G. Kulik, in: Metallography, Physical Chemistry, and Physical Metallurgy of Superconductors, Izd. Nauka, Moscow (1967), p. 161.
217. H. J. Goldschmidt and I. A. Brand, J. Less-Common Metals, 3(1):44 (1961).
218. E. M. Savitskii, M. A. Tylkina, and K. B. Povarova, At. Energ., 7(5):407 (1959).
219. E. Bucher and I. Muller, Helv. Phys. Acta, 34:843 (1961).
220. H. J. Goldschmidt, Research, 10(7):289 (1957).
221. B. T. Matthias and B. S. Chandrasekhar, Rev. Mod. Phys., 36(1):134 (1964).
222 D. Bender, E. Bucher, and J. Muller, Phys. Kondens. Materie, 1:225 (1963).
223. A. G. Knapton, J. Less-Common Metals, 2(2-4):113 (1960).
224. E. A. Bucher, F. Heiniger, and I. Muller, Proc. 8th Internat. Congr. Low-Temperature Phys., London (1962), Vol. 4, p. 8.
225. E. M. Savitskii, V. V. Baron, and A. N. Khotinskaya, Zh. Fiz. Khim., 6(11):2603 (1961).
226. E. M. Savitskii, M. A. Tylkina, and K. B. Povarova, Rhenium Alloys, Izd. Nauka, Moscow (1965), pp. 126 and 129.
227. E. Bucher, E. Heiniger, and J. Muller, Phys. Kondens. Materie, 2(3):210 (1964).
228. V. B. Compton, E. Corenzwit, I. P. Maita, B. T. Matthias, and E. Morin, Phys. Rev., 123:1567 (1961).
229. I. Niemiec, Bull. Acad. Polon. Sci. Ser. Sci. Chim., 11(6):305 (1963).
230. B. T. Matthias and E. Corenzwit, Phys. Rev., 94:1065 (1954).
231. B. T. Matthias, Phys. Rev., 97:74 (1955).
232. S. H. Autler, J. K. Hulm, and R. S. Kemper, Phys. Rev., 140(4A):1177 (1965).
232a. C. W. Chu, T. F. Smith, and E. Gardner, Phys. Rev. Lett., 20(5):198 (1968).
233. V. Ya. Anosov and S. A. Pogodin, Fundamentals of Physico-Chemical Analysis, Izd. AN SSSR, Moscow (1947).
234. V. S. Mikheev and O. K. Belousov, Zh. Neorgan. Khim., 4(8):1905 (1961).
235. T. Doi, F. Ishida, and Y. Tagasi, J. Japan Inst. Metals, 30(2):139 (1966).
236. V. S. Kogan, B. G. Lazarev, and L. F. Yakimenko, Zh. Éksp. Teor. Fiz., Vol. 51, No. 5(11), p. 1328 (1966).
237. N. E. Alekseevskii, O. S. Ivanov, I. I. Raevskii, and N. V. Stepanov, Fiz. Met. Metallov., 23(1):28 (1967).
238. A. G. Grigor'ev, E. M. Sokolovskaya, V. V. Kuprina, M. V. Raevskaya, L. S. Guzen, and I. G. Sokolova, Vestn. MGU, Ser. Khim., 5(4):57 (1966).
239. L. F. Myzenkova, Yu. V. Efimov, V. V. Baron, and E. M. Savitskii, in: Metallography, Physical Chemistry, and Metal Physics of Superconductors, Izd. Nauka, Moscow (1965), p. 39.
240. B. G. Lazarev, O. N. Ovcharenko, A. A. Matsakova, and V. G. Volotskaya, in: Metallography, Physical Chemistry, and Metal Physics of Superconductors, Izd. Nauka, Moscow (1967), pp. 76-78.
241. B. G. Lazarev, O. N. Ovcharenko, and A. A. Matsakova, in: Metallography, Metal Physics, and Physical Chemistry of Superconductors, Izd. Nauka, Moscow (1967), pp. 98-100.
241a. Doi Tosio, Isida Fumihiko, and Kawabe Usio, Hitachi Hyoron, 50(12):1065 (1968); J. Japan Inst. Metals, 32(9):886 (1968).

242. N. E. Alekseevskii, O. S. Ivanov, I. I. Raevskii, and N. V. Stepanov, Dokl. Akad. Nauk SSSR, 176(2):305 (1967).
243. N. E. Alekseevskii, Dokl. Akad. Nauk SSSR, 163(5):1121 (1965).
244. T. Doi, M. Masao, and Y. Tagasi, J. Japan Inst. Metals, 30(2):133 (1966).
245. S. Maeda, T. Doi, and F. Ishida, Transactions of the Tenth International Conference on Low-Temperature Physics (1966), p. 185.
246. T. F. Fedorov, N. M. Popova, and R. V. Skolozdra, Izv. Akad. Nauk SSSR, Metally, No. 2, p. 204 (1967).
247. D. K. Deardorff and H. Kato, The Transformation Temperature of Hafnium, USBM-U-426 (April, 1958), p. 8.
248. E. M. Savitskii, V. V. Baron, and Yu. V. Efimov, Izv. Akad. Nauk SSSR, Metally, No. 3, p. 157 (1966).
249. V. S. Emel'yanov, Yu. G. Godin, and A. I. Evstyukhin, At. Énerg., 4(2):161 (1958).
250. R. M. Rose and J. Wulff, J. Appl. Phys., 33(7):2394 (1962).
251. R. S. Shmulevich, I. A. Baranov, V. R. Karasik, and G. B. Kurganov, Fiz. Met. Metallov., 21(3):379 (1966).
252. I. A. Baranov, R. S. Shmulevich, V. A. Synikov, V. R. Karasik, and N. G. Vasil'ev, in: Metallography, Physical Chemistry, and Metal Physics of Superconductors, Izd. Nauka, Moscow (1967), p. 82.
253. E. M. Savitskii and A. M. Zakharov, in: Study of Nonferrous Metal Alloys, No. 3, Izd. AN SSSR, Moscow (1963), p. 108.
254. J. Murakami, T. Kajyo, and H. Yoshida, Techn. Repts. Engng. Res. Inst. Kyoto Univ., 13(5):65 (1963).
255. I. A. Popov and N. G. Rodionova, Zh. Neorgan. Khim., 9(4):890 (1964).
256. E. M. Savitskii and A. M. Zakharov, Zh. Neorgan. Khim., 7(11):2575 (1962).
257. V. A. Frolov, V. V. Baron, and E. M. Savitskii, in: Metallography, Physical Chemistry, and Metal Physics of Superconductors, Izd. Nauka, Moscow (1967), p. 74.
258. M. I. Bychkova, V. V. Baron, and E. M. Savitskii, in: Metallography, Physical Chemistry, and Metal Physics of Superconductors, Izd. Nauka, Moscow (1967), p. 79.
259. I. I. Kornilov and V. S. Vlasov, Zh. Neorgan. Khim., 4(7):1630 (1959).
260. I. I. Kornilov and V. S. Vlasov, Zh. Neorgan. Khim., 2(12):2762 (1957).
261. I. I. Kornilov and V. S. Vlasov, Zh. Neorgan. Khim., 4(9):2017 (1960).
262. Yu. V. Efimov, V. V. Baron, and E. M. Savitskii, in: Metallography, Physical Chemistry, and Metal Physics of Superconductors, Izd. Nauka, Moscow (1967), p. 86.
262a. E. M. Savitskii, V. V. Baron, Yu. V. Efimov, M. I. Bychkova, and N. D. Kozlova, Dokl. Akad. Nauk SSSR, 196(5):1145 (1971).
262b. H. J. Sheffe, J. Roy. Statistical Soc., Vol. 20, Ser. B, No. 2, p. 344 (1958).
262c. J. M. Gorman and J. E. Henman, Technometrics, 4(4):463 (1962).
262d. M. Suenaga and K. M. Ralls, J. Appl. Phys., 40(11):4457 (1969).
262e. E. M. Savitskii, V. V. Baron, M. I. Bychkova, S. D. Gindina, Yu. V. Efimov, N. D. Kozlova, L. F. Martynova, B. P. Mikheilov, L. F. Myzenkova, and V. A. Frolov, Summaries of Contributions to the Sixteenth All-Union Conference on Low-Temperature Physics, Leningrad (1970), p. 198.

263. I. I. Kornilov and R. S. Polyakova, Zh. Neorgan. Khim., 3(4):879 (1958).
264. M. I. Bychkova, V. V. Baron, and E. M. Savitskii, in: Properties and Uses of Heat-Resistant Alloys, Izd. Nauka, Moscow (1966), p. 30.
264a. E. M. Savitskii, V. V. Baron, and V. A. Frolov, Soviet Patent No. 296,825, Byull. Izobr., No. 9, p. 94 (1971).
264b. M. V. Mal'tsev, G. D. Danilova, and N. P. Druzhinina, Izv. VUZov, Tsvetnaya Met., No. 5, p. 146 (1959).
264c. I. A. Popov and V. I. Rabezova, Zh. Neorgan. Khim., 7(2):436 (1962).
264d. E. M. Savitskii, V. V. Baron, V. A. Frolov, and N. D. Kozlova, in: Superconducting Alloys and Compounds, Izd. Nauka, Moscow (1971).
265. A. Nowikow and H. G. Baer, Z. Metallkunde, 49(4):195 (1958).
266. I. I. Kornilov, V. S. Mikheev, and O. K. Belousov, in: Titanium and Its Alloys, No. 7, Izd. AN SSSR, Moscow (1962), p. 120.
267. I. S. Komjathy, J. Less-Common Metals, 3(6):468 (1961).
267a. E. M. Savitskii, Yu. V. Efimov, and N. D. Kozlova, Izv. Akad. Nauk SSSR, No. 4, 215 (1971).
268. I. I. Kornilov and R. S. Polyakova, Izv. Akad. Nauk SSSR, Otd. Tekh. Nauk, Metallurgiya i Toplivo (1960), No. 1, p. 85.
269. I. I. Kornilov and R. S. Polyakova, Izv. Akad. Nauk SSSR, No. 4, p. 76 (1961).
269a. L. F. Mattheiss, Phys. Rev., 139:A1893 (1965).
269b. A. I. Golovashkin, I. E. Leksina, G. B. Motulevich, and A. A. Shubin, in: The Fifteenth Conference on Low-Temperature Physics, Tiflis (1968).
269c. C. W. Cheng, Phys. Stat. Sol., 40:717 (1970).
269d. M. F. Manning and M. Y. Chodorov, Phys. Rev., 56(15):787 (1939).
270. A. Taylor, M. I. Doyle, and B. J. Kagle, Trans. ASM, No. 56, p. 49 (1963).
271. E. M. Savitskii, Yu. V. Efimov, and E. I. Gladyshevskii, Izv. Akad. Nauk SSSR, Neorgan. Mat., No. 1, p. 354 (1965).
272. E. I. Gladyshevskii and Yu. B. Kuz'ma, Izv. Akad. Nauk SSSR, Metally, No. 3, p. 2101 (1967).
273. E. M. Savitskii, V. V. Baron, and V. A. Frolov, Byull. Izobret., No. 13 (1967).
274. E. I. Gladyshevskii and G. N. Shvets, Izv. Akad. Nauk SSSR, Metally, No. 2, p. 120 (1965).
275. V. V. Baron, K. N. Ivanova, and E. M. Savitskii, Izv. Akad. Nauk SSSR, Otd. Tekh. Nauk, Metallurgiya i Toplivo, No 4, p. 143 (1960).
276. E. M. Savitskii, V. V. Baron, and K. N. Ivanova, Izv. Akad. Nauk SSSR, Otd. Tekhn. Nauk, Metallurgiya i Toplivo, No. 2, p. 126 (1962).
277. O. S. Ivanov and G. I. Terekhov, in: Structure of Alloys in Various Systems Incorporating Uranium and Thorium, Gosatomizdat, Moscow (1961), p. 20.
278. D. A. Prokoshkin and M. I. Zakharova, Izv. Akad. Nauk SSSR, Otd. Tekh. Nauk, No. 5, p. 59 (1961).
279. H. J. Goldschmidt and J. A. Brand, J. Less-Common Metals, 3(1):44 (1961).
280. V. S. Mikheev and T. S. Chernova, in: Titanium and Its Alloys, No. 7, Izd. AN SSSR, Moscow (1962), p. 81.
281. Bi Tsin-Khua and I. I. Kornilov, Zh. Neorgan. Khim., 5(4):902 (1960).
282. Bi Tsin-Khua, Izv. Akad. Nauk SSSR, Metallurgiya i Toplivo, No. 6, p. 100 (1959).
283. Bi Tsin-Khua, Zh. Neorgan. Khim., 6(6):1351 (1961).

284. V. N. Svechnikov, Yu. A. Kocherginskaya, et. al., in: Study of Heat-Resistant Materials and Metals, No. 4, Izd. AN SSSR, Moscow (1959), p. 248.
285. I. I. Kornilov, in: Properties and Use of Heat-Resistant Alloys, Izd. Nauka, Moscow (1966), p. 3.
286. A. M. Zakharov and E. M. Savitskii, Izv. Akad. Nauk SSSR, Neorgan. Mat., 3(4):661 (1967).
287. A. M. Zakharov and E. M. Savitskii, Izv. Akad. Nauk SSSR, Metally, No. 5, p. 159 (1966).
288. A. M. Zakharov and E. M. Savitskii, Izv. Akad. Nauk SSSR, Metally, No. 1, p. 151 (1965).
289. A. M. Zakharov and E. M. Savitskii, Izv. Akad. Nauk SSSR, Metally, No. 6, p. 121 (1966).
290. M. A. Tylkina, K. B. Povarova, and E. M. Savitskii, Zh. Neorgan. Khim., 5(11):2459.
291. C. T. Sims and R. I. Jaffee, Trans. ASM, 52:929 (1960).
292. E. M. Savitskii, M. A. Tylkina, and V. P. Polyakova, Zh. Neorgan. Khim., 8(1):146 (1963).
293. M. A. Tylkina, V. P. Polyakova, and E. M. Savitskii, Zh. Neorgan. Khim., 9(3):671 (1964).
294. W. Rostoker, A Study of Ternary Diagrams of Tungsten and Tantalum, WADC, Techn. Rept., pp. 59-492 (1960).
295. C. Frelin, VIII Seminair de Thermodynamique et Physicochemie Metallurgiques, Grenoble (May 5-6, 1967), p. 15.
296. E. M. Savitskii and A. M. Zakharov, Zh. Neorgan. Khim., 9(10):2422 (1964).
296a. T. Claeson, Phys. Stat. Sol., 42:321 (1970).
296b. W. de Sorbo, P. E. Lawrence, and W. A. Healy, J. Appl. Phys., 38(2):903 (1967).
297. N. E. Alekseevskii, N. V. Ageev, and V. F. Shamrai, Izv. Akad. Nauk SSSR, Neorgan. Mat., No. 12, p. 2150 (1966).
298. E. A. Linton and B. Serin, Phys. Rev., 112:70 (1958).
299. D. F. Seraphim and D. C. Quinn, Acta Metallurg., No. 9, p. 861 (1961).
300. B. T. Matthias, H. Suhl, and E. Corenzwit, Phys. Rev. Lett., 1:92 (1958).
301. Sci. News, 91(20):475 (1967).
302. C. D. Cody, J. J. Hanak, G. T. McConvile, and F. D. Rosi, Rev. RCA, 25(3):338 (1964).
303. E. M. Savitskii, New Metal Alloys, Izd. Znanie, Moscow (1967).
304. E. M. Savitskii, Izv. Akad. Nauk SSSR, Metally, No. 3, p. 100 (1967).
305. E. M. Savitskii, Metal. i Term. Obrabotka Metal., No. 10, p. 12 (1967).
306. F. Heiniger, E. Bucher, and J. Muller, Physik Kondens. Mater., 5:243 (1966).
307. O. S. Ivanov and V. K. Grigorovich, Trans. of the Second Internat. Conf. on the Peaceful Use of Atomic Energy, Contributions of Soviet Scientists, Vol. 3, Geneva (1958).
308. In: Physical Chemistry of Zirconium Alloys, Izd. Nauka, Moscow (1968).
309. S. A. Nemnonov and E. Z. Kurmaev, in: Problems of Superconducting Materials, Izd. Nauka, Moscow (1970), p. 5.

Chapter V

Production of Superconducting Materials

The most promising field of application of superconductors lies in the creation of magnetic systems, for which materials are required in the form of long wires and strips of small diameters and thicknesses; the creation of magnetic fields by means of bulkier superconducting parts (tubes, rings, and so on) is already being studied [1].

The treatment of metals by pressure (forging, extrusion, rolling, etc.) plays a major part in the technological process of producing superconducting materials and parts. The plastic deformation (working) of metal is a complicated physicochemical process, causing changes in not only the geometrical dimensions but also the physical (including superconducting) properties of the metals and alloys so treated [2].

The capacity of a metal to undergo plastic deformation, of course, depends on the nature of the metal (alloy), its composition and structure, the method of deformation, i.e., the character of the stressed state, and the conditions of working (temperature, velocity, degree of deformation, and effect of the external medium).

A real metal or alloy constitutes a physicochemical system of crystals of the original metal with impurities and alloying elements distributed within them and along their boundaries, together with crystallographic defects. These features may be either favorable or unfavorable toward the plastic deformation of the metal crystals [2].

Gubkin, Tselikov, Savitskii, and others have formulated some general principles for choosing the conditions of hot plastic deformation in relation to metallic substances; these are also valid for superconducting metals and alloys [3-5].

In choosing materials for making the windings of superconducting magnets and solenoids, the critical current (I_c, A) is an extremely important characteristic in addition to T_c and H_{c2}. A high critical current enables us to create superconducting magnets and solenoids of small dimensions and low weight; this is extremely important in the majority of practical cases. Moreover, a low critical current or a low critical current density (J_c, typically of the order of 10^3 A/cm^2) makes the use of superconducting materials as solenoid windings undesirable, particularly in a high magnetic field. In this case the dimensions of the solenoid and hence the consumption of liquid helium increase sharply.

The combination of high critical current, magnetic field, and transition temperature is in fact secured in a number of superconducting materials, broadly divided into two groups (alloys and compounds) such as those which we have been discussing in earlier chapters.

EFFECT OF COMPOSITION, DEFORMATION, AND HEAT TREATMENT ON THE CRITICAL CURRENT OF SUPERCONDUCTING ALLOYS

The first group of superconducting materials includes alloys of the Nb–Zr and Nb–Ti systems and a few complex alloys based on these (Nb–Ti–Zr, Nb–Zr–Ta, etc.), as well as alloys of the V–Ti and Ta–Ti systems. At high temperatures these systems are characterized by a continuous series of solid solutions, which decompose on reducing the temperature as a result of the polymorphism of the components (this was discussed earlier). The critical current of such alloys largely depends on the conditions of mechanical and thermal processing (working and heat treatment); appropriate treatment may increase it by two or three orders of magnitude or even more. The critical current may be increased either by cold work or by annealing.

In alloys of systems in which the components are mutually soluble at all temperatures and concentrations (Nb–Ta, V–Nb,

PRODUCTION OF SUPERCONDUCTING MATERIALS 375

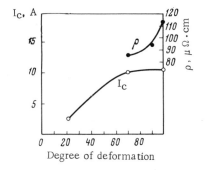

Fig. 203. Critical current and electrical resistance of an alloy of niobium with 75 at.% Ti in relation to the degree of cold working.

etc.), the critical current is only increased by cold working. Thus, for example, in a cold-worked alloy of niobium with 45 at.% Ta the critical current density after cold working is roughly an order of magnitude higher than after annealing [6]. The annealing of worked alloys leads to the removal of stresses arising as a result of working and to a corresponding decrease in the critical current. However, at a certain stage in the prerecrystallization annealing of worked alloys a cellular distribution of the dislocations has sometimes been observed, together with a rise in I_c [6a]. The results of our own investigations [7, 8] into the effect of cold working on the critical current of Nb-Ti alloys shows that the critical current of an alloy containing 75 at.% Ti subjected to various degrees of cold working increases by a factor of five times at the greatest degree of reduction employed, reaching 11 A instead of the 2 A achieved by 20% deformation; the electrical resistance also increases (Fig. 203).

We also studied the effect of annealing on the critical current of worked Nb-Ti alloys containing 50-80 at.% Ti. The alloys were melted in an electric-arc furnace using electron-beam-melted niobium (O_2 – 0.01; N_2 – 0.01; H_2 – 0.0001; C – 0.01; Ta – 0.2 wt.%) and iodide-type titanium of high purity; they were cold-worked so as to obtain a wire of 0.25 mm diameter. Either intermediate or final annealing of the samples was carried out, with subsequent water-quenching. The results showed that the greatest increase in the critical current density of alloys containing 70, 75, and 80% Ti occurred after 1 h annealing at 400-450°C (Fig. 204). For alloys with 70 and 80 at.% Ti, intermediate annealing with subsequent working and final annealing in a number of cases gave very similar results, while the alloy with 75% Ti had the highest values

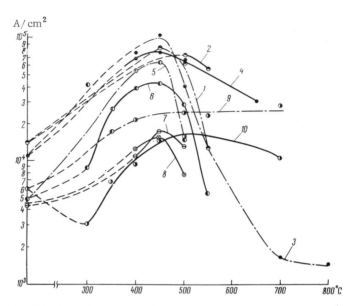

Fig. 204. Dependence of the critical current density on the annealing temperature of a Nb−Ti alloy: titanium content, at.% 1) 80 (final annealing); 2) ditto (intermediate annealing); 3) 75 (final annealing); 4) ditto (intermediate annealing); 5) 70 (final annealing); 6) ditto (intermediate annealing); 7) 65 (final annealing); 8) ditto (intermediate annealing); 9) 50 (final annealing); 10) ditto (intermediate annealing).

of critical current after final annealing: 58 A ($1.2 \cdot 10^5$ A/cm^2). At 500-550°C the annealing of alloys containing less than 70% Ti led to an increase in J_c but only by a factor of two or three. Annealing alloys containing 70-80% Ti at 600-900°C reduced the critical current density, probably as a result of recrystallization and the removal of stresses.

Annealing also changes the strength characteristics of the alloys. The tensile strength rises particularly sharply in alloys with a high titanium content (the same conditions as those under which J_c increases most sharply); after annealing the alloy with 75% Ti the strength reached 105 kg/mm^2.

A study of the structure and properties of a particular alloy subjected in one case to cold working and in another to heat treatment yielded certain information regarding the individual effects of deformation (working) and the decomposition of the solid solution on the critical current of alloys [9]. Samples were prepared in the form of wires from a homogenized cast alloy of niobium with 75 at.% Ti, which according to x-ray analysis had the structure of a bcc solid solution. Samples were cold-worked to various dimensions in such a way as to ensure that, after a recrystallization anneal and further deformation to a final wire size of 0.25 mm, the degree of deformation should total between 20 and 99.9%. As already indicated (Fig. 204), the critical current rose by a factor of five by increasing the cold deformation 20 to 99.9%.

Cold-worked samples of this wire were subjected to a recrystallization anneal at 800°C (the recrystallization temperature was determined metallographically) and then heated to 350-550°C for 1 h. We see from Fig. 205 (curve 1) that in this case J_c rises with increasing temperature, reaches a maximum at 400°C, and then starts falling. For comparison, the same figure (curve 2) shows the effect of the same heating on a cold-worked wire (deformation 99.95%). Although the measurements showed, as might be expected, that the highest values were obtained as a result of a combination of cold work and annealing, there was also a considerable increase in this characteristic (J_c) in the case of previously recrystallized samples, this amounting to an order of magnitude at 450°C. The smaller increment in the critical current of recrystallized wire samples after annealing may be attributed to the slower decomposition of the solid solution in this case as compared with the previously-worked sample. The greater changes in the electrical resistance and tensile strength of the previously-worked samples (the greater reduction in electrical resistance and tensile strength) indicate the same. On subjecting samples annealed at

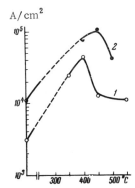

Fig. 205. Critical current in relation to the annealing temperature of a Nb-75-at.% Ti alloy: 1) recrystallized sample; 2) cold-worked sample.

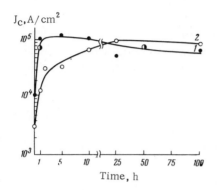

Fig. 206. Critical current density of worked (1) and recrystallized (2) Ni-75-at.%-Ti samples in relation to annealing time at 450°C.

400-550°C to x-ray analysis after a holding period of 1 h, a hexagonal phase was found to have been formed in every case.

In order to increase the degree of decomposition of the solid solution, the annealing time of the alloys was increased (to 100 h) at the optimum temperature established (450°C). It was found that an increase in the annealing period led to an increase in the critical current of the recrystallized samples. The critical current density increased particularly sharply in the range 1-10 h ($7 \cdot 10^4$ A/cm^2) and after 100 h J_c equalled 10^5 A/cm^2 (Fig. 206, curve 2). At the same time, in the cold-worked samples there was a decrease in J_c after holding for more than 10 h (curve 1).

An analysis of the individual effects of plastic deformation and the decomposition of the solid solution on the critical current of a superconducting Nb–Ti alloy showed that the physical and chemical inhomogeneities arising as a result of these processes tended to increase the critical characteristics of the superconductors.

The application of a higher degree of deformation accelerates the decomposition processes, and in this case a high critical current may be achieved after shorter annealing periods than those required for the recrystallized material. The softening of the alloy and an increase in the number and dimensions of the second-phase inclusions beyond the optimum values worsen the characteristics and annul the beneficial influence of plastic deformation.

In order to establish the phase composition of the Nb-75-at.%-Ti alloy we analyzed the structures of these alloys (in collaboration with N. N. Buinov and S. V. Sudareva) under the electron microscope as well as subjecting them to x-ray analysis [8a]. The results were correlated with the critical current measured on the same samples after identical mechanical and heat treatment. The x-ray phase analysis and electron-microscope examination (Fig. 207) showed, that during the decomposition of the β solid solution under the thermal conditions chosen, particles of α solid solution precipitated. The greatest critical current occurred, not when the size and number of these α phase particles were minimal (annealing time 15 min), but at a time when they had reached certain specific optimum values (after 1-5 h): particle size 10^{-5} cm, number 10^{-11} cm^{-2}.

The fundamental correctness of this conclusion was confirmed by a number of subsequent investigations in which electron-microscope analysis was also employed for the phase analysis of alloys containing 78 and ~63 at.% Ti [8b, c, d, e].

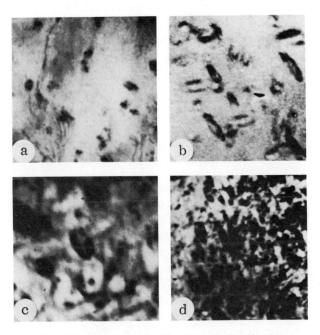

Fig. 207. Microstructure of a Nb-75-at.%-Ti alloy (magnification 10,000): a) recrystallization (800°C) for 1 h, annealing (450°C) for 15 min; b) ditto, annealing 1 h; c) ditto, annealing at 10 h; d) deformation (99.9%) and annealing (450°C) 1 h.

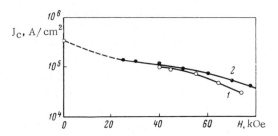

Fig. 208. Dependence of the critical current density of heat-treated Nb alloys with 70 (1) and 75 at.% Ti (2) on the external magnetic field.

All the foregoing data as to the critical current density of Nb−Ti alloys were obtained by measurements in an external magnetic field (created by a superconducting solenoid) of up to 26 kOe. Additional investigations into the characteristics of heat-treated alloys with 70 and 75 at.% Ti were also carried out in higher fields. Good results were obtained in this case also (Fig. 208) [9].

Similar results were also obtained elsewhere [10]. The critical current density of a wire 0.2 mm in diameter made from a niobium alloy with 78 at.% Ti reached $5 \cdot 10^4$ A/cm^2 in a magnetic field of 80 kOe and about $2 \cdot 10^4$ A/cm^2 at H = 90 kOe after heat treatment. The sensitivity of the structure and superconducting properties of Nb−Ti alloys to heat treatment is also discussed in [10a, b].

The dependence of the critical current density on the heat treatment of a Nb−Ti alloy was also studied by Reuter et al. [11]. According to this investigation the heat treatment of an alloy containing 44.7 at.% Nb, 54.3% Ti, and about 1% oxygen had a substantial effect on the value of the critical current. The choice of optimum conditions of heat treatment for the finally-worked 0.25-mm diameter wire of this alloy (annealing at 500°C for 2 h) yielded a J_c of about $5 \cdot 10^4$ A/cm^2 in a field of 80 kOe and about 10^4 A/cm^2 in a field of 90 kOe. In a field of 100 kOe the alloys passed from the superconducting to the normal state. Phase transformations in alloys of niobium with 60% Ti, 0.66-0.72% Er, 0.65-0.82% Sc and traces of oxygen (up to 0.45%) were considered in [11a]. It is to be noted that the precipitation of oxides has a positive influence on the critical current, but not as good as the precipitation of the α and ω phases.

Phase transformations in Ti–V alloys [11b] may also be used in order to raise the critical current. The critical currents of V–Ti alloys are discussed in [11c].

The value of the critical current disrupting the superconductivity of Nb–Zr alloys and its dependence on the external magnetic field has been studied by a number of research workers [12-17]. After cold working with a high degree of deformation, these alloys withstand high current densities without losing superconductivity over a wide concentration range, considerably higher densities, in fact, than cold-worked Nb–Ti alloys. On increasing the degree of deformation from 40 to 99.5% the critical current density in a Nb–25%-Ar alloy increases (according to our data obtained in a magnetic field of 24 kOe) by a factor of five [12].

An increase in the critical current density of the same alloy in high magnetic fields on increasing the degree of cold deformation was also observed elsewhere [17]. Figure 209 shows the dependence of the critical current density on the external magnetic field (from 0 to 70 kOe) for a Nb-25% wire 0.54-0.025 mm in diameter after various degrees of deformation. One notices the high current density for diameters of 0.064 to 0.025 mm, more than 10^5 A/cm^2 in a field of 50 kOe. In a field of 60 kOe the current density is lower, but in this case it still reaches $4-6 \cdot 10^4$ A/cm^2. In a field of 70 kOe the J_c of samples of this alloy de-

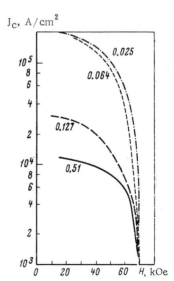

Fig. 209. Critical current density of a Nb–25%-Zr wire in an external magnetic field (numbers of the curves give the diameter of the wire in mm).

creases sharply. It was found in a number of investigations [12, 14, 17, 18, and elsewhere] that heat treatment had a major effect on the critical current of alloys belonging to this system. Heat treatment of a number of such alloys at the intermediate stages of cold working and after the latter had been completed frequently produced a substantial increase in the critical current.

Baron et al. [18] compared the results of measuring the critical current in an external magnetic field at 4.2°K with the structure of several Nb–Zr alloys after various forms of heat treatment. The effect of the structure formed as a result of mechanical and heat treatment on the critical currents of alloys (25 and 50% Zr) was studied in wires 0.2-0.25 mm in diameter. After quenching and working to the sizes indicated, the wire samples had the structure of a bcc β solid solution. After annealing for 1 h at 700-900°C and measuring in an external magnetic field of 22 kOe, the critical current of cold-worked Nb–25-50-at.%–Zr alloys rose sharply in comparison with those not so annealed. The phase composition after heat treatment indicated a nonequilibrium state of the alloys. This was particularly evident when studying the alloy with 50% Zr, in which the decomposition processes took place more rapidly than in the alloy with 25 at.% Zr as a result of the higher diffusion velocity. The critical current of this alloy is shown in relation to the annealing temperature in Fig. 210, together with the changes in phase composition. Samples of the alloy annealed at 700-900°C (for 1 h) withstood a critical current density of up to $1.5 \cdot 10^5$ A/cm^2, the structure consisting of $\beta_1 + \beta_2 + \alpha$ solid solutions.

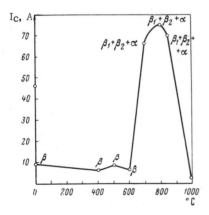

Fig. 210. Dependence of the critical current and structure on the annealing temperature of a Nb–50%–Zr wire.

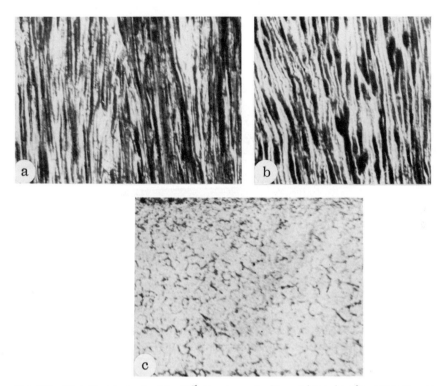

Fig. 211. Microstructure of a Nb–50%–Zr wire: a) after working; b) after annealing at 800°C; c) ditto at 1050°C.

This nonequilibrium state of the alloy may be explained by the inadequate rate of cooling from the $\beta_1 + \beta_2$ region; this was too low to fix the high-temperature state of the alloy. In addition to the bcc phases β_1 and β_2, the hexagonal α phase, with a transition temperature of under 4.2°K, precipitated, and this led to a considerable increase in critical current. Substantial recrystallization of the samples occurred after heating to above 1000°C (Fig. 211).

Figure 212 shows the change in the critical current of coldworked samples of the alloy containing 25% Zr after various periods at 800°C. The phase composition is indicated on the same curve.

We see from the data presented that the critical current rises even after holding the samples at 800°C for 10 min (the greatest value of the critical current is obtained after holding for

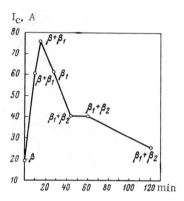

Fig. 212. Dependence of the critical current and structure on the holding time at 800°C for cold-worked Nb-25%-Zr samples (H = 22 kOe; wire diameter 0.25 mm).

15 min). By considering the intensity of the (211) lines on the x-ray diffraction patterns we may conclude that these samples have an identical phase composition ($\beta_1 + \beta_2$) but differ simply in the quantities of the phases. The amount of β_1 phase precipitating after 15 min is greater. The critical current is in this case 75 A instead of 60 A after 10 min. Increasing the holding time leads to a further decomposition of the β phase. The critical current thereupon diminishes. On approaching the equilibrium state the critical current continues to decrease and assumes a value close to the original.

Thus the maximum rise in the critical current taking place as a result of heat treatment involving the decomposition of the solid solution occurs at the onset of this process, although not at the very earliest stages.

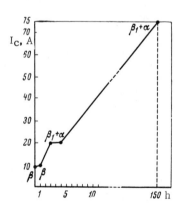

Fig. 213. Dependence of the critical current and structure on the holding time at 500°C for cold-worked Nb-50%-Zr samples (H = 22 kOe; wire diameter 0.25 mm).

An increase in the holding time at lower temperatures in the $\alpha + \beta$ phase region had a beneficial effect on the critical current of the alloys. Thus in the alloy with 25% Zr after annealing at 600°C for 100 h a wire 0.25 mm in diameter withstood 100 A without losing superconductivity ($J_c = 2 \cdot 10^5$ A/cm^2).

Annealing samples of the alloy with 50% Zr at 500°C for an hour led to neither structural transformations nor increased current (Fig. 213). Annealing for 3 and 5 h led to the precipitation of the α phase and a rise in the critical current of cold-worked alloys from 10 to 20-21 A. A substantial rise only occurred however after 150 h annealing, reaching 75 A ($1.5 \cdot 10^5$ A/cm^2) when the proportion of α phase became greater.

Annealing cold-worked Nb–Zr samples and changing their phase composition as a result of the decomposition of the solid solution leads to a still greater rise in critical current density, the maximum value of this being reached immediately after the onset of decomposition (the critical current density may rise by more than an order of magnitude). On increasing the annealing time and temperature without producing recrystallization of the alloys the critical current remains higher than in the original cold-worked samples, but falls gradually on approaching the equilibrium state. In the case of heat treatment under conditions leading to the substantial softening or recrystallization of the alloys, J_c decreases rapidly, independently of the phase composition.

A high current density may also be achieved by intermediate heat treatment. Thus in a Nb–25%-Zr alloy the value reaches $3 \cdot 10^5$ A/cm^2 in this way [12]. The extremely beneficial influence of this kind of treatment was also demonstrated elsewhere [19] when studying the effect of the structural state on the critical current density of alloys with higher zirconium contents.

It is well known that on quenching and annealing zirconium alloys containing a certain amount of β stabilizing elements (Nb, Mo, Cr, Re, V, etc.) there may be a precipitation of the metastable ω phase, which sharply changes the properties of the alloys. It was found in one case [19] when studying alloys with 15-25% Nb (balance zirconium) that the formation of the ω phase in these alloys after intermediate annealing (at 400-450°C) corresponded to a sharp rise in J_c if cold working were applied after this annealing period. In

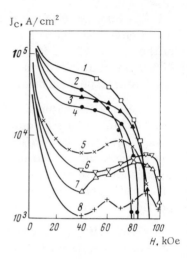

Fig. 214. Critical current density as a function of external magnetic field for Nb-Zr-Ti alloys. Content, at.%: 1) 40 Zr, 10.1 Ti; 2) 20 Zr, 10 Ti; 3) 24 Zr, 28 Ti; 4) 27 Zr, 10 Ti; 5) 11.1 Zr, 38.6 Ti; 6) 5,9 Zr, 52.4 Ti; 7) 5.03 Zr, 62.0 Ti; 8) 13.0 Zr, 48.0 Ti.

Zr-rich alloys subjected to other forms of thermomechanical treatment the value of J_c in an external magnetic field was quite low (under 10^4 A/cm^2); the application of the thermomechanical treatment indicated raised the critical current density by over an order of magnitude.

We have discovered an analogous positive effect of certain forms of thermomechanical treatment on J_c in alloys of the Nb−Hf and V−Ti systems [20, 21]. Other information indicates a rise in the critical current densities of Nb−Zr alloys with interstitial impurities (O, N, C) [15, 22].

The effect of heat treatment and cold working on the critical current has also been studied for ternary alloys of a number of systems [23-29]. In one case [23] it was shown that alloys of the

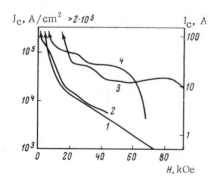

Fig. 215. Dependence of the critical current on the field for Nb−Zr−Ta samples. Content, at.%: 1) 25 Zr, 2 Ta; 2) 25 Zr, 10 Ta; 3) 25 Zr, 5 Ta; 4) 25 Zr, 10 Ta (intermediate annealing at 860°C, 15 min).

Nb−Ti−Zr system belonging to sections with a constant content of 30, 50, and 70 wt.% Nb might after heat treatment (550°C, 1-3 h) withstand a high critical current density of up to $6\text{-}8 \cdot 10^4$ A/cm^2 in a field of about 40 kOe. A rise in the J_c of these alloys occurred when the α ternary solid solution precipitated.

Elsewhere [24, 25] it was found as a result of an analysis of alloys of the same system that the greatest J_c occurred for an alloy of niobium with 40% Zr and 10% Ti; in magnetic fields of 60, 70, and 80 kOe J_c equalled $4 \cdot 10^4$, $2.7 \cdot 10^4$, and $1.5 \cdot 10^4$ A/cm^2, respectively (Fig. 214). Phase transformations in an alloy of niobium with 40% zirconium and 10% titanium are further considered in [25a].

Superconducting alloys of the Nb−Zr−Ta system were studied in another paper [26]. The best results in fields up to 80 kOe were obtained for an alloy with 10% Ta and 25% Zr (Fig. 215) in the heat-treated state.

The foregoing values of the critical current of various alloy wires after heat treatment relates only to individual short samples. In order to retain these high values of J_c in actual superconducting solenoids further research and experimental work is required, mainly in connection with discovering the nature of the so-called "degradation" or "degeneracy" effect in which the critical current of a wire is much lower when used in a solenoid than when tested as a short sample.

TECHNOLOGY OF THE PRODUCTION OF SUPERCONDUCTING ALLOYS

The limited information so far accumulated in relation to the effect of plastic deformation and heat treatment on the critical current of alloys with particularly promising superconducting characteristics illustrates the importance of choosing the best conditions for manufacturing parts from superconducting materials. The question of the technology of processing these materials has as yet received scant attention. We consider that the pressure treatment (working) of superconducting materials is to become one of the most important means of creating materials with high and stable superconducting characteristics.

After studying the superconducting characteristics of Nb–Zr alloys (up to 50 at.% Zr) prepared and processed in various ways in 1961-1962 we developed a technology for producing superconducting wires having good characteristics and secured a patent in respect of this [30]. Using this technology, Nb–Zr alloys were prepared by the following operations: vacuum casting of the original bars in an electric-arc or electron-beam furnace, hot pressing (extrusion) or forging in vacuum or a protective medium, or else in protective sheaths of other metals at 1000-1500°C, heat treatment of the extruded or forged billets, rolling in shaped rolls or rotational forging, cold-drawing, and annealing. In order to obtain a high total degree of cold deformation (about 99.97%), cold treatment was applied from a diameter of 6-4 mm.

Subsequently, in view of the necessity of shunting and insulating the wire used for the manufacture of magnets and solenoids, we developed electrolytic copper-plating and the insulation of the wire with paint or lacquer (type VL). The technology so developed was adopted by the Ministry of Nonferrous Metallurgy of the USSR. Niobium–zirconium alloys of various compositions were assigned the nomenclature RNS-1, RNS-2 (T_c = 10.1°K, H_{c2} ~ 100 kOe, J_c = $2 \cdot 10^4$ A/cm^2), and RNS-3 (T_c = 10.8°K, H_{c2} ~ 80 kOe, J_c = $8 \cdot 10^4$ A/cm^2).

In order to improve the superconducting characteristics of the wire, attempts have been made in the Soviet Union and elsewhere at depositing coatings (Cu, Al, brass, Cu–Ni, etc.) during the processing of the castings. Positive results have been obtained in a number of cases.

As a result of studying the superconducting properties of Nb–Ti alloys and the effect of working and heat treatment on these properties, we developed a technology for making wire and strip of three compositions which were given the names of NT-1, NT-2, and NT-3. After appropriate mechanical and heat treatment, wire made from NT-1 and NT-2 has a critical current density of $4.8 \cdot 10^4$ A/cm^2, H_{c2} = 120-140 kOe, and T_c = 7.6-8.6°K. The strip prepared from these alloys is being successfully used for magnetic-field screening. We have developed a technology for making superconducting wire of ternary Nb–Ti–Zr alloys types 65BT, 50BT, 35BT [30a] under industrial conditions (Central Scientific-Research Institute of Ferrous Metallurgy).

In recent years the creation of multiple-filament composition-type superconducting materials has been vigorously developed [30b, 30c]. The use of multiple-filament superconductors substantially reduces the inductance of a solenoid; this reduces the time required for the setting up of the system, ensures the creation of a structure with good thermal characteristics, and eases the fitting of protective shunts for preventing the system from being damaged on suddenly passing into the normal state. The problem of winding the solenoids is made much simpler, since fewer turns are required in order to create a specific magnetic field.

The A. A. Baikov Institute of Metallurgy, Academy of Sciences of the USSR [30b], has developed a method of producing internally-stabilized multiple-filament-composition superconductors of the NTB-1 type consisting of very fine superconducting filaments of a Nb–Ti alloy 10-90 μ in diameter incorporated in a matrix of normal metal (high-purity copper). The characteristics of the multiple-filament superconductors are presented below:

Number of superconducting filaments and diameter of conductors, mm			Occupation factor relative to the superconductor	Critical current density at 4.2°K in a transverse magnetic field 50 kOe (A/cm^2)	
7fil.	19fil.	37fil.		superconducting filaments	whole conductor
0.3	0.5	0.8	0.25-0.30	up to $2 \cdot 10^5$	up to $4 \cdot 10^4$

The foregoing multiple-filament-composition superconductors possess a reasonable mechanical strength and ductility and are suitable for winding on cylinders of any small diameter.

Various superconducting cables obtained by twisting superconducting and copper wires together have been developed and are being manufactured [30d]. Key questions of the technology of producing hard superconductors are considered in [30e]; these include the effect of technological processing operations, the structure of the alloys, and the proportion of interstitial impurities within them on the superconducting parameters. The phenomena of the degradation, conditioning, and stabilization of the superconductors are discussed, and data relating to the development of multiple-filament superconducting materials are presented [30e, 30f]. Superconducting materials are also produced by a number of companies in other countries. In the United States the largest producer of semifinished superconducting materials is Supercon (wire 0.08 mm in diameter, strip, multiple-filament superconductors) [30g]. General Electric produces superconducting wire consisting of Nb_3Sn with the trade name Cryostrand, giving a current density of over $5 \cdot 10^4$ A/cm^2 in a magnetic field of 100 kG. The production of a multiple-filament superconductor of the Supercon VSK-type is indicated by Nortan International Ind. [30h]; the conductor incorporates 400 filaments of a Nb–Ti alloy in a copper matrix, and the corresponding solenoid allows a current density of $4 \cdot 10^3 A/cm^2$ at 60 kG and 4.2°K. It should be noted that the majority of producers manufacturing superconducting materials are developing multiple-filament composition conductors. In West Germany Vakuumschmelze GMBH, in France Thomson Houston–Hotchkiss Brandt, in Japan Vacuum Metallurgical, and in the United States Airco Kryoconductor indicate the production of multiple-filament (7-10 to 420 filaments) superconductors of various profiles, the development of these materials being chiefly based on alloys of the Nb–Ti system enclosed in a copper matrix.

In order to use superconducting materials in pulse-type or low-frequency ac devices the losses must be reduced to a minimum. To this end a so-called three-component composition superconductor has been developed: Each of 55 Nb–Ti filaments is enclosed in a Cu–Ni sheath, then 19 such groupings are placed in a copper matrix, so that the total number of superconducting filaments is 285. Tests on this material (manufactured by Imperial Metals Industries) have shown that on working under ac conditions its losses are analogous to those in a material with a matrix entirely consisting of the Cu–Ni alloy, i.e., to the losses in an individual filament [30i].

With the development of Nb–Ti/Cu composites (capable of being used in much higher magnetic fields) the use of Nb–Zr alloys has been considerably reduced. However, for certain fields of application some of the characteristics of Nb–Zr alloys are preferable to those of the Nb–Ti type. An attempt was made in one case [30j] to prepare composites of Nb–25% Zr and copper under laboratory conditions by the bimetal method (simultaneous processing). Wire of a total diameter of 0.4 mm was made, the copper being metallurgically linked to the superconductor (diameter of the superconducting filament 0.24 mm). The critical current density of cold-worked samples at 4.2°K and 30 kOe was 1.3 kA/mm^2 and after heat treatment (600°, 1 h) 2.7 kA/mm^2. A composite was also made with a thin intermediate niobium layer between the copper and the Nb–Zr alloy. This structure of the composite allowed heat treatment at higher temperatures and for longer periods without mutual diffusion between the zirconium and copper.

PROPERTIES AND PRODUCTION TECHNOLOGY OF PARTS MADE FROM SUPERCONDUCTING COMPOUNDS

The practical application of the superconducting compounds Nb_3Sn, Nb_3Al, V_3Ga, V_3Si, and others with high superconducting parameters is impeded by their great brittleness. Ordinary methods of working fail to yield reasonably ductile wire or other thin semifinished materials as required in the manufacture of superconducting magnets and solenoids.

According to the Savitskii effect [1, 2], at high temperatures a considerable ductility appears in brittle metallic compounds. In principle it is possible to work brittle materials at high temperatures, particularly under conditions of all-round (hydrostatic) nonuniform compression. However, the range of temperatures corresponding to the appearance of ductility in the fairly refractory compounds is very high, and the process is therefore technologically difficult (heat-resistance of the tools must be great, etc.).

Savitskii also proposed a method of replacing the eutectic (for example, in alloys based on compounds) by other superconducting constituents in order to control their mechanical and electrical (including superconducting) properties [2].

In the case of the formation of a compound from solid solution (for example, V_3Ga, see Chap. IV), the production of a wire is in principle quite possible by working alloys having the composition of the compound after quenching from temperatures above that at which the compound is formed. Subsequent heat treatment of the resultant wire may convert the latter into the superconducting state. However, there are considerable difficulties even in this method.

The production of wire from brittle compounds may be achieved by various indirect methods, chiefly those based on diffusion processes.

1. Production of Vanadium-Gallium Wire by Working the Quenched Solid Solution

Thin wire made from vanadium alloys containing 20-30 at.% Ga should have high superconducting characteristics. According to the phase diagram (Fig. 136) the quenching of these alloys should be carried out from temperatures above 1525°C. Lower quenching temperatures lead to the formation of β and β' phases, in the presence of which the alloys cannot be worked in the cold or "warm" states.* In order to fix the V-base solid solution at ordinary temperatures a high cooling rate on quenching is essential [31, 32].

Our investigations showed that, after cold and warm (400-700°C) rolling in calibers, cast V–Ga alloys in the concentration range of V_3Ga ruptured under fairly low degrees of deformation as a result of thermal stresses, the partial decomposition of the V-base solid solution, and the inhomogeneity of the alloy. Wire 1 mm in diameter could only be prepared from alloys containing under 14 at.% Ga. The nonsuperconducting cold-worked V–10-14-at.%-Ga wire became superconducting after appropriate heat treatment. However, the critical superconducting temperature was rather low (5.6-6.4°K).

The hardness of alloys containing 20-30 at.% Ga is greater than 500 kg/mm^2 after sharp quenching. On cold or warm rolling in calibers the maximum possible degree of deformation (without rupture) is no greater than 15%. By alternately rolling and annealing at 1525-1600°C followed by quenching from these temperatures, wire 1 mm in diameter may be obtained. After heat treatment the structure of these alloys comprises alternating plates consisting of two phases with a bcc lattice and a structure of the Cr_3Si type. The number and size of the superconducting particles depends on the annealing time and temperature and also on the composition of

---
*"Warm" (not "hot") plastic deformation is carried out at a temperature such as to ensure merely the removal of the internal stresses so arising.

the alloy. A two-phase decomposition of the vanadium solid solution on heat treatment was also found elsewhere [32]. The superconducting phase first precipitates at the grain boundaries, then inside the grains. The width of the superconducting plates is 1000-2000 Å after annealing at 900°C and a few microns after annealing at 1050°C. The critical temperature of this wire after heat treatment at 860-1150°C is only 8-10°K.

Our experiments indicate the fundamental possibility of producing V–Ga wire with the composition of the compound by working quenched alloys (or by high-temperature working) with subsequent annealing. The production of wire from other superconducting compounds (Nb_3Sn, Nb_3Al, V_3Si, etc.) by this method is impossible owing to the conditions of their formation (Chap. IV). However, this method has some serious disadvantages; its technological difficulty lies in the necessity of repeated annealing and quenching at a very high cooling rate (over 500 deg/sec). The high-temperature quenching of thin wire with a diameter of under 1 mm is particularly difficult.

Another serious disadvantage of this method is the comparatively low set of superconducting characteristics of the resultant wire. The critical temperature of the samples falls well below that of bulk samples of the same composition. The reason for this is evidently the precipitation of the superconducting phase in an intermediate, nonequilibrium state arising as a result of the heat treatment of the wire. Prolonged heat treatment sufficient to produce equilibrium phases leads to the embrittlement of the wire owing to the brittleness of the precipitating phases themselves.

2. Production of Superconducting Coatings

The production of a coating with the composition of a compound on a soft, ductile base from the gaseous or liquid phase is widely employed in various fields of technology [33]. An example is the preparation of a pigment (TiO_2) from $TiCl_4$ and the production of graphite layers with special characteristics (pyrolytic graphite). In nuclear technology the decomposition of volatile compounds on a hot surface is used in order to coat uranium with zirconium or nickel or parts made from uranium oxide with thick layers of alumina. Other examples involve the creation of coatings by the reduction of volatile hydrogen halides.

The production of superconducting Nb–Sn, V–Ga, and V–Si coatings on a thin ductile wire is achieved by diffusive interaction between the vapor or liquid of the volatile component and the solid refractory component [34-38].

At high temperatures (over 800°C) the vapor (or liquid) of gallium, silicon, or tin interact actively in vacuum with vanadium or niobium, with the formation of thin, continuous coatings of uniform thickness, identical in composition after identical heat treatment. The thickness of the coatings is affected by the annealing time and temperature, and by the amount of interacting volatile metal per unit surface of the vanadium or niobium wire. The thickness of the coating varies from a few microns to millimeters, depending on the conditions of production.

Coatings obtained by the diffusion method are usually multi-phased.

In the vanadium–silicon coatings obtained by annealing at temperatures up to 1200°C one constituent is V_5Si_3 [36]. The compound V_3Si is formed in coatings obtained by annealing above 1000°C; after annealing at 1100-1200°C this is the main phase in the coatings. The proportions of V_5Si_3 and VSi_2 depend on the annealing period and the amount of silicon reacting. Small quantities of V_3Si appear in the coatings after annealing at 800-1000°C, while after annealing above 1200°C the coating consists mainly of this compound. However, in any coating obtained in this way the compound V_3Si is always accompanied by small amounts of the other V–Si phases, which is entirely reasonable, since all the compounds contributing to the V–Si system exist at the diffusion-annealing temperatures (800-1300°C) (Fig. 135).

The V–Ga [34-36] and Nb–Sn [35-37] coatings are also many-phased. In samples of Nb–Sn and V–Ga obtained by diffusion from the liquid or gaseous phases with a large excess of tin or gallium, the upper layer is soft and consists almost exclusively of tin or gallium, respectively. The boundary surface between the layer of Nb_3Sn and the soft surface layer is severely cracked. In samples obtained from the gas phase the soft surface layer is very thin (under 1μ), and there are far fewer cracks at the boundary of the Nb_3Sn layer [35]. In the diffusion process the Nb_3Sn layer acts as a barrier to tin atoms [38a].

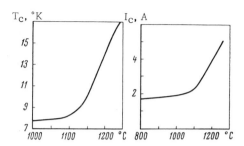

Fig. 216. Superconducting transition temperature and critical current of V–Si diffusion coatings in relation to the diffusion-annealing temperature.

The individual layers in these coatings can only be distinguished by color etching (Chap. I).

The effect of the diffusion conditions on the superconducting properties of the V–Si coatings was studied in detail by Efimov et al. [36]. On increasing the temperature of diffusion annealing the T_c of the V–Si coatings increases (Fig. 216). Coatings obtained at 1220°C and consisting almost entirely of V_3Si become superconducting at 17°K. An increase in the annealing period, tending to produce more equilibrium phases in the coating, raises the critical temperature (Fig. 217).

Vanadium–silicon coatings obtained at temperatures up to 1200°C pass into the superconducting state over a certain temper-

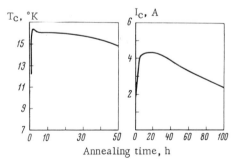

Fig. 217. Effect of the period of diffusion annealing on the superconducting transition temperature and critical current of diffusion V–Si coatings.

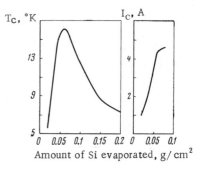

Fig. 218. Superconducting transition temperature and critical current of V–Si coatings produced at 1220°C in relation to the amount of reacting silicon per unit surface of the vanadium wire.

ature range (diffuse transition). A possible reason for this is the nonequilibrium nature of the phases formed in the coating, due to the actual diffusion method of producing the samples. After annealing above 1220°C, when the coating consists mainly of V_3Si, there is a sharp transition into the superconducting state.

Slightly different results were obtained elsewhere [34, 39]. Maximum critical temperatures and sharp superconducting transitions were observed in V–Si coatings after diffusion annealing at 1000°C (20 h) and in V–Ga coatings at 1200°C (20 h). For higher or lower temperatures the T_c of the coatings diminished.

The form of the transition curve and the value of the critical temperature of the coatings are substantially affected by the amount of reacting silicon per unit surface of the vanadium wire (Fig. 218). For each temperature of diffusion annealing a specific amount of silicon per unit surface of the vanadium wire is required in order to achieve the maximum T_c and a sharp superconducting transition. At 1220°C the maximum T_c (17°K) is reached for 0.054 g/cm² Si.

The effect of annealing time and temperature and also the amount of evaporating silicon (per unit surface of the vanadium wire) on the critical current (at 26.6 kOe) of the V–Si coatings obtained at temperatures above 1200°C is illustrated in Figs. 216-218 [36]. The multiphase coatings mainly containing the Si-rich compounds have low critical currents at 26 kOe. The critical current of coatings containing solely or almost solely the compound V_3Si reach values of 5-6 A at 26 kOe. These values are comparable with those obtained earlier [40]. For the same thickness of the superconducting coating (10 μ = 100 A), the critical current is directly proportional to the circumference of the wire.

In the first case [36] a wire 0.2 mm in diameter was used and in the second [40] one of 0.5 mm. The critical current density ($4 \cdot 10^4$-$1 \cdot 10^5$ A/cm^2), referred to the coating, in these investigations is quite comparable.

The laws governing the superconducting properties of Nb–Sn and V–Ga coatings are analogous to those just discussed [34, 35, 40-43]. The maximum critical temperature of V–Ga diffusion coatings is 15.05°K [34, 39] and that of Nb–Sn coatings 18.2°K [44].

The dependence of the critical temperature of the diffusion layers on their composition was studied earlier [44] in Nb–Sn coatings deposited from the gas phase. The T_c of the coatings decreases linearly with increasing excess of niobium in the coating relative to the stoichiometric content. The highest T_c of Nb–Sn coatings (18.2°K) is achieved after prolonged diffusion annealing at 1000°C. There is an excess of niobium in this coating.

Figure 75 illustrates the transition curves (plotted from resistance measurements) for diffusion layers obtained on niobium at 700-1250°C [45]. The superconducting transition temperature for samples obtained at 990 and 1250°C is 18.0 and 18.1°K respectively, in agreement with published data relating to the T_c of Nb$_3$Sn [46, 47]. The width of the superconducting transition is $\Delta T = 0.02$°K. For diffusion layers formed at temperatures of under 850°C the superconducting transition temperature diminishes, T_c being the lower, the lower the temperature of formation of the layer (for example, $T_c = 17$°K if the layer is formed at 800°C). The width of the superconducting transition increases substantially in this case ($\Delta T = 3$°K at 800°C).

An analogous dependence of T_c on the processing temperature was found in bulk samples of Nb$_3$Sn [48].

The effect of temperature on the critical magnetic field of diffusion layers treated at 800-1250°C [45] is shown in Fig. 76. The value of the critical magnetic field at 0°K was determined by extrapolating in accordance with a quadratic or linear law

$$H_c = H_0 \left[1 - \left(\frac{T}{T_c}\right)^2\right], \qquad H_0 = \frac{dH_c}{dT}\left(T - T_c\right).$$

For samples obtained at 990 and 1250°C $dH_c/dT = -17$ kOe/deg; for samples obtained at 800-850°C the ratio lies between -12

and −15 kOe/deg. Extrapolation of the critical magnetic field to 0°K by a quadratic temperature/field law gives $H_0 = 155$ kOe for diffusion layers formed above 850°C and $H_0 = 100\text{-}140$ kOe for those formed below this temperature. Extrapolation by the linear law gives 308 and 200-270 kOe in the two cases respectively. The value of H_0 for the diffusion layers obtained by Bychkova et al. [21] at temperatures above 850°C agrees closely with the value of H_0 for Nb_3Sn obtained by Kunzler [49, 50].

X-ray analysis confirmed that at temperatures of sample preparation above 850°C only the compound Nb_3Sn was formed [45]. At diffusion temperatures below this a phase containing more tin than Nb_3Sn appeared on the surface of the niobium wire.

The variation in the critical current of V−Ga and Nb−Sn diffusion coatings in fields up to 150 kOe is shown in Fig. 219 [35, 40]. In fields of over 30 kOe higher critical currents occur in coatings subjected to more stringent diffusion treatment. The critical current curves of diffusion samples obtained from the gas phase under various conditions lie parallel to one another (if plotted in semilogarithmic coordinates) for high magnetic fields (over 30 kOe) and are almost straight. The absolute values of the critical currents depend on the production conditions. The properties of the current-carrying regions of the diffusion layers in strong magnetic fields

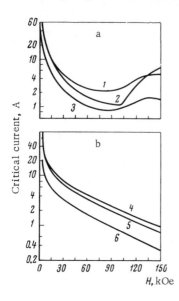

Fig. 219. Critical current at 4.2°C of diffusion-type V−Ga (a) and Nb−Sn (b) wires 0.5 mm in diameter in relation to the external magnetic field: 1) diffusion from the gas phase at 1200°C for 20 h; 2) and 3) ditto for 10 and 0.5 h; 4) annealing at 1110°C (60 min) in vacuum to clean the wire, diffusion annealing in tin vapor for 15 min; 5) annealing at 1100°C (15 min) in vacuum, diffusion annealing in tin vapor for 15 min; 6) annealing at 900°C in vacuum (60 min); diffusion annealing in tin vapor for 60 min and then heating at the same temperature in the absence of tin vapor for 15 min.

are independent of the conditions of preparation and subsequent heat treatment. However, the production conditions affect the number of such regions and hence the absolute value of the critical current.

The critical currents of coatings obtained by immersion in molten tin or gallium are higher than those of coatings obtained from the gas phase [34, 35, 39, 40]. For niobium–tin coatings the critical current reaches a maximum under the following conditions of production: annealing at 1100°C for 60 min *in vacuo*, immersion in molten tin for 4 min, and subsequent heat treatment at the same temperature for 15 min [35].

Under optimum conditions of diffusion the critical current of a wire 0.5 mm in diameter with a Nb–Sn diffusion layer some 10μ thick reaches 10-11 A in a field of 90 kOe [35, 37, 42]. A slight deviation from the conditions indicated produces a sharp drop in critical current. For example, the critical current of layers obtained at 1100°C in pure tin is 11 A, but in a bath of Sn + Nb with ratios of the components equal to 2:1 and 1:4 it is only 5.3 and 4.8 A [35], respectively.

The strict proportionality of the critical current of the diffusion samples to the perimeter of their cross section shows that the current-carrying regions only occur in the thin zone of the diffusion layer. The reduction in the critical current on etching away the soft surface layer indicates that the latter contains a large number of inclusions of superconducting phase; this is confirmed by anodic etching [35]. Raising or lowering the diffusion temperature relative to the optimum value, the use of contaminated tin, and also the dissolution (on etching) of the outer tin coating lead to a decrease in the critical current.

The parallel shift of the critical-current curves on passing from a transferse to a longitudinal magnetic field indicates the presence of current-conducting zones of any orientation in the diffusion layers, only the number of zones oriented along the axis of the wire is greater [35].

The mechanical deformation of V–Ga, Nb–Sn, and V–Si diffusion coatings reduces the critical current and widens the range of the transition from the superconducting to the normal state [35, 51]. We may suppose that on deforming the layers no additional superconducting filaments are formed during the low-temperature working, but that existing superconducting filaments of the zone are partly disrupted, so that the material passes into a state of normal conductivity at low critical currents [51].

The critical current of V–Ga diffusion layers rises in a high magnetic field (above 90 kOe). The critical current of Nb–Sn and particularly V–Si coatings decreases in fields over 90 kOe and passes below the critical current of the V–Ga coatings. The reason for the "peak" effect in V–Ga coatings is still not clear.

The introduction of copper into the Nb–Sn coating leads to an increase in the J_c value, reaching $4.5 \cdot 10^5$ A/cm^2 at 100 kOe [51a]. High critical current densities are reached in thin films of niobium nitride (50 A for 8μ): $2 \cdot 10^7$ and $5 \cdot 10^5$ A/cm^2 at 0 and 210 kOe [51b]. The production of complex coatings with high critical currents and magnetic fields by the interaction of tin, aluminum, silicon, and gallium with a substrate of ternary or quaternary alloys based on niobium, vanadium, tantalum, and titanium, with traces of other elements (zirconium, vanadium, molybdenum, uranium, rhenium) was mentioned in [51c].

Certain other methods for producing diffusion coatings have also been described. In one case [52] a gas–chemical method was described for producing Nb$_3$Sn layers on a metal substrate of refractory ductile material by conveying the sintered compound Nb$_3$Sn from a low- to a high-temperature zone with the help of a gas-transport reaction based on HBr or HCl as transporting agent. The critical temperature of such coatings depends on the distance between the source of evaporating material and the substrate surface. For large distances a nonuniform coating is obtained owing to the different reactivities of Nb and Sn to HBr (or HCl) and the different reduction characteristics of NbBr(Cl$_4$) and SnBr$_2$(Cl$_2$).

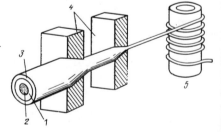

Fig. 220. Arrangement for producing a solenoid consisting of a superconducting compound by the Kunzler method: 1) niobium tube; 2) tin and niobium powder (composition Nb$_3$Sn); 3) Monel sheath; 4) draw plates; 5) solenoid.

For small distances between the source and substrate T_c reaches 18°K and the critical current density $6 \cdot 10^4$ A/cm² in fields up to 50 kOe. The deposition of Nb_3Sn from the gas phase has also been described elsewhere [50].

3. Production of Superconducting Wire from Compounds by Working a Mixture of the Original Components in a Soft Sheath with Subsequent Heat Treatment (Kunzler Method)

In 1961 Kunzler et al. [53] described a method of producing a superconducting wire with a core of niobium and tin. A wire of this kind withstands a current density of over 10^5 A/cm² in magnetic fields of over 90 kOe.

Wire produced in the manner indicated constitutes a tube of ductile material with a core consisting of a superconducting compound. The method of preparation is illustrated in Fig. 220. A niobium tube with external and internal diameters of 6 and 3 mm respectively is fed with a homogeneous mixture of pure tin and niobium powders (in a ratio corresponding to the compound Nb_3Sn) or a mixture of Nb_3Sn and pure tin powders (9:1). The ends of the tube are firmly closed with niobium stoppers. In this form the tube is drawn through a draw-plate into a fine wire with an external diameter of up to 0.38 mm. A solenoid is wound from the ductile wire and the entire solenoid is annealed at 970-1400°C. In the course of the heat treatment the superconducting compound Nb_3Sn is formed in the core of the wire. The wire thus becomes brittle. Depending on the composition of the core and the annealing time and temperature, a variety of superconducting properties of the resultant wire may be obtained (Fig. 221) [53]. The best results were obtained after annealing at 970°C for 16 h, using a wire with a core formed by a powder mixture of (3Nb + Sn) + 10% Sn (T_c = 17.8°K, $J_c = 1.5 \cdot 10^5$ A/cm² at 4.2°K and 88 kOe) [54]. An estimate of the critical magnetic field gives a value of the order of 200 kOe. This is confirmed by measurements in a pulsed field, using a wire with an Nb_3Sn core, heat-treated at 1000°C for 16 h [55, 56].

The critical current of such a wire exceeds that of ordinary Nb_3Sn samples, usually by about a factor of 1000 [48]. This difference is due to the special "filamental" structure of the core mate-

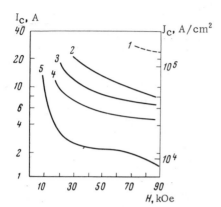

Fig. 221. Dependence of the critical current on the applied magnetic field for samples with a core of $Nb_3Sn + 10\%$ Sn in a niobium sheath (magnetic field normal to the direction of the current): 1) sintering for 16 h at 970°C, test at 4.2°K; 2) ditto at 1.5°K; 3) core 3Nb + Sn, sintering for 16 h at 1200°C, test at 1.5°K; 4) ditto, at 4.2°K; 5) sintering for 14 h at 1200°C, test at 1.5°K.

rial, which consists of very thin regions of composition Nb_3Sn and thin intermediate regions of a different composition. This kind of structure is attributable to the degree of diffusion of the molten tin into the niobium grains during heat treatment. The filamental structure is more strongly expressed for lower annealing temperatures owing to the incomplete diffusion [48, 54, 57]. The optimum temperature for annealing is about 1000°C [34]. Lower annealing temperatures (under 900°C) reduce the critical temperature of the samples.

Hauser [58] mentioned observing a filamental structure in the compound V_3Si.

Several solenoids have now been made from wire with a Nb_3Sn core (Chap. VI). The effect of annealing temperature on the maximum magnetic fields created in these solenoids (length 19.8 m, wire diameter with copper coating 0.381 mm, 612 turns) for various currents was studied by Kolbe [59]. On increasing the annealing temperature from 875 to 945°C the magnetic field thus created increased from 12.6 to 23.8 kOe. On further raising the annealing temperature to 960° the magnetic field decreased to 19.3 kOe. The change in the magnetic fields of these solenoids with time shows that, owing to the poor insulation between the turns and layers in the solenoid, the time required to obtain the maximum field was quite prolonged.

An analogous method was used to obtain a wire in a vanadium sheath containing V_3Si or V_3Ga [39]. Increasing the annealing temperature from 800 to 900° (20 h) raises the critical temperature of the V–Si wire from 16.75 to 16.9°K. Further raising the annealing temperature has little effect on the critical temperature. The critical temperature of the V–Ga wire increases, on increasing the annealing temperature from 1000 to 1300°, from 15.5 to 16.4°K. These values are higher than the maximum critical temperatures of the corresponding diffusion coatings. The width of the superconducting transition in the V–Si samples is 0.08°K and in the V–Ga wire 0.2°K.

The changes in the critical current of the wire with the V_3Ga core are indicated in Fig. 222 in relation to the conditions of production [40]. In fields above 90 kOe a peak effect of the same kind as before occurs. The critical current of the wire with the V_3Si core is smaller than that of the analogous V–Ga wire.

The effect of extension under loads of up to 6000 kg/mm² on the T_c and I_c in a magnetic field of 80 kOe at 4.2°K was studied for a wire with an Nb_3Sn core in a niobium sheath or in a double sheath of niobium and Monel [60]. In the case of the wire in the niobium sheath there was a decrease in T_c, reaching 0.16° at 6000 kg/mm². In the case of the wire with the double sheath, T_c increased for loads up to 2000 kg/mm² and then decreased sharply. The rise in T_c under small loads may be due to the preliminary compression of the wire by the double sheath. The critical current in both wires first rose and then fell sharply on further increasing the load. The reversible decrease in the critical current (for a stress of 6000 kg/mm²) was approximately 10%.

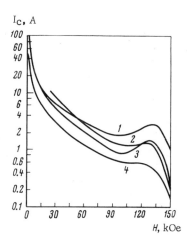

Fig. 222. Critical current at 4.2°K for a wire with a superconducting core of V_3Ga in relation to the external magnetic field. Annealing conditions: 1) 1000°, 100 h (d = 1 mm); 2) ditto (d = 0.5 mm); 3) 1300°, 2 h (d = 1 mm); 4) ditto, 10 h (d = 1 mm).

Saur [61] described the production of a superconducting strip of a ductile refractory metal (Nb, Ta, V, and their alloys) with an inner layer of a superconducting compound formed between the material of the sheath and a low-m.p. element (Sn, Ga, Si, Al, Gr, In, etc.). On filling the sheath the temperature of both the sheath and the metal fed into it should be below the temperature of formation of the superconducting compound. After filling, the tube is cold-worked and then flattened into a strip no more than 0.05 mm thick. After the strip has been wound into a spiral, annealing is carried out in order to form the compound. After heat treatment a layer of superconducting compound is obtained on the inner surface of the sheath. This procedure as it were combines both the foregoing methods of producing a superconducting wire from brittle compounds.

The production of a complex combined superconducting wire of niobium with tin was mentioned elsewhere [62, 63]. The formation of a filamental structure in the cored wire [48, 54, 57] and the associated creation of high critical currents led to the idea of regulating this kind of filamental structure by artificial means. Thus 195 niobium wires 0.05 mm in diameter covered with a layer of tin 1μ thick were placed in an annealed niobium tube with an outer diameter of 10 mm and a wall thickness of 1 mm. The gaps between

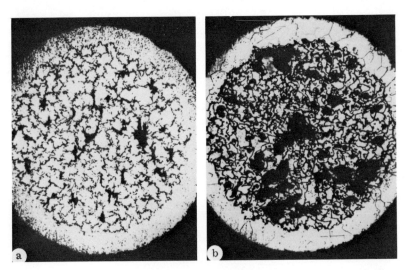

Fig. 223. Cross section of a combined multifilament Nb−Sn wire: a) cold-drawn to 0.5 mm; b) after annealing (1100°C, 20 h).

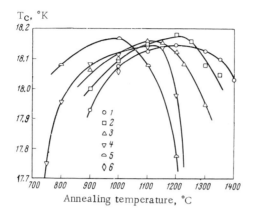

Fig. 224. Variation in the T_c of a combined multifilament Nb-Sn wire in relation to the diffusion annealing time and temperature. Annealing time, h: 1) 0.5; 2) 2; 3) 10; 4) 20; 5) 100; 6) 200.

the wires were filled with molten tin in an oxygen-free atmosphere (e.g., in vacuum). The resultant combined material was cold-worked to a diameter of 0.3-0.5 mm and then annealed (also in an oxygen-free atmosphere) at 700-1400°C for up to 500 h. In this way the formation of a multitude of internal diffusion layers of the compound Nb_3Sn artificially simulated the filamental structure [62]. Figure 223 shows a combined wire consisting of 195 niobium filaments covered with tin after cold-drawing in a niobium tube.

The effect of the conditions of heat treatment on the critical temperature of a combined multifilament wire (195 filaments) and the effect of an external magnetic field on the critical current are shown in Figs. 224 [52] and 225 [40]. At temperatures of 800 to

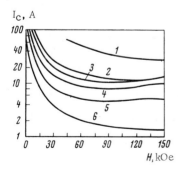

Fig. 225. Critical current at 4.2°K of a combined multifilament Nb-Sn wire (total diameter 0.5 mm) subjected to diffusion annealing at 1000°C for 20 h (1), 1100°, 20 h (2), 1150°, 10 h (3), 1250°, 2 h (4), 1300°, 2 h (5), and 200°, 20 h (6), in relation to the external magnetic field.

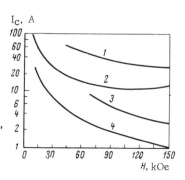

Fig. 226. Critical current at 4.2°K of a combined multifilament Nb-Sn wire with a total diameter of 0.3 and 0.5 mm in relation to the external magnetic field: 1) diameter 0.5 mm, diffusion anneal at 1000°C; 2) ditto at 1100°C; 3) diameter 0.3 mm, diffusion anneal at 1000°C; 4) ditto at 1100°C (annealing time of all samples 20 h).

1050°C, increasing the annealing time from 5 to 100 h increases the T_c of the wire (for example, at 1000°C from 18.06 to 18.17°K). Only for a very long anneal (500 h at 1000°C) is there a decrease in critical temperature. For higher annealing temperature the maximum T_c are achieved with shorter holding periods (18.18 and 18.21°K in 5 h at 1100 and 1200°C). Long anneals at high temperatures reduce the critical temperature, evidently as a result of a change in the composition of the diffusion layers.

The critical-current curves of the combined wire differ from the corresponding curves of ordinary diffusion coatings (Fig. 219) and wires with an Nb_3Sn core (Fig. 221); they have a plateau at 100

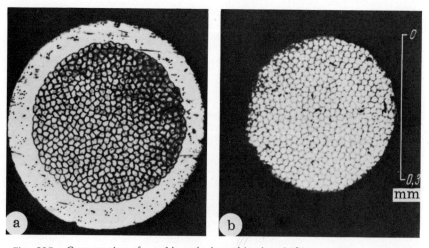

Fig. 227. Cross section of a cold-worked combined multifilament Nb—Sn wire (522 filaments) before heat treatment, with a copper sheath (a), and after removing the sheath chemically (b).

kOe. In certain samples the critical current rises smoothly in higher fields [40, 64]. In general, annealing at a lower temperature for a shorter time raises the critical current of a combined wire.

The critical currents of a combined multifilament wire 0.3 mm in diameter are lower than those of a 0.5 mm diameter wire (Fig. 226) [40]. The critical-current curves of the 0.3-mm multifilament wire have no plateau, but the slope of the curves is not as sharp as that of simple Nb_3Sn wire. Evidently the reason for this is that some of the interlayers within the sheath are ruptured when the latter is drawn into a thinner wire (0.3 instead of 0.5 mm). The approximate equality of the critical temperatures supports this view.

The ductility of the combined multifilament wire is considerably reduced on heat treatment, as a result of the diffusion of tin into the niobium sheath. In order to eliminate this disadvantage, the niobium sheath was replaced by one made of copper, and then this was removed chemically after drawing (Fig. 227) [62]. A wire obtained in this way has a much greater flexibility after heat treatment.

Solenoids made from combined multifilament superconducting wire create a magnetic field of 100-130 kOe [65, 66].

4. Production of Superconducting Coatings by Hydrogen Reduction

Methods of producing superconducting coatings of brittle compounds based on the classical hydrogen-reduction technique have been described in the literature.

Hanak and colleagues [67] developed a chemical method for the rapid and continuous growth of a Nb—Sn alloy on Nb, Pt, or Hastelloy substrates. The proposed arrangement of the apparatus for the chemical vapor-phase coating of a niobium wire with a crystalline film of a Nb—Sn alloy is shown in Fig. 228 [67, 68, 68a].

In a single technological cycle, the original Nb_3Sn powder prepared metalloceramically is chlorinated, and the chlorides $NbCl_5$ and $SnCl_4$ and also HCl vapor are fed into the hot zone of the apparatus. A niobium wire is drawn through this zone, and at 730-

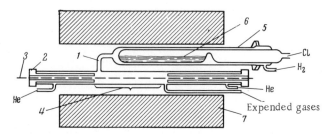

Fig. 228. Apparatus for the vapor-phase coating of a wire with a film of Nb-Sn: 1) inlet tube for HCl, $NbCl_5$, and $SnCl_4$; 2) graphite stopper; 3) wire; 4) chamber for coating the conductor with Nb_3Sn; 5) quartz tube for chlorinating the Nb_3Sn; 6) sintered Nb_3Sn; 7) electric furnace.

1600°C (preferentially 900-1200°C) the chlorides are reduced on this simultaneously with the formation of a crystalline layer of Nb-Sn alloy ($3NbCl_4 + SnCl_2 + 7H_2 \rightarrow Nb_3Sn + 14HCl$). The composition of the crystallizing layer depends on the concentration of the components in the gas phase and on the substrate temperature. Under laboratory conditions (in the optimum mode) the wire is coated with an alloy having the composition of the compound Nb_3Sn at a velocity of 9.1 m/h. The wire with its crystallized layer of Nb_3Sn possesses excellent superconducting characteristics. A solenoid 1.8 mm in diameter and consisting of 90 turns was made from the wire. The critical magnetic field of this wire is over 94 kOe and the critical current about 7 A. This corresponds to a current density in the film of superconducting compound of the order of $1 \cdot 10^5$ A/cm^2. The maximum critical temperature is 17.7°K and the width of the transition 0.8°K. The usual T_c of such a wire is 14-16°K.

The method of depositing Nb_3Sn by the thermal dissociation of niobium and tin iodides was described elsewhere [68b].

A plasmochemical method of synthesizing superconducting compounds on a plastic base was described by Bashkirov [69]. In this method the hydrogen and the original components are passed in the form of their volatile chlorides into the plasma of an electrodeless high-frequency discharge in argon. The chlorides dissociate in the stream of low-temperature plasma [70]. As a result of the reaction taking place in a reducing atmosphere at normal pressure, an alloy with the composition of the compound crystallizes in the region of the discharge on the hot substrate (niobium wire), the material being derived from the elements formed by the dissociation of the chlorides. The chlorides are obtained in a single technological cycle immediately before passing into the discharge. The rate of flow of chlorine determines the dosing of the chlorides. The composition of the end product is identical with the composition of the original gas mixture. The purity of the process is determined by the purity of the argon, hydrogen, original metals, and chlorine and the purity (cleanness) of the reaction vessel, which is ensured by the electrodeless discharge and the airtight chamber. The method is suitable for producing compounds of any metals. The reducing medium is required to prevent the formation of strong lower compounds of certain metals with chlorine (for example, vanadium dichloride and so on). Under optimum conditions a hundred-percent reduction of the chlorides to the pure metals is achieved. The process is regulated by the flow of gases through the burner, the power of the discharge, and the amount of hydrogen in the argon—hydrogen mixture. The substrate temperature is 1000°C.

The critical temperature of V—Si coatings obtained by the plasmochemical method is 17°K and that of V—Ge coatings 6.5°K [69].

5. Production of Composite Superconductors from Compounds of the Cr_3Si-Type

The ability of copper to accelerate the formation of the compound V_3Ga during heat treatment served as a basis for creating a technology of producing V—Ga composite superconductors [70a-70f].

Vanadium strip (very pure or containing 0.2% Zr) 50μ thick and up to 13 mm broad is passed through a bath containing molten gallium at 700°C. After this treatment layers of V_3Ga_2 and VGa_2 4μ thick are produced on the two sides of the strip. The gallium-treated strip is covered with an electrolytic layer of copper 5μ thick. Then the result is annealed in a quartz ampoule in an argon atmosphere at a pressure of 500 mm Hg.

The optimum annealing temperature is 650-675° (up to 100 h). After heat treatment the core of the strip ($\sim 30\mu$) consists of vanadium, the intermediate layer ($8-10\mu$) of V_3Ga, and the surface

TABLE 44. Superconducting Properties of Composite Vanadium–Gallium Superconductors after Various Types of Heat Treatment

Conditions of heat treatment		H_{c2}, kOe	I_c, A (at 200 kOe)
Temp., °C	Annealing time, h		
650	100	215	30
650*	100	215	51
675	40	212	41
700	15	212†	34
750	15	206†	13
800	15	200	0
900	15	150	0

*Sample containing 0.2% zirconium
†Extrapolation

layer of an alloy of copper with gallium (~15% Ga) containing less than 1% V.

After heat treatment the strip is again coated with copper to a total thickness of 90 μ. During the heat treatment the copper acts as a catalyst, increasing the diffusion between the vanadium and the V_3Ga_2, and also increases the rate of formation of the V_3Ga layer by an order of magnitude. The copper hardly penetrates at all into the layer of V_3Ga and has no effect on the properties of this layer under the heat treatment in question.

TABLE 45. Properties of a Composite Vanadium–Gallium Superconductor Obtained on a Short Strip 12.7 mm Wide at 4.2°K

H, kOe	I_c, A
50	500
100	280
150	175

Raising the annealing temperature to 900° results in the almost complete disappearance of the copper layer and a worsening of the properties of the superconductor as a result of the substantial dissolution of copper in the superconducting phase.

The properties of short samples of strip ~ 13 mm thick at 4.2°K in a transverse magnetic field of up to 215 kOe are shown in Tables 44 [70d] and 45 [70g].

The critical magnetic field of this composite superconductor is over 210 kOe [70g].

A solenoid with an internal diameter of 20-25 mm made of this kind of strip with a total thickness (including insulation) of $120 \pm 10 \mu$: a width of 12.7 ± 0.2 mm, and a length of 500 m creates a magnetic field of 100 kG [70g, 70h].

Another technology for producing a vanadium–gallium superconducting cable was described in [70i]. Cables of various configurations (as regards the number of filaments and the pitch of the twist) were made from annealed vanadium wire. The cable was coated with gallium by hot metallization in a molten gallium bath at 750-1000°C with a pulling speed of 1.5-6 m/min. After coating, the cable was annealed at 950-1000°C. The thickness of the resultant layer of V_3Ga on the vanadium filaments was $2-10 \mu$. The critical current density of a seven-filament cable at 4.2°K in fields of 40-50 kOe reached 10^5 A/cm^2, while T_c = 14.4-14.6°K.

The production of a multiple-filament superconductor by the tinning of a niobium wire *in vacuo* was described in [70j]. The critical current density of a cable 1.45 mm in diameter at 4.2°K in a field of 55 kOe reaches $1 \cdot 10^4$ A/cm^2.

The construction of a vacuum installation for the continuous production of the compound Nb_3Sn on a niobium profile (strip, wire, cable, etc.) of considerable length by drawing this through a bath containing molten tin (950-1050°C) was described in [70k].

6. Production of Large Superconducting Parts

In certain constructions of superconducting solenoids, relatively bulky cylinders, tubes, plates, and rings of superconducting materials may be used (Chap. VI). In view of the high supercon-

Fig. 229. Metalloceramic Nb_3Sn rings.

ducting properties of compounds, their use for manufacturing such large objects is very promising [71].

Large parts made of brittle compounds may be prepared by metalloceramic techniques or by casting in an arc furnace. Large metalloceramic rings which we made from Nb_3Sn are illustrated in Fig. 229, and cast V_3Ga rings in Fig. 230.

The brittleness of the compounds at ordinary temperatures, of course, impedes the processing of such parts. However, large samples of such compounds may be successfully ground. The possibility of subjecting such brittle compounds to plastic deformation under a high hydrostatic pressure at room temperature has been demonstrated [72]. Single-phase samples of V_3Ga ruptured under a pressure of 18,000 kg/mm² after 28% plastic deformation. An alloy of vanadium with 21% Ga was extremely ductile (E_{st} = 64%) under a hydrostatic pressure of 15,500 kg/mm². With increasing degree of deformation the strength became greater as a result of the precipitation of a second phase in the course of working.

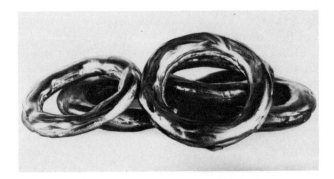

Fig. 230. Cast rings made from V–Ga superconducting alloys.

In samples of V_3Si a certain ductility was observed under a hydrostatic pressure of 23,000 kg/mm^2 (E_{st} = 7% before rupture), despite the presence of visible cracks formed by thermal stresses during the coolings of the castings. Raising the temperature of processing to a level close to the melting point also facilitated the plastic deformation of compounds which were brittle at room temperature [2].

Thus by using special methods of metallurgical processing large parts may be made from brittle compounds with high superconducting characteristics. The problem is now to make the most reasonable use of large superconducting parts in magnets and other installations.

The addition of traces of uranium or boron during the melting of an A_3B compound with the Cr_3Si structure substantially raises the critical current density of the resultant materials after irradiation by large doses of neutrons (from $1.1 \cdot 10^{17}$ to $1.7 \cdot 10^{18}$ cm^{-2}) [73]. It is interesting to note that the whole effect is produced by an extremely small number of decaying atoms, no more than one in $75 \cdot 10^6$ of the matrix atoms.

The creation of scientific bases for the plastic deformation of superconducting alloys has only just started, experimental data being sparse and approximate. The pressure treatment of metals and alloys is governed by specific laws: shear stresses, constancy of volume, least resistance, similarity, additional stresses, and nonuniformity of plastic deformation. The process of plastic deformation in the working of a polycrystalline material is characterized by the following phenomena: change of shape and grain orientation, the formation of texture, the accumulation of potential energy (residual stresses), intra- and intercrystallite damage, and the healing of the latter.

The optimum conditions of hot working (above the recrystallization temperature) are determined by the effect of the temperature, velocity, and degree of deformation and also the form of the stressed state on the structure, mechanical, and physical properties of the material being processed [2-5]. The fundamental information as to the effects of these factors may be elicited from the phase, ductility, and recrystallization diagrams [2].

In order to create a scientific basis for the pressure treatment of superconducting materials, it is equally important to develop all the foregoing aspects of the case.

One method of preparing bulk materials based on superconducting compounds, ensuring a combination of high superconducting characteristics, together with a reasonable ductility, sufficient to facilitate the production of finished articles from these, is that developed in the Institute of Metallurgy, Academy of Sciences of the USSR, in which a substitution is made for the low-m.p. constituent in the alloys [74-77]. This method enables the mechanical properties of the alloys to be varied by creating composition materials containing specified proportions of brittle and ductile phases, as well as varying other properties such as the superconducting characteristics by creating a specified combination of the superconducting phase with other superconductors, nonsuperconducting phases, semiconductors, or insulators. For example, in an alloy based on Nb_3Sn (T_c = 17.5-18°K) the replacement of the tin (which has a T_c = 3.72°K) by a solid solution of bismuth in lead (T_c = 8.2°K) raises the critical temperature of the second phase [77].

LITERATURE CITED

1. E. M. Savitskii, in: Metallography and Metal Physics of Superconductors, Izd. Nauka, Moscow (1965), p. 3.
2. E. M. Savitskii, Effect of Temperature on the Mechanical Properties of Metals and Alloys, Izd. AN SSSR, Moscow (1957).
3. S. P. Gubkin, Plastic Deformation of Metals, Metallurgizdat, Moscow (1960).
4. E. M. Savitskii and G. S. Burkhanov, Metallography of Refractory Metals and Alloys, Izd. Nauka, Moscow (1967).
5. A. I. Tselikov, Fundamentals of the Theory of Rolling, Izd. Metallurgiya, Moscow (1965).
6. I. W. Heaton and A. C. Rose-Innes, Phys. Letts., 9:112 (1964).
6a. D. Dew-Hughes and A. Narlicar, Transactions of the Tenth International Conference on Low-Temperature Physics [Russian translation], Izd. VINITI, Moscow (1967), p. 84.
7. M. I. Bychkova, V. V. Baron, and E. M. Savitskii, in: Metallography, Physical Chemistry, and Metal Physics of Superconductors, Izd. Nauka, Moscow (1967), p. 48.
8. V. V. Baron and M. I. Bychkova, in: Metallography, Physical Chemistry, and Metal Physics of Superconductors, Izd. Nauka, Moscow (1968).
8a. M. I. Bychkova, V. V. Baron, E. M. Savitskii, S. V. Sudareva, and N. N. Buinov, in: Physical Chemistry, Metallography, and Physical Metallurgy of Superconductors, Izd. Nauka, Moscow (1969), p. 76.
8b. W. G. Brammer and C. G. Rodes, Phil. Mag., 16:477 (1967).
8c. D. Kramer and C. G. Rodes, Trans. Met. Soc. AIME, 236:1612 (1967).
8d. V. A. Vozilkin, N. N. Buinov, et al., Fiz. Met. Metallov., 26(4):655 (1968).

8e. V. A. Vozilkin, N. N. Buinov, Yu. F. Bychkov, V. G. Vereshchagina, V. R. Karasik, G. B. Kurganov, and V. A. Mal'tsev, Fiz. Met. Metallov., 30(4):253 (1970).
9. E. M. Savitskii, V. V. Baron, I. I. Goncharov, M. I. Bychkova, I. S. Khukhareva, and L. V. Petrova, in: Metallography, Physical Chemistry, and Metal Physics of Superconductors, Izd. Nauka, Moscow (1968).
10. A. D. McInturff, G. G. Chase, C. N. Whetstone, R. W. Boom, H. Brechna, and W. Haldemann, J. Appl. Phys., 38(2):524 (1967).
10a. B. A. Hatt and V. G. Rivlin, Brit. J. Appl. Phys., DI(9):1145 (1968).
10b. Yu. F. Bychkov, V. G. Vereshchagin, V. R. Karasik, G. B. Kurganov, and V. A. Mal'tsev, Zh. Éksp. Teor. Fiz., 56:505 (1969).
11. F. W. Reuter, K. M. Ralls, and I. Wulff, Trans. Metallurg. Soc. AIME, 236(8):1143 (1966).
11a. R. L. Ricketts, T. H. Courtney, L. A. Shepard, and J. Wulff, Met. Trans., 1(6):1537 (1970).
11b. B. S. Hickman, J. Inst. Metals, 96(11):330 (1968).
11c. J. B. Vetrano, J. Appl. Phys., 39(6):2524 (1968).
12. E. M. Savitskii and V. V. Baron, Izv. Akad. Nauk SSSR, Metallurgiya i Gornoe Delo, No. 5, p. 3 (1963).
13. K. M. Rolls, A. L. Donieva, R. M. Rose, and D. M. Wulff, in: New Materials for Electronics [Russian translation], Izd. Metallurgiya, Moscow (1967), p. 37.
14. M. S. Walker, R. Stichler, and F. E. Werner, in: New Materials for Electronics [Russian translation], Izd. Metallurgiya, Moscow (1967), p. 48.
15. D. Betterton, J. Kneip, D. Easton, and D. Scarborough, in Superconducting Materials [Russian translation], Izd. Mir, Moscow (1965), p. 102.
16. D. Young, in: Superconducting Materials [Russian translation], Izd. Mir, Moscow (1965), p. 137.
17. K. Olsen, R. Jak, E. Fusch, and F. Hsu, in: Superconducting Materials [Russian translation], Izd. Mir, Moscow (1965), p. 194.
18. V. V. Baron, in: Metallography, Physical Chemistry, and Metal Physics of Superconductors, Izd. Nauka, Moscow (1965), p. 29.
19. Yu. F. Bychkov, I. N. Goncharov, and I. S. Khukhareva, in: Metallography and Metal Physics of Superconductors, Izd. Nauka, Moscow (1965), p. 44.
20. Yu. V. Efimov, V. V. Baron, and E. M. Savitskii, in: Metallography and Metal Physics of Superconductors, Izd. Nauka, Moscow (1965), p. 59.
21. M. I. Bychkova, V. V. Baron, and E. M. Savitskii, in: Metallography, Physical Chemistry, and Metal Physics of Superconductors, Izd. Nauka, Moscow (1967), p. 68.
22. D. E. Anderson, L. E. Toth, L. Rosner, and C. N. Yen, Appl. Phys. Lett., 7:90 (1965).
23. N. E. Alekseevskii, O. S. Ivanov, I. I. Raevskii, and N. V. Stepanov, Dokl. Akad. Nauk SSSR, 176(2):305 (1967).
24. T. Doi, M. Masao, and Y. Tagasi, J. Japan. Inst. Metals, 30, No. 2, 133 (1966).
25. S. Maeda, T. Doi, and F. Yoshida, Transactions of the Tenth International Conference on Low-Temperature Physics (1966), p. 185.
25a. Kitada Masihiro and Doi Tosio, J. Jap. Inst. Metals, 34(4):369 (1970).

26. R. M. Rose and J. Wulff, J. Appl. Phys., 33(7):2394 (1962).
27. E. M. Savitskii and V. V. Baron, Izv. Akad. Nauk SSSR, Metally, No. 3, p. 157 (1966).
28. M. I. Bychkova, V. V. Baron, and E. M. Savitskii, in: Metallography, Physical Chemistry, and Metal Physics of Superconductors, Izd. Nauka, Moscow (1968), 00-00.
29. Yu. V. Efimov, V. V. Baron, and E. M. Savitskii, in: Metallography, Physical Chemistry, and Metal Physics of Superconductors, Izd. Nauka, Moscow (1968), 00-00.
30. E. M. Savitskii, V. V. Baron, M. I. Bychkova, V. Ya. Pakhomov, and V. R. Karasik, Soviet Patent No. 168,250, Byull. Izobret., No. 4, published November 18, 1965.
30a. N. V. Polyakova, A. F. Prokoshin, I. P. Sergeev, and L. N. Fedotov, in: Collection of Works of the Central Scientific-Research Institute of Ferrous Metallurgy, No. 71 (1969), p. 158.
30b. E. M. Savitskii, V. V. Baron, N. I. Bychkova, S. D. Gindina, Yu. V. Efimov, N. D. Koslova, L. F. Martynova, B. P. Mikhailov, L. F. Myzenkova, and V. A. Frolov, in: Summaries of Contributions to the Sixteenth All-Union Conference on Low-Temperature Physics, Leningrad (1970), p. 198.
30c. V. E. Keilin, E. Yu. Klimenko, and B. N. Samoilov, Pribory i Tekh. Éksperim., No. 1, p. 216 (1971).
30d. L. Z. Kamskii and G. G. Svanov, Pribory i Tekh. Éksperim., No. 4, p. 243 (1970).
30e. H. Hillman, Z. Metallkunde, 60(3):157 (1969); Preprint from the Laboratory of Vakuumschmelze GMBH, Hanau (1970).
30f. M. V. Wilson, C. R. Walters, J. D. Lewin, P. F. Smith, and A. H. Spurway, J. Appl. Phys., 3(11):1517 (1970).
30g. Cryogenics, Vol. 10, No. 5 (1970).
30h. Cryogenics, 11(1):75 (1971).
30i. A. C. Barber and P. F. Smith, Cryogenics, 9(6):483 (1969).
30j. M. T. Taylor, C. Graeme-Barber, A. C. Barber, and R. B. Reed, Cryogenics, 11(3):224 (1971).
31. E. M. Savitskii, P. I. Kripyakevich, V. V. Baron, and Yu. V. Efimov, Izv. Akad. Nauk SSSR, Neorgan. Mat., No. 1, p. 45 (1967).
32. H. J. Levinstein, J. H. Wernik, and C. D. Capio, Phys. and Chem. Solids, 26(7):1111 (1965).
33. H. E. Schmid, BBC-Nachr. 46(9):476 (1964).
34. C. B. Muller, H. Otto, and E. Saur, Z. Naturforsch., 19a(5):539 (1964).
35. L. Rinderer and E. Saur, Z. Physik, 176:464 (1963).
36. Yu. V. Efimov, V. V. Baron, E. M. Savitskii, and S. N. Sokolov, in: Metallography, Physical Chemistry, and Metal Physics of Superconductors, Izd. Nauka, Moscow (1967), p. 122.
37. L. Rinderer, E. Saur, and J. Wurm, Z. Physik, 174:405 (1963).
38. R. Engstrom, T. Courtney, G. Pearsall, and J. Wulff, Proc. Conf. Advanced Electron. Materials, Philadelphia, Metallurg. Soc. AIME (August 1962), p. 121.
39. D. Koch, G. Otto, and E. Saur, Phys. Letts., 4(5):292 (1963).
40. J. Babiskin, P. G. Siebermann, G. Otto, and E. Saur, Z. Physik, 180(5):483 (1964).

41. G. Meyer and H. Wizgall, Z. Physik, 183:412 (1965).
42. L. Rinderer and E. Saur, Z. Naturforsch., 18a:771 (1963).
43. C. Muller and E. Saur, Z. Physik, 179:111 (1963).
44. H. J. Bode and Y. Uzel, Phys. Letts., A24(3):141 (1967).
45. V. S. Kogan, A. I. Krivko, B. G. Lazarev, L. S. Lazareva, A. A. Matsakova, and O. N. Ovcharenko, in: Metallography and Metal Physics of Superconductors, Izd. Nauka, Moscow (1965), p. 76.
46. B. T. Matthias, T. H. Geballe, S. Geller, and E. Corenzwit, Phys. Rev., 95:1435 (1954).
47. S. Geller, J. Amer. Chem. Soc., 77:1502 (1955).
48. E. Saur and P. Schult, Z. Physik, 167:170 (1962).
49. J. E. Kunzler, J. Appl. Phys., 33:1042 (1962).
50. I. I. Hanak, G. D. Cody, P. R. Aron, and H. C. Hitchcock, High Magnetic Fields, Technol. Press, New York–London–Cambridge (1962), p. 592.
51. C. Muller and E. Saur, in: Advances in Cryogenic Engineering, Vol. 9 (K. D. Timmerhaus, ed.), Plenum Press, New York (1964), p. 338.
51a. J. S. Caslaw, Cryogenics, 11(1):57 (1971).
51b. J. R. Gavaler, M. A. Janocko, A. Patterson, and C. K. Jones, J. Appl. Phys., 42(1):54 (1971).
51c. W. de Sorbo, U. S. Patent, class 75-174, No. 3416817, Dec. 17, 1968.
52. T. L. Chu and J. R. Gaveler, J. Electrochem. Soc., 113(12):1289 (1966).
53. J. E. Kunzler, E. Buchler, F. S. L. Hsu, and J. H. Wernick, Phys. Rev. Lett., 6:89 (1961).
54. J. E. Kunzler, Rev. Mod. Phys., 33:501 (1961).
55. V. D. Arp, R. H. Kropschot, J. H. Wilson, W. F. Love, and R. Phelan, Phys. Rev. Lett., 6:452 (1961).
56. G. D. Kneip, R. E. Worsham, J. O. Betterton, R. W. Boom, and C. E. Roos, Phys. Rev. Lett., 6:532 (1961).
57. E. Saur, Metall, 16(5):380 (1962).
58. I. I. Hauser and H. C. Theurerer, Phys. Rev., 129(1):103 (1963).
59. K. L. Kolbe and K. H. Rosner, in: New Materials for Electronics [Russian translation], Izd. Metallurgiya, Moscow (1967), p. 21.
60. E. Buchler and H. J. Levinstein, J. Appl. Phys., 36(12):3856 (1965).
61. E. J. Saur, U. S. Patent, No. 3243871, published April 5, 1966.
62. D. Koch, H. Speidel, G. Otto, and E. Saur, Z. Physik, 180:476 (1964).
63. C. P. Bean, M. V. Doyle, and A. G. Pincus, Phys. Rev. Lett., 9:93 (1962).
64. H. E. Cline, R. M. Rose, and J. Wulff, J. Appl. Phys., 34:1771 (1963).
65. D. L. Martin, M. G. Benz, C. A. Buch, and C. H. Rosner, Cryogenics, 114(3):161 (1963).
66. C. H. Rosner and M. G. Benz, Metallurgy and Ceram. Lab. Rept., N65-C-0-062 (1965).
67. J. J. Hanak, K. Strater, and G. W. Gullen, RCA Rev., 25(3):342 (1964).
68. K. A. Kapustinskaya and B. I. Kogan, in: Metallography and Metal Physics of Superconductors, Izd. Nauka, Moscow (1965), p. 132.
68a. Kawano Masasi and Fukube Yosichito, Newer Metal Ind., 14(9):251 (1969).
68b. V. I. Statsenko, Yu. F. Bychkov, et al., Metallurgy and Metallography of Pure Metals, Atomizdat, Moscow (1967), p. 30.

69. Yu. A. Bashkirov and S. A. Medvedev, in: Metallography, Physical Chemistry, and Metal Physics of Superconductors, Izd. Nauka, Moscow (1967), p. 201.
70. Kinetics and Thermodynamics of Chemical Reactions in a Low-Temperature Plasma, Izd. Nauka, Moscow (1965), p. 223.
70a. K. Tachikawa, International Cryogenic Engineering Conference, Berlin (1970).
70b. K. Tachikawa and Y. Tanaka, Japan J. Appl. Phys., 5:834 (1966).
70c. E. Nambach and K. Tachikawa, J. Less-Common Metals, 19:359 (1969).
70d. K. Tachikawa and Y. Iwasa, Phys. Letts., 16(6):230 (1970).
70e. K. Inoue and K. Tachikawa, J. Japan Inst. Metals, 34:202 (1969).
70f. N. S. Vorob'eva, V. I. Sokolov, V. Ya. Pakhomov, and S. M. Kuznetsova, Summaries of Contributions to the Sixteenth Conference on Low-Temperature Physics, Leningrad (1970), p. 61.
70g. Prospectus of Sumimoto Electrical Industries Ltd. (Osaka, Japan), April 1971.
70h. Cryogenics, 10(5):358 (1970).
70i. N. S. Vorob'eva and Ya. N. Kunakov, in: Problems of Superconducting Materials, Izd. Nauka, Moscow (1970), p. 106.
70j. I. A. Baranov, N. T. Konovalov, Ya. N. Kunakov, and L. Z. Kamskii, in: Problems of Superconducting Materials, Izd. Nauka, Moscow (1970), p. 120.
70k. I. S. Krainskii, S. S. Mazokhin, V. I. Sokolov, I. F. Shchegolev, and V. K. Énman, in: Problems of Superconducting Materials, Izd. Nauka, Moscow (1970), p. 124.
71. P. S. Schwartz and C. H. Rosner, J. Appl. Phys., 33(7):2292 (1962).
72. E. D. Martynov, B. I. Beresnev, I. A. Baranov, V. Ya. Mezis, A. E. Fokin, S. P. Chizhik, and Yu. N. Ryabinin, Fiz. Met. Metallov., 24(3):522 (1967).
73. P. S. Swartz and R. L. Fleischer, U. S. Patent, No. 3310395, published March 21, 1967.
74. E. M. Savitskii, Soviet Patent No. 69308, Byull. Izobret., No. 9 (1947).
75. E. M. Savitskii and V. V. Baron, Transactions of the A. A. Baikov Institute of Metallurgy, Academy of Sciences of the USSR, Vol. 1, Izd. AN SSSR, Moscow (1958), p. 148.
76. E. M. Savitskii and T. A. Kim, Izv. Akad. Nauk SSSR, Metallurgiya i Gornoe Delo, No. 4, p. 89 (1963).
77. E. M. Savitskii, V. V. Baron, and B. P. Mikhailov, in: Problems of Superconducting Materials, Izd. Nauka, Moscow (1970), p. 112.

Chapter VI

Applications

All the possibilities of using superconducting materials are difficult to predict at the present time. However, even the very fact that the remarkable properties of these materials are known offers wide prospects for their use in various fields of science and technology.

Thus almost all the elements and machines of electrical technology may be invented anew (or, more precisely, seen in a new light) using one or other of the special effects characterizing the phenomenon of superconductivity. Superconducting materials may be used to create superconducting transformers, electrical transmission lines, generators, motors, switches, and rectifiers. There are immense possibilities of using superconductors in accelerators, computers, electronics, and measuring devices. Whereas for some branches of science and technology the use of superconductors enables us to improve the characteristics of instruments, their accuracy, rapidity of action, and compactness and to make them economically more viable, for other branches such as the control of thermonuclear reactions, magnetohydrodynamic generators, and shielding from cosmic radiation the use of superconductors is clearly the only possible direction of development.

Three main properties of superconductors are used in instruments and apparatus: the absence of electrical resistance at low frequencies and hence the absence of energy losses, the Meissner effect (complete repulsion of a magnetic field from a superconductor), and the controllable transition from the normal to the superconducting state.

The extensive introduction of devices using superconducting materials started soon after 1961, when a group of materials entitled "hard" superconductors and possessing high critical magnetic fields in combination with large critical current densities was discovered. From this period more and more work was done in connection with the factors affecting the properties of this group of materials. A large number of new alloys was developed. However, further progress in the application of superconductors largely depends on metallurgists, metal scientists, and engineers. The present problem is the creation of superconducting materials with high and stable characteristics, the development of mass-production methods of producing objects of any desired shape with good uniformity of structure, electrophysical, and mechanical properties, and the reduction of the net cost involved in the manufacture of these. Of fundamental significance for the rational use of superconductors is the creation of low-temperature cryostats, compact standardized magnetic systems ensuring low thermal losses, and effective cryogenic installations of various powers, in particular those of the closed-cycle type; this offers a wide scope for specialists in low-temperature technology, electrical power, electrical technology, and technical physics. A radical advance would be the discovery of superconductors working at hydrogen, neon, nitrogen, and higher temperatures, thus avoiding the problem of using the expensive and comparatively rare helium.

The scientific endeavors of the specialists in question should be combined with the work of instrument makers in those branches of science and technology particularly concerned with superconductivity. Thus the practical use of superconductors is quite a complicated problem.

In this Chapter we shall give some examples of the use of superconductors in the following respects: magnetic fields, computer technology, electronics and measuring instruments, nuclear power, and electrical technology.

SUPERCONDUCTING MAGNETS

Immediately after the discovery of superconductivity Kamerlingh-Onnes [1] conceived the idea of making magnets from these materials. However, it was soon found that a magnetic field of only a few hundred oersteds destroyed the superconductivity of the materials under consideration (silver, lead, and tin). A little later, in 1930, it was found that a lead–bismuth alloy remained superconducting in moderate fields, up to 20 kOe; however, it still proved impossible to make the desired magnet [2]. The real advance in the creation of solenoids started after 1961, when the unique capacity of the compound Nb_3Sn to pass high currents in strong magnetic fields was discovered [3]; the high superconducting properties of Nb–Zr alloys were found later [4].

The reason for the interest in superconducting solenoids lies in the difficulty of creating magnetic fields using ordinary materials. Under ordinary conditions even the best conductors have such a high resistance that the creation of magnetic fields of the order of 100 kOe is an extremely difficult and costly task. Whereas for fields of 20 (or at best 24) kOe copper wire and iron cores may be employed, for fields above 24 kOe an iron core is entirely useless, this field constituting the saturation limit of the material in question. In order to obtain higher fields, solenoids with a large number of ampere-turns are required, but the power required for this increases more rapidly than the square of the field. A rise in power means an increase in thermal losses. The release of the thermal losses and a number of additional difficulties (large radial forces) demand a fairly complex engineering solution.

According to [5] there are perhaps twenty magnets with coils made from ordinary wire capable of giving fields of the order of 100 kOe. Each of these needs a power of about 2000 kW, cooling with 220 liters of water per minute, and huge capital expenses. In the National Magnet Laboratory at Cambridge steady fields of 250 kOe are obtained by means of an electromagnet requiring some 16 million watts of electrical power, equivalent to that of a town with a population of 15,000 [5a]. One of the most powerful magnets of this type has been made in the Physical Institute of the Academy of Sciences of the USSR; it is used for studying the properties of solids in high magnetic fields [6].

The use of superconducting materials in solenoids is very attractive; they eliminate ohmic losses, reduce the power required by several orders of magnitude, substantially reduce the expenditure of materials in short supply, and avoid the use of cooling water. The running costs of superconducting magnets are thus very low even on allowing for the cost of the liquid helium required for cooling.

A positive feature of superconducting magnets is their small weight (a solenoid producing 40-50 kOe weighs 300-500 g as compared with 15-20 tons for ordinary magnets), and the small supply sources required to feed them (after the magnetic field has been created no sources are needed to maintain it). Hence the efficiency of superconducting magnets tends to 100% rather than zero as in ordinary magnets. Thus, for example, in the superconducting magnet developed in the Argonne National Laboratory a saving of 3-5 million watts of electrical power is expected as compared with ordinary electromagnets of the same power; the magnet is intended for experiments in high-energy physics [7]. In addition to this, these features make superconducting magnets extremely portable and facilitate their use in aircraft.

One of the first superconducting magnets with a field of 15 kOe was made of Mo–Re wire [8]. Perhaps the very first superconducting magnets were a magnet with a niobium winding giving 4.3 kOe [9] and a magnet with an iron core and a superconducting coil [10]. Still earlier (in 1942) Justi used a nonattenuating current excited in a superconducting ring for magnetizing iron [11].

Later superconducting magnets were frequently based on Nb–Zr wire. One of the first magnets made by Westinghouse from this material had a magnetic field of 43 kOe, an internal diameter of 12 mm, an external diameter of 50.8 mm, a length of 38 mm, and a weight of 453 g; a current of 20 A flowed through the 0.25 diameter wire [12]. A second solenoid was made by the same company from wire 190μ in diameter; this consisted of five sections, the total length of the wire being 3000 m, the internal diameter 3.8 mm, the external diameter 60 mm; the maximum magnetic field was 58 kOe with a current of 9.7 A.

In the Soviet Union a Nb–Zr superconducting solenoid was first made in the Physical Institute of the Academy of Sciences of the USSR, using wire developed in 1961-1962 in the Institute of Metallurgy of the Academy of Sciences of the USSR [13]. In 1963 this solenoid with a magnetic field of 45 kOe and a weight of 300 g was demonstrated in the Pavilion of the Academy of Sciences of the USSR in the VDNKh. Later a solenoid with a maximum magnetic field of 46.5 kOe was made in the Physical Institute of the Academy of Sciences of the USSR [14]. The same group also made a solenoid using Nb–Zr and Nb–Zr–Ti wire; the solenoid with dysprosium tips 3 mm in diameter and a gap of 1 mm gave a magnetic field of 94 kOe. A superconducting magnet for 250 kOe has been made in Canada [15].

The creation of superconducting magnets developed along the lines of making systems with higher magnetic fields and a larger working space. At the same time work proceeded in connection with perfecting the construction of the magnets so as to give more stable working conditions with better arrangements for cooling the coils and stabilizing the wire. The chief organizations engaged in the production of solenoids from Nb–Zr wire outside the USSR are Westinghouse, RCA, Magnion, Avco-Everet (USA), C. A. Parsons (UK), Siemens (West Germany), Cohern CSF (France), Hitachi and Er (Japan).

In the Physicotechnical Institute of the Academy of Sciences of the Ukrainian SSR a solenoid has been made of Nb–Zr–Ti wire giving a maximum field of 112 kOe in an aperture of 5 mm at 2°K. Insertion pieces made of dysprosium raise the magnetic field to 137 kOe [16].

A high magnetic field has been obtained in a solenoid of Zr–Nb wire in Dubna. A magnetic field of 103 kOe is obtained at a temperature of 1.5°K in a gap of 0.6 mm using pole tips of Permandur 1.0 mm in diameter [17].

The external appearance of the Nb–Zr-wire solenoids is shown in Fig. 231 [18].

Fig. 231. Superconducting solenoids made by Magnion.

The creation of solenoids with still higher magnetic fields involves the development of wire capable of withstanding such fields. Of the alloys based on solid solutions the highest-field materials are Nb–Ti alloys, their field limit being 120-140 kOe [19]. Further increase in field requires the use of materials with still higher H_{c2} values such as the compounds Nb_3Sn, V_3Si (200 kOe at 4.2°K), V_3Ga (196 kOe at 4.2°K), Nb_3Al (250 kOe at 0°K), and $Nb_{0.79}[Al_{0.75}Ge_{0.25}]_{0.21}$ (410 kOe) [20, 20a-e].

The first solenoids using Nb_3Sn wire were made by Bell Telephone in America [21], a field of 70 kOe being obtained in a solenoid with an internal diameter of 8.5 mm and a length of 100 mm. Later a field of 101 kOe was obtained with the same solenoid at 1.5°K in an internal diameter of 6 mm. The Nb_3Sn wire was made by the Kunzler method. At about the same time workers in the Research Laboratory of General Electric made an Nb_3Sn magnet with a field of 101 kOe [22]. In addition to the Kunzler wire-making technique, the principles of depositing Nb_3Sn from the liquid and gas phases on wire and strip were also employed (see Chap. V).

The greatest success in using strip made from Hastelloy stainless steel (width 2.3 mm, thickness 0.05 mm) and covered with Nb_3Sn deposited from the gas phase was achieved by RCA, who used this material to create a solenoid with a maximum field of 107 kOe. The diameter of the working region of the magnet was 25 mm. The weight of the solenoid was 12 kg. The layer of Nb_3Sn on the surface of the strip was firmly bonded to the latter and was quite uniform. The current density reached 200-300 A/cm^2 in a field of 100 kOe. The strip retained these characteristics when bent, even at loads exceeding 8000 kg/cm^2. In order to improve the electrical properties of the strip as well as its ductility a layer of copper or silver was deposited on the strip in the manufactured state [23].

The original construction of the solenoid was described earlier [24]; it was made of round discs prepared from Nb_3Sn and coated with copper. A solenoid of this construction gives a magnetic field of 100 kOe in a 1-inch gap. The current is fed in along copper busbars.

A field of 132 kOe was obtained in a gap of 6 mm in the General Electric Laboratory, using a four-section solenoid (Fig. 232) wound from Nb–Sn wire (Cryostrand) [25].

Fig. 232. Superconducting solenoid with a field of 132 kOe made of Cryostrand.

A solenoid based on superconducting compounds with a maximum induction of 180 kOe [26] has also been described. One of the chief difficulties in creating high-field magnets, particularly those based on superconducting compounds, is the mechanical stress in the wire or strip due to the field; this increases as the square of the field [5a].

The possibility of creating a magnetic field over a comparatively large volume (internal diameter 276 mm) was demonstrated later [27]. The solenoid consists of six individual units (Fig. 233) and gives a field of up to 32.8 kOe for a current of 10 A, corresponding to a total magnetic energy of 300 kJ. Each coil is wound from wire 0.25 mm in diameter consisting of Nb–Zr alloy electroplated with copper.

A solenoid with a field of variable intensity used for studying magnetic susceptibility was described in [27a]. The characteristics of a combined sectional magnet with internal sections made of Nb–Sn strip (by RCA) and an outer section wound from Nb–Ti wire ("Niomaos 5") 50/80 with a total field of 114 kOe at 1.5°K are given in [27b].

Existing methods of producing strip and wire coated with a layer of Nb_3Sn or containing this compound on the inside are very diffi-

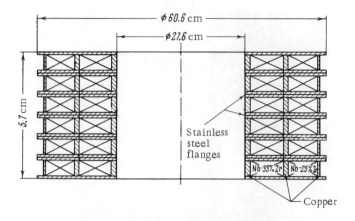

Fig. 233. Superconducting magnet with an internal diameter of 276 mm: a) arrangement of coil system; b) general appearance of sectioned solenoid.

cult and not always successful. It is thus highly desirable to use Nb_3Sn or other compounds (V_3Ga, V_3Si) in compact form for creating magnetic fields. In order to achieve this, the method of magnetic field pumping was proposed. According to this method, strong magnetic fields are first obtained by creating a magnetic field inside a hollow cylinder made from a superconducting material. The aperture in the cylinder has the cross-sectional shape of a figure 8, one loop being much larger than the other. After the magnetic field has been created in the hollow cylinder, a piston of superconducting material is inserted into the large aperture so as to fill it. Since the magnetic field cannot penetrate through the superconductor, the magnetic lines of force are driven out of the larger opening and create a more concentrated magnetic field in the small one. This arrangement for magnetic pumping is illustrated in Fig. 234. In this manner fields of up to 25 kOe were produced in an Nb_3Sn cylinder [28].

The Oxford Instrument Company (UK) used Hastelloy coated with Nb_3Sn and then silver on the outside to make a solenoid with a field of 100 kOe for low-temperature experiments. The length of the winding was 6 in., the external diameter 5 in., and the main gap 1 in. The electromagnet required a current of 92 A. A further winding of Nb–Zr wire is to be added to this later [29].

The transition of a magnet from the superconducting to the normal state is accompanied by the evolution of a great deal of heat so that the solenoid may be put out of action. In order to dissipate the heat so evolved special measures are required. In order to protect the installations combined superconductor–copper windings are employed in many practical cases.

Avco developed a cable consisting of several Nb–Ti wires mounted in a sheath of pure copper. Depending on practical requirements the shape of the cable might be round, square (0.04–0.15 in.), or rectangular (width 0.7 in.). The ratio between the Nb–Ti conductor and the copper sheath varied from 1 to 5 or over. At the present time this cable has yielded a critical current density of $2 \cdot 10^4$ A/cm^2 in a field of 45 kOe [30].

A superconducting strip made of Nb_3Sn foil coated with copper and then rolled has been made in the United States; the width of the strip is 12.7 mm and the thickness 0.4 mm. The strip has high mechanical and electrical properties. In view of the layered structure, brittleness is largely overcome. A helical magnet 160

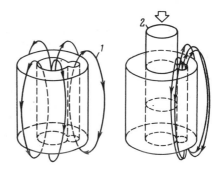

Fig. 234. Schematic representation of magnetic pumping: 1) distribution of magnetic lines of force in a cylinder with an aperture in the shape of a figure 8; 2) piston inserted in the large loop.

mm in diameter and 110 mm long made from this strip gives a magnetic field of 100 kOe [31].

Specialists of the Argonne National Laboratory have proposed a new form of cable consisting of a Nb–Zr wire, a large number of aluminum wires, and some of copper and Nichrome (for mechanical strength). Good contact between the aluminum and Nb–Zr wires is achieved by coating the aluminum with indium and the superconductor with copper. The advantage of aluminum over copper lies in the absence of mechanical stresses from this metal (annealing is carried out at a low temperature), which increases the electrical conductivity. In addition to this, aluminum has a lower density, a greater specific heat, and is much cheaper than copper.

Tests on an electromagnet made from a cable of this construction with the winding completely immersed in liquid helium showed that a larger current could be passed through it than through a cable using copper. Using resistance measurements carried out on the winding in its operational state, together with tables giving the rate of heat transfer from the winding to the liquid helium, the stability of the current in the winding over and above that which would be regarded as normal for the superconductor was calculated. The calculated current was 745 A, while in actual fact the electromagnet operated stably at a current of 850 A. These results show that the stable operation of electromagnets incorporating a winding containing aluminum may be determined fairly exactly by calculation, while in view of the small weight of the winding there is a clear possibility of making a wider use of large superconducting magnets [32]. Using special equipment, the manufacture of stabilized cable may be made extremely efficient by drawing an aggregate of wires or strip [33] inside an aluminum or copper sheath [34].

In recent years the main efforts in the development of superconducting materials have been directed at a reduction in their degradation in large magnetic systems. On the basis of theoretical treatments of Hancocks, Chester, and others, criteria for the stability of superconductors have been formulated [34a-b]. The main factors which arise when selecting a superconducting composite material reduce to three parameters: the diameter of the filament, the diameter of the composite, and the twist pitch. For dc magnets the diameter of the superconducting filaments should be of the order of 50 μ, the matrix should have a low specific electrical resistance (usually pure copper), and twisting is not obligatory. For ac magnets the diameter of the filament should be smaller than that corresponding to a ratio of 1500/11, the matrix should have a high resistance (for example, a Ni–Cu alloy), and the material must certainly be given a twist with a pitch of $\ll l_c$, where

$l_c^2 \approx 10^8 \lambda J_c\, d\rho/H$,
λ is the occupation factor,
J_c is the current density,
ρ is the specific resistance of the normal metal,
H is the alternating field,
d is the diameter of the filament.

The manufacture of composite materials (Nb–Ti in Cu) of various cross sections and with various numbers of filaments has been undertaken by a number of companies; in the United States by Wire Reduction Company, Inc., in Britain by Imperial Metals Industries, in France by Thomson–Houston–Brandt, in West Germany by Vakuumschmelze GMBH, and others. Imperial Metals Industries together with the Rutherford Laboratory has adopted a triple-composite material. The composite comprises niobium-titanium filaments in a matrix of Cu–Ni, placed in an outer matrix of pure copper [34c].

As regards compounds, there is considerable interest in materials based on V_3Ga; a solenoid has already been made from this [34d]. Work on Nb_3Sn superconductors is proceeding in the direction of producing multiple-filament materials with adequate strength and reasonably stable properties. Mention has been made of the creation of composite materials in which a copper matrix contains a large number of Nb–Sn filaments (Whittaker Corporation) [34e]. Composite materials are now being used for the winding of solenoids [34f].

Considerable advances have been made over the past few years in the construction and manufacturing technology of superconducting magnets. In many cases superconducting magnets have become far more efficient and convenient and even cheaper than ordinary magnets. It would appear likely that in the near future the demands of physics and modern technology for very strong magnetic fields will be entirely met by means of superconducting materials and magnets.

COMPUTING TECHNOLOGY

1. Cryotrons

In computer technology, work is being carried out in connection with the use of superconductors as fast-acting switches (cryotrons) for memory systems [35, 36].

The first cryotron [35] consisted of a thin layer (0.003 in.) of niobium wire wound on a thicker (0.009 in.) tantalum conductor (Fig. 235). A fairly large current passing through the niobium wire, called the control coil, disrupts the superconductivity of the tantalum wire, called the valve.

Other materials have also been used [37], the valve being made of a quartz filament coated with a tin film and the coil of a lead wire.

Using such mutual-control cryotrons a great variety of logical systems may be made. Complex logical systems of this kind have been described in several places [35, 38-40].

The rate at which resistance develops (i.e., at which a particular valve is able to pass into the normal state) depends on the time of the fundamental phase transformation; this is quite short (around 10^{-10} sec) and in no way restricts the field of use of the cryotron [41, 42].

A positive aspect of wire cryotrons is their ease of manufacture. However, despite all their advantages, wire cryotrons are hardly ever used in view of the complexity involved in making reliable connections between units and the comparatively long time required for switching the current from one branch to another (at least 10^{-5} sec). The switching time is determined by the L/R ratio, where L is the inductance of the superconducting loop created by the current paths and R is the resistance arising when the valve opens.

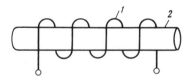

Fig. 235. Wire cryotron: 1) control coil; 2) valve.

APPLICATIONS

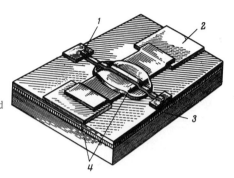

Fig. 236. Thin-film cryotron: 1) lead strip (control); 2) tin strip (valve); 3) lead screening plane; 4) insulation.

In order to eliminate this kind of defect, film cryotrons have been proposed (Fig. 236); these use basically the same principle as wire cryotrons [43-45]. The time constant for film cryotrons equals $4.46 \cdot 10^{-9}$ sec, which represents an extremely good switching time [36]. The technique of vacuum film deposition enables the units and the connections between them to be created in a very small number of technological stages. The reliability of the interconnections is greatly increased.

The greatest rapidity of action is achieved in the so-called longitudinal cryotrons [46, 47]. The rapidity of action of these in no way loses the other advantages characterizing the transverse cryotrons just described. The use of a longitudinal cryotron opens the possibility of developing a memory device acting extremely rapidly although not having a very great volume.

2. Memory Devices

Superconductors are also used in computer technology as memory units [48-50]. Figure 237 illustrates a basic memory

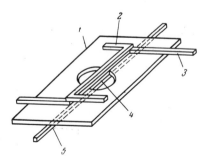

Fig. 237. Arrangement of the Crowe cell: 1) plane with aperture; 2) control bar A; 3) control bar B; 4) computing bar.

cell proposed by Crowe [50]. The cell consists of a superconducting film (usually lead) with a small hole and a narrow cross-piece.

There are also other forms of memory cells: the Persistor [49, 51], the Persistatron [48], and a memory device based on a continuous film [52]. Apart from lead, tin is also used for depositing superconducting films [53]. The whole circuit is usually deposited on a glass substrate.

A positive feature of superconducting memory devices is the absence of "delta noise," a lack of inertia, compactness, probably a low cost [36], and a high signal-noise ratio.

The chief problem to be solved in creating a large superconducting memory system is that of making a matrix containing a thousand cells in a single substrate plate with a satisfactory rate of production and an acceptable variation of parameters. The positive features of superconducting memory devices should exceed those of magnetic cores sufficiently to justify the use of a liquid helium bath. According to existing data [54] the flow of liquid helium for cooling a circuit comprising 10^6 memory cells is 2 liters per hour. This kind of flow may easily be achieved with closed-cycle refrigerators [55]. Superconducting memory systems are most promising in cases in which helium is already used for cooling purposes (for example, when the equipment incorporates cryotrons).

Memory units made from superconductors are at present the only ones suitable for making fast-acting devices containing billions of information units. Superconducting memory devices are the most capacious with respect to the amount of information. Other advantages include their miniaturization and the prospects of using mass-production techniques for film systems.

Special apparatus has been developed for making superconducting film memory devices; this comprises a vacuum-deposition installation enabling circuits of 100 × 100 mm in size to be produced to a manufacturing accuracy of $2.5\,\mu$ [56]. Smaller circuits are made automatically [57].

IBM has developed a fast-acting superconducting logical element with a switching speed of 800 psec [58].

A memory device based on superconductors with a memory of 14,000 binary digits is being made by RCA. The memory device is arranged on four lead plates 50×50 mm in size with a memory-cell density of 10 mm^{-2}. The access time is 4 μsec and the working temperature 3.5°K. The memory cell of the memory device is made in the form of a closed loop formed from a tin film. The decoders work at ordinary temperatures [59].

ELECTRONICS AND MEASURING TECHNOLOGY

1. Bolometers — Receivers of Thermal Radiation

The creation of bolometers (instruments for measuring thermal radiation) from superconductors was proposed in 1939 by Goetz [60]. Others started developing these instruments at the same time [61, 62]. The operating principle of these devices lies in the change in the electrical resistance of a superconducting film when radiation is incident upon it. The advantage of superconducting bolometers over other forms of such instruments is their high sensitivity (about $3 \cdot 10^{-12}$ W) and low noise voltage, resulting from the low temperature. It is considered that the sensitivity of superconducting bolometers might be extended to $6 \cdot 10^{-16}$ W at 4.2°K [63].

Both hard (V, Ta, Nb, NbN, etc.) and soft (Pb, Sn) superconductors are used for making bolometers. Instruments based on hard superconductors have the greater noise voltage [36, 64]. In order to remove the noise voltages and realize all the potentialities of these radiation receivers, methods of measuring the resistance without passing a current through the receiver must be devised. A receiver of this kind was used to study α particles emitted by a polonium source [65]. The development of nuclear-particle detectors using thin superconducting tin films has also been mentioned; these may prove extremely sensitive to α particles [66].

Bolometers find applications in aviation and cosmic technology [67].

2. Superconducting Magnetic Lenses

In the Laboratory of the United States Atomic Energy Commission new superconducting magnetic quadrupole lenses have been constructed and tested. Despite their relatively small sizes, the lenses create magnetic-field gradients up to 10 kOe/cm.

The lenses are made of Nb—Sn strips and are cooled with liquid helium to 4.0°K. Laboratory tests show that lenses with still higher field gradients are quite feasible. Such lenses offer the possibility of creating charged-particle beams and focusing these at far shorter distances than is possible at the present time. The second advantage of the new lenses is the small amount of electrical power required [68].

Magnetic lenses may also be used for precisely regulating particle beams of high energy in giant accelerators [69].

Two superconducting magnetic quadrupole lenses are being made in CERN [69a] (Geneva).

3. Masers

Masers employ magnetic fields of great uniformity (0.02%) created by solenoids made of superconducting Nb—Zr wire and Nb—Ti strip.

The positive aspect of this method of producing magnetic fields lies in the small weight of the magnetic systems. For example, the solenoid which we ourselves have developed for these purposes weighs 0.25 kg (instead of 40-50 kg ordinarily) and the magnet 2 kg (instead of 100 kg) (Fig. 238). The superconductor

Fig. 238. Solenoids and magnet for KPU.

has no resistance and absorbs no energy; thus if the solenoid terminals are short-circuited (by a superconductor) the current induced in the coil will circulate forever (while the helium supply lasts) and the magnetic field will remain absolutely stable. The supply source is therefore only used when the magnet is first put into commission. The system requires no thermal compensation, as the temperature of the liquid helium remains stable [70].

Low-noise amplifiers are used in America for communications with the artificial satellites Telestar, Relay, Synch, and the commercial communications satellite Early Bird. All these satellites are of small dimensions and the power of their converters is no greater than a few watts. Thus, an important problem is that of separating the weak signal from the noise. The usual receiver has a high noise level and a weak signal will be suppressed. The advantage of masers is that they enable such a signal to be distinguished [71].

Other examples of the use of magnetic fields created by means of superconductors in various instruments include photodetectors [68], lasers [77], and devices for microwave spectroscopy [72].

Lead superconducting resonators are used in "field" diodes [73].

For measuring voltages to a high accuracy, RCA and Pennsylvania State University have constructed a device based on the Josephson effect in superconductors. In this process one of the most fundamental physical constants, the charge/Planck's-constant ratio, is measured [74].

At the present time extensive investigations are being conducted into various kinds of contacts possessing the properties of a Josephson junction. For example there are some six kinds of construction classified as Josephson devices: a tunnel system made of two superconducting films a few hundred ångstroms thick, separated by a layer formed by the oxide of one of these materials 10-20 Å thick; a Diem bridge including one superconducting film of rectangular shape with a contraction or cross-piece in the center, $1 \times 1 \mu$ in dimensions; a noncontact bridge (a version of the previous device made by Ford Research Laboratories), distinguished by the presence of a break filled with a strip of ordinary metal in the cross-piece; a point device consisting of a sharpened wire made of a superconducting material (radius of tip $\sim 1\mu$) in mechanical contact with a superconducting plate or rod; a superconductor–normal metal–superconductor system (Cambridge, England), constructionally reminiscent of the tunnel version, in which the oxide layer is replaced by a film of ordinary metal 10,000 Å thick; a rod device (Cambridge, England) constituting an "excrescence" of hardening solder in rod form on an oxidized niobium wire [74a]. Josephson tunnel devices are used in the United States for preparing a high-accuracy standard voltage (a few parts in 10^8 as against the present norm of 10^{-6}), for measuring physical constants (for example, the 2e/h constant in Pennsylvania State University) to an accuracy twenty times better than existing measuring methods, and for low-temperature thermometers with a scale down to 0.023°K [74a].

Josephson devices may be used in magnetocardiography for measuring the magnetic fields associated with human heart activity and in magnetometers for measuring magnetic induction to an accuracy of 10^{-11} G, voltages to 10^{-11} V, and currents to 10^{-8} A, as well as in the guise of logic and memory systems in computers, and also high-frequency electromagnetic-radiation generators and detectors [74a-74c]. Niobium and its alloys are used as contacts [36, 75].

Sperry Gyroscope is developing microwave switches, inexpensive wide-band microwave devices acting in a frequency range extending to centimeter wavelengths. Attempts are being made at studying switching in the 90 Hz range. Very thin films (20-100 Å) are used in the experiments with a view to determining the possibility of employing these for repeated switching at 9 kc/sec under the action of a microwave field [76].

The possibility of using superconductors as cryogenic tunnel diodes is also being considered [77, 78].

NUCLEAR POWER AND SPACE

1. Magnets for Thermonuclear Reactions

The development of thermonuclear power is currently retarded by the absence of necessary insulating materials. Thus, according to the calculations of Academician E. K. Zavoiskii the plasma required in order to obtain a thermonuclear reaction of

APPLICATIONS

practical interest should have a temperature of about one billion degrees and a particle density of $10^{15}/cm^3$, the pressure being 120 atm. On coming into contact with solids the plasma loses its temperature and the reaction ceases. In order to control a plasma filament heated to millions or billions of degrees, it is proposed that extremely strong magnetic fields should be employed. Such fields may be created in fairly small volumes by means of superconducting materials [79].

An experimental system for studying plasma physics was described earlier [80]. The apparatus consists of two spaced superconducting coils (with an adjustable distance between them) having a common horizontal axis. Both coils are enclosed in Dewars; these provide a diameter of the working part of the field in the magnetic stopper (mirror) equal to 17.8 cm. Each coil is capable of creating a magnetic field of up to 25 kOe on the axis. As a result of the working characteristics of the coils and Dewars this arrangement is more economical and convenient in operation than ordinary magnets and gives a greater flexibility (Fig. 239).

An experimental superconducting system has been developed in the United States for use in a controlled thermonuclear reaction [81]. In Japan an apparatus has been developed for creating and maintaining hot plasma by means of superconducting magnets [81a].

Superconducting magnets have been made by RCA for use in high-energy physics and plasma research. The National Aeronautics and Space-Research Board intends using these magnets in the development of plasma motors for space ships.

It is considered that thermonuclear reactors will have great advantages over rocket motors using chemical energy as power sources for space-ship motors.

The coils of the magnets are made of strip based on stainless steel coated with a layer of Nb_3Sn. The coils are fed from a battery source (6 V, 100 A). The range of magnetic fields which may be created by these magnets fluctuates from 60 to 125 kOe. Constructionally the coils are made in the form of demountable and nondemountable solenoids. The latter create magnetic lines of force directed along the longitudinal axis and the former magnetic lines of force directed along either the longitudinal or the transverse axis [82].

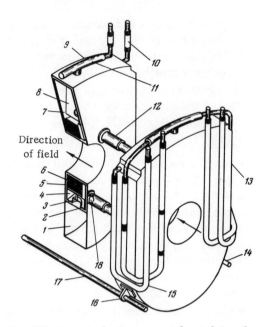

Fig. 239. Superconducting magnet for studying plasma physics: 1) reservoir containing liquid nitrogen; 2) vacuum space; 3) gas absorber; 4) reservoir containing liquid helium; 5) coil; 6) stainless steel support plate; 7) reservoir containing liquid helium; 8) reflector; 9) stationary Dewar; 10) line for evacuating gaseous nitrogen; 11) line for supplying and removing liquid helium; 12) screen of support rod (at liquid-nitrogen temperature); 13) mobile Dewar; 14) guide rod; 15) liquid-nitrogen supply line; 16) roller bearing; 17) guide rod; 18) support rod (at liquid-helium temperature).

Cryogenic devices for experimental nuclear physics are considered in an earlier review [82a].

In the Lawrence Radiation Laboratory (Livermore, USA) initial tests have been carried out on a superconducting magnet six feet in diameter and 13 tons in weight (designed magnet field 20 kG, achieved 13.5 kG), intended for the creation of a "magnetic bottle" in which plasma is to be held by the magnetic field and heated to a temperature of $3 \cdot 10^8 \, °K$ [82b].

2. Elementary-Particle Accelerators

The accelerators used for studying the properties of elementary particles are of large dimensions [85]. Thus the recently-commissioned 70 GeV Serpukhovsk proton accelerator has a diameter of over 1.5 km and the weight of the whole magnet system exceeds 20,000 tons. Plans for 500-1000 GeV accelerators provide for the creation of magnets lying around a circle 2-5 km in diameter. The size of the accelerators can only be reduced by increasing the magnetic field strength. Calculations show that if the magnetic field in the accelerators were increased from 16 to 300 kOe, the giant Berkeley accelerator could be put on a dinner table. The possibility of creating such magnets is perfectly feasible on using superconducting materials.

Apart from using superconductors in the magnet systems of large particle accelerators, they may also be employed in order to improve the characteristics of accelerating tubes [84, 84a].

In the Stanford center (USA) opportunity is being taken at the present time to reconstruct a 20-GeV 2.5-km electron linear accelerator so as to give a superconducting accelerator capable of accelerating electrons to an energy of 100 GeV [84c]. The proposed length is 165 m.

Work is being carried out on replacing the deflecting and focusing magnets used for directing the particle beams created by an accelerator into the experimental zone with superconducting versions [85]. The strong fields created by superconducting magnets and the high gradient of these fields facilitate the analysis and focusing of such beams at shorter distances; this is particularly important for beams of short-lived particles.

At the present time the use of superconducting magnets for the construction of an accelerator in Weston (NAL Laboratory) is being considered. It is considered that 250 superconducting magnets will be required for the equipment of this device, these occupying an area 300 m wide and 600 m long. An estimate of the cost is $25,000,000 [85a].

The prospects of making proton accelerators for pulsing at a frequency of 0.5 Hz [85a] are also being examined. The fundamental possibility of creating ac solenoids was indicated in [34c].

3. Bubble Chambers

Superconducting magnets have started playing an increasing part in high-energy physics.

One of the fields in which the use of superconducting magnets is particularly promising is that of the bubble chambers used in studying the properties of elementary particles. Strong magnetic fields are needed in these in order to change the trajectories of the elementary particles from which their properties are calculated. The use of superconducting solenoids is reasonable in view of the low temperatures at which the chambers operate. Usually bubble chambers employ either liquid helium or liquid hydrogen (20°K). If liquid helium is employed, then the cooling systems of the windings and the chamber may be partly coupled together.

In the Argonne National Laboratory construction of a large bubble chamber was started in 1967 and completed in 1969. The magnetic field in the chamber is created by a superconducting cylindrical solenoid with a winding made from Nb−Ti wire. The magnetic field amounts to 20 kOe. The diameter of the chamber is over 3.6 m, the weight 2000 tons, and the volume 27,000 liters; it is to be filled with liquid hydrogen or deuterium. Strong walls around the chamber 1200-1500 tons in weight protect it from unwanted radiation. Detailed technical data relating to this magnet and a number of others are presented in [35a].

In the Brookhaven National Laboratory a bubble chamber 4.3 m in diameter with a volume of 47,000 liters has been developed. The magnet of this chamber gives a magnetic field of 30 kOe [86].

For these purposes a solenoid with an internal diameter of 25 cm was employed. The field created by the solenoid reached 31.8 kOe [87].

For the American bubble chamber a superconducting magnet 25.4 cm in diameter with a magnetic field of 32.8 kOe was also constructed and prepared. This consists of six discs 61 cm in diameter; each disc has outer and inner superconducting coils wound with Nb−Zr wire 0.25 mm in diameter; the weight of the magnet is 382 kg [88]. Press releases indicate that the bubble chamber of Cambridge University (England) will incorporate a 70 kOe superconducting magnet. Recent data relating to the use of superconducting magnets in accelerator technology appear in a review article [50].

4. Resonance Pump

Superconductors are also employed in cryogenic devices used in cosmic technology. Thus General Electric has constructed a resonance pump for feeding liquid helium into a cryostat [89] in the Space-Flight Scientific-Research Center.

5. Gyroscopes

Gyroscopes are extremely important navigational devices used in sea, air, and space navigation. In principle the gyroscope is a rapidly rotating disc seated on an axle. The direction of the gyroscope axle (axis) always remains constant, irrespective of the position of the ship in space. Strictly speaking, there are always changes in the direction of the axle as a result of friction in the supports. Research is therefore aimed at reducing the friction so as to eliminate errors in the readings. Gyroscopes using superconductors aid the solution of this problem. The operating principle lies in the fact that a sphere of superconducting material (rotor) is suspended in a magnetic field and rotated (Fig. 240). In this way friction is reduced to a minimum. Friction only occurs between the rotor and the gaseous helium in the cryostat. Gyroscopes constructed on this principle have a much lower friction

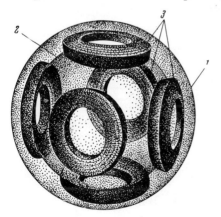

Fig. 240. Arrangement of a cryogenic gyroscope: 1) Superconducting cell; 2) rotor; 3) sustaining coil.

than all other known forms [90]. The rotor is made of niobium and its alloys or the compound Nb_3Sn; it is coated on top with a layer of superconductor having a lower T_c; this enables it to be turned by the magnetic field. The pairs of superconductors used in this manner are Nb–Ta, Pb–Sn, Pb–In, and Nb_3Sn–Nb [79].

6. "Zero" Magnetic Field

The capacity of superconductors to screen a magnetic field may be used in order to screen the magnetic field of the Earth. Workers at Stanford University have attempted to create a system of superconducting cylinders giving a "zero" magnetic field, with the intention of installing a nuclear gyroscope, which is only able to operate successfully in the complete absence of a magnetic field [91, 91a].

7. Magnetohydrodynamic (MHD) Generators

Superconductors may be used in order to create strong magnetic fields in MHD generators. As already indicated, without compact and powerful superconducting magnets MHD generators hold little promise of practical use. The principle of operation of MHD generators has been known for a long time: A flow of ionized gas moves between the poles of a magnet; the motion of the electrically-charged particles through the lines of force creates an electromotive force (Faraday's law); as the gaseous medium, plasma at a temperature of 2000-2500°K is employed.

If the magnets are made of nonsuperconducting materials, the energy losses are high, and the prospects of generating electrical power by this method are unpromising. Calculation of an experimental MHD generator designed to produce 2000 kW with a "normal" magnet creating a field of 40 kOe in the channel showed that the power required by such a system on using a magnet with copper windings and water cooling was 15,000 kW; *in toto*, the equipment required for creating the field would cost $600,000. On using a superconducting solenoid, however, the power required (including cooling) is 50-100 kW; the size of the whole system is also far less [92]. The use of superconductors facilitates the construction of economically viable MHD generators.

APPLICATIONS

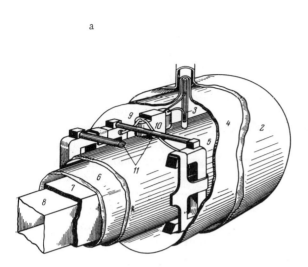

Fig. 241. MHD generators with superconducting magnets. a) section of an MHD generator with superconducting excitation coils: 1) inflow of liquid helium; 2-6) outer windings; 3) current lead to transformer; 4) liquid nitrogen; 5) liquid helium; 7) thermal insulation of the channel; 8) channel with plasma; 9) current leads to winding; 10) dc transformer; 11) support structures; b) external view of an experimental MHD generator.

Figure 241 illustrates the arrangement of a generator with a superconducting magnet and the general appearance of the installation created by the Westinghouse laboratory. The power required is no more than 3-5 W. The plasma is created by the combustion of ethylene and oxygen in a tube 25 mm in diameter [93, 94]. The power of the MHD generator depends on the diameter of the channel for the ionized gas: If instead of a channel 25 mm in diameter we take one of 100 mm, the power increases by a factor of 100. The channel diameter is in turn limited by the diameter of the superconducting magnet [95]. The insulation between the central channel and the magnet includes a ceramic casing, a water jacket, vacuum (the magnet is enclosed in a Dewar vessel), and a jacket of liquid nitrogen surrounding the helium bath.

In the Tokyo Electrotechnical Laboratory an experimental 1000 kW MHD generator is being constructed; this will work on petroleum fuel. The generator will be furnished with a 40-45 kOe superconducting magnet [97].

MHD generators may in turn be used in various fields of science and technology: thermonuclear physics, electrical power, chemistry, metallurgy, atomic reactors, space ships, and supersonic aircraft.

Avco has developed and made a saddle-shaped magnet with a magnetic field of 40 kOe suitable for use in MHD generators. The length of the magnet is 300 mm. Inside a cylinder 150 mm long and 30 mm in diameter the magnetic field is held in a steady state continuously. As it is essential for the magnetic field to be perpendicular to the axis of the MHD generator channel, the windings of the magnet are given a saddle-shaped form. The magnet windings are made from nine rows of Nb–Zr wire placed in a copper sheath of high purity. The copper serves as a shunt [96]. It has been proposed that a powerful (1 MW) MHD generator be used as a source of electrical power on board an aircraft [96a].

8. Protection of Astronauts from Radiation

Perhaps one of the most important fields of future use for superconducting magnets will lie in the field of cosmic research. Lockheed is making a many-turned superconducting solenoid 1.8 m in diameter. The field on the inner surface of the toroid is up to 15 kOe and in the center 1000 kOe (Fig. 242). Such a solenoid is capable of protecting a ship from particles with energies of hundreds of millions of electron volts. The weight of the solenoid to-

Fig. 242. Solenoid for protecting astronauts from radiation.

gether with the cryostat and the cooling system is 85 kg. Cooling is achieved by means of a liquid-helium supply.

The use of a superconducting magnet as a radiation shield gives a twenty-fold reduction in weight as compared with ordinary shielding formed from massive, armored walls. The effect of strong magnetic fields on the human organism is not yet sufficiently clear. Investigations are proceeding in this direction.

9. Hydromagnetic Braking

Superconductors may also be used in the return of a space ship to the Earth's atmosphere. A strong magnetic field creates hydromagnetic braking in the cloud of ionized air formed by the entry of the space ship into the atmosphere. The kinetic energy of the ship will be absorbed by the magnetic field, preventing the ship from overheating and breaking up.

10. Energy Stores

A superconducting magnet is an effective device for storing energy, comparable with an electrical condenser, a Mach ring, or even TNT. The energy of the magnet may be rapidly and efficiently reconverted to electricity [85, 98].

ELECTRICAL MACHINES

Existing electrical machines may be entirely reconstructed on the basis of superconducting materials.

Superconductors are even now being used in ac and dc generators [99, 100, 101] and other electrical machines (rotor and stator windings are being made from superconductors). The efficiency in motors of this kind is almost 100% [90, 102]. In the United States a 50 h.p. motor has been made from superconducting materials and a 3000 h.p. motor is being constructed.

International Research and Development is making the excitation winding of a unipolar dc electric motor using superconducting material. The motor power is 200 h.p. at 200 rpm [102a]. The manufacture of motors of 3000 h.p. or over has also been mentioned [102b]. Tests have already been made on an experimental electric motor with superconducting windings, giving a power of 50 h.p. [102c].

It is intended to use the motors in the steel-melting industry, in paper factories, electrical power stations, ships, compressors, and blowers [102d]. Work is being carried out on making transformers with superconducting coils [103, 104, 105]. This mode of transformer manufacture should eliminate thermal losses.

It has been proposed to use superconducting magnets in high-speed trains. Thus Japanese National Railways Corporation intends to introduce an express train in 1981 developing a speed of over 260 mph with linear induction motors and a magnetic suspension system produced by interaction between magnetic fields in superconducting windings on board the train and in the guide rail [105a].

At the Stanford National Research Institute a "magnetic-cushion" train is being designed for a speed of over 360 km/h [106a].

Current amplifiers based on superconductors are being developed [90].

The property of superconducting materials to expel a magnetic field is being used to create frictionless bearings [106].

The most direct use of superconductivity is that of transmitting power losslessly over long distances. According to the formula $J_c = 2.5 dH_c$ [107], a lead cable 1 cm in diameter can transmit a current of 2000 A without resistive losses. For introducing the power into the low-temperature medium and extracting it at the ends of the cable a special transformer may be used, one winding being superconducting and being held at a low temperature while the other is made of copper and is thermally insulated from the superconducting turns. A disadvantage of superconducting cables is the great expense of maintaining them at 4.2°K, which exceeds ordinary resistive losses.

As yet this method of transmitting energy is uneconomical for long distances; however, it is to be expected that the development of new materials with a higher critical superconducting temperature and better critical field and current will enable such a project to be realized some time in the future.

APPLICATIONS 447

At the present time the possibilities of using composite materials for this purpose are being considered [85a].

Coaxial transmission lines made from nonideal superconductors are extremely promising for the transmission of direct current. A line made of Nb_3Sn with a diameter of a handsbreadth could transmit a power equal to the peak load required by all users in the United States [103, 105, 108, 109, 110, 111]. Under laboratory conditions an ac current of 2080 A has been transmitted along a coaxial cable 10 m long consisting of layers of niobium foil. With an aluminum sheath the cable withstood a current density of $2 \cdot 10^4$ A/cm^2 [112].

Conclusion

In conclusion we shall attempt a brief assessment of the present state of the problem of introducing superconducting metals, alloys, and compounds into science and industry.

Superconductors occupy an ever-increasing place in the new technology and are beginning to exert a decisive influence on leading aspects of scientific and technological progress. Advances in the development and use of new superconducting materials are largely associated with the general state and level in the development of the physical theory of superconductivity, the physical chemistry of metals, metallography, metal physics, and technical physics.

At the present stage of research it is inevitable that there should be disagreements in relation to the effects of fine structure, lattice distortion, lattice defects, dislocations, the degree of cold plastic deformation, and the heat treatment of superconductors on their superconducting characteristics. It would be highly desirable to achieve a more or less unified view among specialists as to the effect of structure, impurities, defects, and defect distribution on superconducting properties. It is a matter of some urgency to secure a proper level of understanding as to the interrelationship between the macrostructure, microstructure, and fine structure, on the one hand, and the mechanical properties of metals and alloys on the other.

It would be particularly valuable to conduct further research into the properties of single crystals of precisely-specified composition and to analyze the effects of impurities (particularly those of the interstitial type) to which the rare and refractory metals are especially sensitive.

It is desirable to seek out new superconducting elements and compounds in a state of high purity and to study their behavior under high pressures, the effects of penetrating radiations, and so on.

The application of scientific ideas and the laws governing the physicochemical analysis of metallic systems to the field of low temperatures achieved in our own laboratory has already yielded important scientific and practical results in searching for superconducting alloys and compounds based on niobium, vanadium, and other superconducting metals.

All alloys of the metallic systems formed by superconducting components possess superconductivity, independently of the phase constitution and crystal structure. The superconducting transition temperature of alloys constituting continuous substitutional solid solutions of two superconducting metals often varies almost additively with composition. In the presence of polymorphism in one or both of the superconducting constituents, a maximum appears on the T_c-composition curve.

In alloys constituting a structural mixture of superconducting phases the transition temperature varies linearly with composition. For a textured structure comprising mixtures of superconducting materials (composite superconductors) the superconducting parameters increase.

Superconducting compounds may be found in many metallic systems formed by both superconducting and nonsuperconducting components; as a rule in the T_c-composition curve such compounds are distinguished by singular points. The highest superconducting transition temperature occurs for superconducting metal compounds based on niobium, vanadium, and certain other metals with a crystal structure of the Cr_3Si type with a high electron specific heat and a high density of electron states on the Fermi surface. Recently some success has been achieved in predicting new superconducting compounds of the A_3B type and discovering methods of forming these by analyzing the electron structure of the components on an electronic computer. Taking due account of the positive results so far achieved, the way is open for a broader approach to the physicochemical analysis and cybernetics of superconducting systems.

In the Soviet Union and elsewhere attention is now passing from binary to ternary and more complicated systems of superconducting alloys and compounds.

It is very important to emphasize the necessity of studying not only the superconducting characteristics but also as many as possible of the other physical properties of superconductors (thermal expansion coefficient, specific heat, elastic modulus, hardness, electrical resistance, and so on) over a wide range of temperatures and compositions with a view to establishing a mutual relationship between structure and properties.

It is extremely attractive to seek for "high-temperature" superconductors with a critical temperature of over 20°K. It must always be remembered, however, that the superconducting state is disrupted not only by the heating of the superconductor above the critical temperature but also by the imposition of a fairly strong (critical) magnetic field or the passage of a large (critical) electric current. All these three critical characteristics of superconductors are mutually related and depend on the electron and crystal structure of the metals and alloys, their composition and purity, and also on the metallurgical history of their preparation and methods of melting and thermomechanical treatment.

There is no doubt that the laws relating the critical characteristics of superconductors to their electron structure and the type of interatomic bond have as yet been insufficiently studied. This aspect of the problem of superconductivity must always be borne in mind when pursuing physicochemical investigations.

The number of the superconducting elements and the values of their superconducting constants cannot be regarded as exactly established, owing to the fact that impurities and structural defects have as yet not been entirely eliminated from these. It is therefore important to advance chemical, physical, and metallurgical methods of purifying these elements as far as possible from unwanted atoms, producing perfect crystals, estimating the purity and degree of perfection, and determining the quantitative values of the superconducting characteristics (T_c, H_c, and I_c). The potentialities of the physicochemical analysis of superconducting systems in providing a deeper insight into the nature and properties of superconducting phases have as yet not been adequately realized; the number of completely studied systems and composition–superconducting-property diagrams can be counted on the fingers of one hand. Frequently the superconducting properties of alloys are determined without ensuring reproducible conditions (purity, degree of equilibrium, conditions of measurement, and so on). The attention of research workers has to be drawn to this aspect of the problem of superconductivity. The number of superconducting compounds has passed into the thousands, some of these having extreme characteristics and are already being used in superconducting magnets and other installations, yet the progress of this field of activity is still empirically controlled. The correlation of the superconducting properties with the fundamental parameters of compounds has been insufficiently developed and awaits the attention of research workers.

The extensive potentialities of the use of superconductors in the new technology evoke an ever-growing demand for superconducting materials with specific and stable superconducting and technological parameters; this requires the development of experimental and research work on the technology of manufacturing superconductors and superconducting parts. However, as yet little attention has been paid to the theory of the pressure treatment (rolling, extrusion, etc.) of superconducting materials.

The stability and the general level of the properties of these materials depend on the methods employed in their processing, the thermal conditions, and so forth, so that it is essential to establish a special branch of this discipline: the pressure treatment (working) of superconducting materials.

A review of the principal possible fields of application of superconducting materials illustrates the importance and necessity of a further-accelerated development of this work.

As we have already mentioned, the problem of superconductors is a complex one. In order to estimate their properties, develop alloys, ensure a successful application of superconducting materials, use these to create magnetic systems, and incorporate superconductors in various installations and equipment, physicists, metallurgists, metal physicists, and specialists in cryogenic technology, electronics, and other related aspects must all work together.

At the moment the center of gravity of the problem is beginning to move in the direction of the technological applications of superconductivity. The importance of superconductivity research will increase still further as the practical use of superconducting materials in the popular economy and new technology advances.

LITERATURE CITED

1. H. Kamerlingh-Onnes, Communs Phys. Lab. Univ. Leiden, p. 139 (1914).
2. W. J. de Haas and J. Voogd, Communs Phys. Lab. Univ. Leiden, 208 (1930) and 214 (1931).
3. J. E. Kunzler, E. Buchler, F. S. L. Hsu, and J. H. Wernick, Phys. Rev. Lett., 6:89 (1961).
4. J. E. Kunzler, Bull. Amer. Phys. Soc., Ser. II, 6:298 (1961).
5. D. Kunzler, Uspekhi Fiz. Nauk, 86(1):125 (1965).
5a. W. Samson, P. Craig, and M. Strongin, Uspekhi Fiz. Nauk, Vol. 93, No. 12(4), p. 707 (1967).
6. Nedelya, No. 44 (1967) (TASS communication).
7. Chem. Eng. News, No. 14, p. 44 (1966).
8. J. E. Kunzler, E. Buchler, F. S. L. Hsu, B. T. Matthias, and C. J. Wahl, J. Appl. Phys., 32:325 (1961).
9. S. H. Autler, Rev. Scient. Instrum., 31:369 (1960).
10. G. B. Yntema, Phys. Rev., 98:1197 (1960).
11. E. Justi, Electrotechn. Z., 63(49/50):578 (1942).
12. J. K. Hulm, M. J. Fraser, H. Riemersma, A. J. Venturino, and R. E. Wein, Bull. Amer. Phys., 6:501 (1961); R. E. Wein, High Magnetic Fields (1961), p. 332.
13. E. M. Savitskii, V. V. Baron, V. R. Karasik, V. Ya. Pakhomov, and M. I. Bychkova, Pribory i Tekh. Éksperim., No. 2, p. 152 (1963).
14. V. R. Karasik, Pribory i Tekh. Éksperim., No. 6, p. 5 (1962).
15. Science News, 93:124 (1968).
16. B. G. Lazarev, L. S. Lazareva, V. R. Golik, and S. I. Goridov, Contributions to the 15th All-Union Conference on Low-Temperature Physics, Tiflis (1968), p. 65.
17. O. I. Goncharov, Author's abstract of dissertation "Study of critical currents in niobium alloys with 65-80% Zr for superconducting magnet systems based on these alloys," Joint Institute for Nuclear Research, Dubna (1967).
18. Physics Today, 16(4):57 (1963).
19. C. K. Jones, J. K. Hulm, and B. S. Chandrasekhar, Rev. Mod. Phys., 36(1):74 (1964).
20. E. Saur and H. Wizgall, Les Champs Magnetiques Intenses, Colloque Internat., Grenoble (1966), p. 223.
20a. D. B. Montgomery, Bull. Amer. Phys. Soc., 10(3):359 (1965).

20b. H. R. Hart, J. S. Jacobs, C. L. Kolbe, and P. E. Lawrence, High Magnetic Fields, New York (1962), p. 584.
20c. D. B. Montgomery and H. Wizgall, Phys. Letts., 22(1):48 (1966).
20d. K. Hechler, E. Saur, and H. Wizgall, Z. Phys., 205(4):400 (1967).
20e. B. T. Matthias, Science, 168(3927):103 (1970).
21. S. H. Autler, High Magnetic Fields (1962), p. 326.
22. D. L. Martin, M. G. Benz, C. A. Bruch, and C. H. Rosner, Cryogenics, No. 3, p. 114 (1963).
23. Electronics, 36(9):18 (1963).
24. M. G. Benz, Metallurgy and Ceram. Lab. Rept., No. 66-K-19 (1966).
25. C. H. Rosner, M. G. Benz, Metallurgy and Ceram. Lab. Rept., No. 65-C-062 (1965).
26. JEEE, No. 3, 338 (1966).
27. Z. J. J. Stekly, E. J. Lucas, and T. A. Winter, Rev. Scient. Instrum., 39(9):1291 (1965).
27a. J. P. Scott and J. R. Laning, Cryogenics, 10(3):208 (1970).
27b. J. W. Metselaar, H. A. Jordaan, J. W. Schutter, and D. de Klerk, Cryogenics, 10(3):220 (1970).
28. P. S. Swartz and C. H. Rosner, J. Appl. Phys., 33(7):2292 (1962).
29. New Scientist, 31(507):257 (1966).
30. Electronics, 40(11):192 (1967).
31. Usine Nouvelle, 42:137 (1966).
32. New Scientist, 31(508):312 (1966).
33. P. F. Chester, Roy. Soc. Meeting on Advanc. Metals for MHD Power Generation, London (1965).
34. Electrical Times, p. 984 (Dec. 11, 1966).
34a. H. Hillman, Laboratory of Vakuumschmelze GMBH, Hanau, Preprint L-Hi/Spä, April 4, 1970.
34b. M. N. Wilson, C. R. Walters, J. D. Lewin, P. F. Smith, and A. H. Spurway, J. Phys. D. Appl. Phys., 3(11):1517 (1970).
34c. A. C. Barber and P. F. Smith, Cryogenics, 9(6):483 (1969).
34d. Cryogenics, 10(4):358 and 10(5):456 (1970).
34e. Electrical World, 174(5):49 (1970).
34f. V. E. Keilin, E. Yu. Klimenko, and B. N. Samoilov, Pribory i Tekh. Éksperim., No. 1, p. 216 (1971).
35. D. A. Buck, Proc. IRE, 44:482 (1956).
36. J. Bremer, Superconducting Devices [Russian translation], Izd. Mir, Moscow (1964).
37. General Electric Review, 91(1):41 (1958).
38. D. R. Young, in: Progress in Cryogenics (K. Mendelssohn, ed.), Vol. 1 (1958).
39. M. K. Haynes, Proc. Sympos. Superconduct. Techniques, Washington (1960), p. 399.
40. I. M. Lock, Cryogenics, 2:65 (1961).
41. C. Nebell and C. P. S. Cichter, Phys. Rev., 113:1504 (1959).
42. D. L. Fench and I. B. Woodford, J. Appl. Phys., 32:1881 (1961).
43. J. Bremer, Superconducting Devices [Russian translation], Izd. Mir, Moscow (1964), p. 71.

44. J. W. Bremer, Electr. Manufact., 61:78 (1958).
45. V. L. Newhanse and J. W. Bremer, J. Appl. Phys., 30:1458 (1958).
46. D. Young, Brit. J. Appl. Phys., 12:359 (1961).
47. J. Bremer, Superconducting Devices [Russian translation], Izd. Mir., Moscow (1964), p. 208.
48. M. I. Buckingham, Transactions of the Fifth International Conference on Low-Temperature Physics (J. R. Dillinger, ed.), Madison (1958), p. 229.
49. E. C. Crittenden, Transactions of the Fifth International Conference on Low-Temperature Physics (J. R. Dillinger, ed.), Madison (1958), p. 232.
50. J. W. Crowe, IBM J. Res. Develop., 1(4):294 (1957).
51. E. C. Crittenden, J. N. Cooper, and F. W. Schmidlin, Proc. IRE, 48:1233 (1960).
52. L. L. Burns, G. W. Leck, V. A. A. Caphouse, and R. W. Katz, Solid-State Electr., 1(4):343 (1960).
53. D. H. Parkinson, Solid-State Electr., 1(4):306 (1960).
54. A. C. Rose-Innes, Brit. J. Appl. Phys., 10:452 (1959).
55. H. O. McMahon and W. E. Gifford, Solid-State Electr., 1:273 (1960).
56. Electr. Engng., 82:150 (1963);
57. Automat. Control, No. 14, p. 52 (1961).
58. Sc. News, No. 11, p. 250 (1967).
59. Élektronika, 40(7):39 (1967).
60. A. Goetz, Phys. Rev., 55:1270 (1939).
61. D. H. Andruos, R. M. Milton, and W. de Sorbo, J. Opt. Soc. Amer., 36:518 (1946).
62. D. H. Andruos and C. W. Clark, Nature, 158:945 (1946).
63. New Scientist, 15(295):68 (1962).
64. B. Lavelic, J. Appl. Phys., 24:19 (1953).
65. D. H. Andruos, R. D. Fowler, and M. C. Williams, Phys. Rev., 76:154 (1949).
66. Aviat. Week and Space Technol., 79(21):96 (1963).
67. Aviat. Week, Vol. 72 (1960).
68. Chem. and Engng. News, 43(37):55 (1965).
69. K. A. Kapustinskaya and B. I. Kogan, in: Metallography and Metal Physics of Superconductors, Izd. Nauka, Moscow (1965), p. 132.
69a. CERN Courier, No. 9, p. 281 (1970).
69b. V. V. Baron, T. F. Demidenko, S. I. Klimov, E. M. Savitskii, and V. M. Turevskii, in: Problems of Superconducting Materials, Izd. Nauka, Moscow (1970), p. 209.
70. V. V. Baron, T. F. Demidenko, S. I. Klimov, E. M. Savitskii, and V. M. Turevskii, in: Physical Chemistry, Metallography, and Metal Physics of Superconductors, Izd. Nauka, Moscow (1968).
71. New Scientist, 73(503):24 (1966).
72. New Scientist, 27(453):218 (1965).
73. New Scientist, 26(448):781 (1965).
74. Electronics, 40(5):52 (1967).
74a. Electronics, 44(5):38 (1971).
74b. E. A. Combet, Rev. Phys. Appl., 4(4):557 (1969, 1970).
74c. S. I. Bondarenko, E. I. Bulanov, L. E. Kolin'ko, and T. P. Narbut, Pribory i Tekh. Éksperim., No. 1, p. 235 (1970).

75. V. R. Karasik, Physics and Technique of Strong Magnetic Fields, Izd. Nauka, Moscow (1964).
76. J. Pinkham, Interavia, 16(1):1833 (1961).
77. H. Meissner, Phys. Rev., 109:686 (1958).
78. H. Meissner, Phys. Rev., 117:672 (1960).
79. E. M. Savitskii, New Metallic Alloys, Izd. Znanie, Moscow (1967).
80. J. R. Routh, D. C. Freemann, and D. A. Haid, Rev. Scient. Instrum., 36(10):1481 (1965).
81. Sci. News, 91(7):163 (1967).
81a. New Scientist, 38(600):502 (1968).
82. Electronics, 39(24):140 (1967).
82a. E. F. Hammel, J. D. Rogers, and W. F. Hassenzahl, Cryogenics, Vol. 10, No. 5 (1970).
82b. Nuclear News, 14(2):51 (1971).
83. Conference on High-Energy Acceleration, Dubna (1963), p. 61.
84. New Scientist, 28(475):868 (1965).
84a. New Scientist, 36(575):663 (1967).
84b. Chem. and Engng. News, 46(20):18 (1968).
84c. New Scientist, 48(728):365 (1970).
85. M. G. Kremlev, Uspekhi Fiz. Nauk, 93(4):675 (1967).
85a. E. F. Hammel, J. D. Rogers, and W. F. Hassenzahl, Cryogenics, 10(3):186 (1970).
85b. Neue Züricher Zeitung, No. 306, p. 76 (1968).
86. Chem. and Engng. News, 45(27):11 (1967).
87. Z. P. Kartsev, Superconductors in Physics and Technology [in Russian], Izd. Znanie, Moscow (1965), p. 30.
88. New Scientist, 24(416):366 (1964).
89. Design News, 22(13):24 (1964).
90. T. A. Buchhold, Scient. Amer., 202:74 (1960).
91. Mond, August 18 (1966).
91a. Science News, 93(23):540 (1968).
92. C. W. Wilson and D. C. Roberts, Sympos. Magnetoplasmodynamic Electrical Power Generation, Newcastle (1962), pp. 8, 9.
93. D. K. Fox and W. J. Reichenecker, Mater. in Design Engng., Vol. 57, No. 492 (1963).
94. New Scientist, 18(338):26 (1963).
95. Khimicheskaya Tekhnologiya, No. 2, 25 (1963).
96. Mech. Engng., 88(8):41 (1966).
96a. New Scientist, 41(634):233 (1969).
97. New Scientist, 34(543):284 (1967).
98. V. L. Newhouse, Applied Superconductivity, John Wiley and Sons (1964).
99. D. H. Douglass, Jr. and R. H. Blumberg, Phys. Letts., 1:78 (1966).
100. Design News, 26:247 (1964).
101. New Scientist, 26(439):163 (1965).
102. K. F. Schoch, in: Advances in Cryogenic Engineering, Vol. 6 (K. D. Timmerhaus, ed.), Plenum Press, New York (1961), p. 65.

102a. Financial Times, No. 24700, p. 11 (Nov. 19, 1971).
102b. New Scientist, 37(582):242 (1968).
102c. Spectrum, No. 49, p. 8 (1968).
103. R. McFee, Rev. Scient. Instrum., 30(2):98 (1959).
104. Electr. World, 156(11):54 (1961).
105. New Scientist, 23(406):500 (1964).
105a. New Scientist and Science Journal, 49(740):424 (1971).
106. T. A. Buchhold, Scient. Amer., 202:74 (1960).
106a. Financial Times, No. 25232, p. 9 (Aug. 20, 1970).
107. R. McFee, Electr. Engng., 2:122 (1962).
108. New Scientist, 29(488):753 (1966).
109. New Scientist, 34(548):557 (1967).
110. New Scientist, 34(544):337 (1967).
111. Chem. and Engng. News, 44(11):35 (1966).

Index

Accelerators
  Elementary-particle, 439
Alloying
  Effects on critical temperature, 29, 159, 170
  Effects on structures and properties, 158, 168, 179, 184
Anderson Model, 22
Anisotropy of superconducting properties, 100
Applications of superconductivity, 4, 419, 430, 431, 433, 434, 436, 439, 440, 441, 442, 445

BCS theory, 10, 14, 31, 215
Bolometer, 433
Bubble chamber, 440

Composite superconductors
  Production of, 409
Composition effect
  Critical current, 374
Cooper pairs, 15
Critical current, 11
  Composition effects on, 374
  Heat treatment effects on, 198, 374
  Measurement of, 44
Critical magnetic field, 6, 9, 13
  Heat treatment effect on, 197
Critical temperature, 9, 30
  Alloying effect on, 29, 159, 170

Critical temperature (cont.)
  Interatomic distance effect on, 35
  Heat treatment effects on, 191, 243, 289
  Measurement of, 37
Cryostats
  Measurement of superconductivity, 56
Cryotrons, 430
Crystal structure of superconductors, 108, 116, 128, 136, 140, 146, 151, 183, 289

Discovery of superconductivity, 9
Dislocation effects, 9, 12

Electrical machines, 445
Electron characteristics
  Cooper pairs, 15
  Specific heat, 16
  Valence density relationships, 84, 114, 118, 137, 139, 142, 145, 147, 153
Elementary particle
  Accelerators, 439
Elements
  Lattice-types of superconducting, 82, 85
  Properties of superconducting, 81
Empirical rules, 27, 215
Energy storage devices, 445
Eutectic superconducting systems, 252

Fluxoids, 22
Filament (sponge) model, 25
Filamentary superconducting material, 389, 404

Generators, 442
GLAG theory, 13, 18
Gyroscope, 441

Hard superconductors, 13, 25
Heat treatment effects on
  Critical current, 198, 374
  Critical magnetic field, 197
  Critical temperature, 191, 243, 289
  Superconductivity, 190, 231, 240, 249
Helium-temperature techniques, 53

Interatomic distance
  Effect on critical temperature, 35

Lattice-type of superconducting elements, 82, 85
London theory, 17
Lorentz forces, 22

Magnetic lenses, 433
Magnets, 420
Manufacture of
  Multi-filament superconducting materials, 389, 404
  Superconducting components, 411
  Superconducting wire, 401
Masers, 343
Matthias rules, 215
Meissner effect, 9, 12
Memory units, 431
Metallography of superconducting alloys, 59
Microscopic theory of superconductivity, 10, 14
Microsection
  Etching, 62
  Preparation of, 59
Microstructure studies, 66

Phase diagrams for superconducting systems
  Indium, 257, 261
  Niobium, 222, 224, 225, 229, 235, 241, 244, 245, 253, 283, 284, 286, 294, 297, 298
  Tin, 261
  Tantalum, 227, 247, 248, 280
  Titanium, 229, 245, 247, 249, 251, 266, 267, 270
  Vanadium, 227, 244, 246, 253, 274, 276, 280
  Zirconium, 251
Phases
  Superconducting laves, 138
  Superconducting sigma, 135
Physiochemical properties
  Binary systems, 221
  Pseudoquaternary systems, 358
  Pseudoternary systems, 353
  Quaternary systems, 351
  Ternary systems, 307
  Systems with intermediate phases, 262
  Systems with nontransition metals, 252
  Systems with unlimited solubility, 228
Plastic deformation effects, 70, 90, 374
Pressure effects on superconductivity, 86, 200
Properties (see also Physiochemical properties)
  Alloying effects on, 158, 168, 179, 184
  Anisotropy, 100
  Interstitial impurity effect, 90, 179
  Superconducting elements, 81

Recrystallization, 70
Resonance pump, 441

Silsbee rule, 11
Specific heat of electrons, 16

INDEX

Structure of superconducting compounds
   Alloying effects, 158, 168, 179, 184
   Cr_3Si, 108
   Cubic, 152
   Hexagonal, 155
   NaCl, 124
   Orthorhombic, 157
   Tetragonal, 157
Superconducting coatings, 393, 407
Superconducting components
   Computers, 430
   Measuring devices, 433
   Nuclear physics devices, 436
   Production of, 411
Superconducting materials
   Compounds, 107
   Compounds of metals and nonmetals, 120
   Elements, 81
   Multi-filament, 389, 404
Superconducting systems
   Eutectic, 252
   Niobium, 221, 282, 307, 316, 326, 329, 342
   Niobium–tin, 287
   Peritectic, 255
   Tantalum, 222, 246, 280, 317, 329, 339, 342
   Titanium, 229, 245, 266, 307, 326, 329, 337, 339
   Vanadium, 226, 244, 254, 271, 316, 326, 337, 339, 342
   Zirconium, 234, 251, 271, 307, 316, 337
Superconductor type
   Type I, 11
   Type II, 12
   Criteria for, 20
Superconductors
   Hard, 13, 25
   Production of composites, 409
   Type I, 11
   Type II, 12

Tensile strength of superconducting alloys, 376
Theories
   BCS, 10, 14, 31, 215
   GLAG, 13, 18
   London, 17
   Microscopic, 10, 14

Valence electron density relationship, 84, 114, 118, 137, 139, 142, 145, 147, 153
Valence laws, 28